Perspectives in European History

No. 17

TRIUMPH OF THE FACTORY SYSTEM
IN ENGLAND

THE TRIUMPH
OF THE
FACTORY SYSTEM
IN ENGLAND

HSIEN-T'ING FANG

PORCUPINE PRESS

Philadelphia

First edition 1930
(Tientsin: Chihli Press, 1930)

Reprinted 1978 by
PORCUPINE PRESS, INC.
Philadelphia, Pennsylvania 19107

Library of Congress Cataloging in Publication Data

Fang, Hsien-t'ing, 1902-
The triumph of the factory system in England.

(Perspectives in European history ; no. 17)
Originally presented as the author's thesis, Yale, 1928.
Reprint of the 1930 ed. printed by the Chihli Press, Tientsin, China, which was issued as no. 2 of Nankai University Committee on Social and Economic Research publications, Economic history series.
1. Factory system—Great Britain—History.
2. Great Britain—Manufactures—History. I. Title.
II. Series: Nan k'ai ta hsëh, Tientsin, Ching chi yen chiu so. Economic history series ; 2.
HD2356.G7F3 1978 338.4 0941 78-15111
ISBN 0-87991-624-9

Manufactured in the United States of America

PREFACE

That the triumph of factory system is the chief outcome of the industrial revolution in England has been commonly accepted among students of economic history. No attempt, however, has yet been made to examine the facts of a large number of English manufactures, so as to determine the validity of such acceptance. Again, to speak of the triumph of one system pre-supposes an understanding of the relative importance of other systems. It is for these two reasons that I have covered in the present study on the "Triumph of Factory System in England" a comprehensive survey of the industrial organization in English manufactures, at the close of the first phase of the industrial revolution and at the beginning of the second about 1840. As a result of this survey, a new scheme of classifying the industrial systems in English manufactures, with a new terminology, is also formulated and presented.

The present study was completed as a doctoral dissertation for the Graduate School of Yale University in the summer of 1928. In this connection, I wish to avail myself of the opportunity to acknowledge my indebtedness to all who have so kindly and generously assisted me in its preparation. I wish to tender my thanks to the staffs of the Yale University Library, the Harvard University Library, the Columbia University Library, the New York Public Library, and the Boston Public Library for their permission to use the parliamentary papers and whatever secondary literature available on the subject. My indebtedness to my teacher, Professor Clive Day, cannot be over-emphasized. Professor Day not only suggests the topic for the present study, but also helps me in securing the necessary financial aids to complete it. Aside from his inspiring scholarship, his penetrating criticisms and suggestions on many phases of the study will long remain as my cherished memory. Finally, I wish to acknowledge my gratitude to one who, even unconsciously himself, must be held responsible for what I have been and shall be. Mr. H. Y. Moh, a great industrialist and now vice-minister of the Ministry of Industry, Commerce and Labor, gave me a new vision of life and a new career when he parted with one of the apprentices in his cotton mill and gave him an education as he did to many other youths.

H. D. Fong

Committee on Social and Economic Research
Nankai University, Tientsin
August, 1930

To

One who has trained many apprentices

PROFESOSOR CLIVE DAY

CONTENT

ERRATA

Page	Paragraph	Line	Table	Error	Correction
5			2	Nail	Nail — 1 2
32		1	18	236.0	236.1
60		1	47	71 (a)	71
60		2	47	53	53 (a)
78	3	14		amount.”	amount.”[84]
78	3	18		pound.	pound.[85]
93	1	4		2,021	2,011
104		1	104	spindles	power looms
106		2	108	1840	1841
116		1	20	20	120
162	2	19		.”	.”[52]
171	1	25		.”	.”[101]
214	The first line should be removed to the top of page 218.				
274		28		Leads	Leeds
282		35		more	lore
284		23		Gambridge	Cambridge
285		35		deplorably	are deplorably
297		32		Yeoil, 187	

Chapter I. English Industrial Organization About 1840

Introduction

A study of the industrial organization of England about 1840 calls for an explanation of method as well as of scope. The method employed is a combination of description and analysis. The collection and arrangement of facts has been the primary concern of the present study, but no pain has been spared in an attempt to differentiate the systems of industrial organization and to determine their extent. The scope covered relates to the choice of date, manufactures and topics for analysis. As regards the date, the year 1840 is preferred. This year "is chosen by Mrs. Knowles to mark the completion in England of the first phase of the industrial revolution, in which the most significant feature was the spread of manufacture by machinery under the factory system. The railroad era had begun ten years earlier, with the opening of the line from Liverpool to Manchester, but in 1840 railways were still in the experimental stage..... 1840 was a year of active construction, and may be taken to mark the beginning of a general system of railroad communication".[1] Furthermore, the year 1840 falls within the thirty year period (1820-1850) during which the growth of population (England and Wales) and of staple industries (United Kingdom) has been most rapid and revolutionary throughout the whole of the eighteenth and nineteenth centuries[2]. From the viewpoint of the present study, the year 1840 is taken on account of the materials available. In the history of the British Parliamentary Papers, perhaps no other period has been so rich in the materials available as that around 1840, for the study of social and economic history in general, and of industrial organization in particular. The Census in the following year, 1841, attempted for the first time to enumerate in detail the occupations of the people. The factory returns were first published in the year 1835, giving statistics on the number of textile factories, employees and motive power in the United Kingdom, but it was not until the year 1838 that such returns became a reliable source of information. The various parliamentary committees and commissions on the employment of children in textile factories (1831-1834) and non-textile factories (1843), on the conditions of the handloom weavers (1839-1841) and the framework knitters (1844), on the exportation of machinery (1841), on the arts and manufactures (1835-1836) and on many other subjects that had a direct or indirect bearing on the industrial organization of English manufactures around 1840, gave a mine of information not to be found elsewhere or at any other year.

It is plain at the outset that our study of the industrial organization cannot be extended to all the English manufactures about 1840. The materials available for such a study are uneven in nature for different manufactures, and are wholly absent for some of them.[3] The difficulty of research in a foreign land[4] is another factor. Several manufactures, however, offered typical systems of industrial organization, which were largely representative of those in other manufactures. An analysis of the former, therefore, would indirectly throw light on the latter. In the present study the term 'manufacture' in 1840 was confined, in the definition of Professor Day, "to all those working up raw materials for sale outside the household"[5]. According to this definition, the English manufactures of 1841 employed a total of about two million persons, of whom about one million persons, or fifty percent., are included in the present study[6]. These one million persons are classified into four different groups in Table 1, in which the relative importance of different manufactures is expressed in terms of the number of persons employed[7].

Table 1. Relative importance of selected manufactures according to number of persons employed, England, 1841

Name of manufacture	Number of persons employed	
Textiles		505,075
Cotton	263,154	
Woolen and worsted	144,527	
Silk	66,630	
Flax and hemp	30,764	
Clothing		328,671
Wearing apparel	209,891	
Hosiery	44,183	
Lace	32,585	
Hat	16,665	
Straw plait	9,775	
Glove	8,746	
Ribbon	6,826	
Metals		141,167
Engine and machine	32,667	
Iron	30,173	
Cutlery	25,099	
Nail	18,040	
Clock and watch	13,118	
Button, etc.	9,305	
Lock and key	5,408	
Gun	5,385	
Anchor and chain	1,672	
Paper, etc.		35,793
Pottery	23,468	
Glass	6,799	
Paper	5,526	
Total		1,010,706

The topics for analysis in treating each manufacture include definition, localization and industrial organization. The definition of a manufacture, which determines its extent, is oftentimes a difficult task, a task that was poorly executed by even the highest authority, the English Census[8]. In our treatment of each manufacture, the Census definition has been either accepted or modified where modification seems necessary on the basis of information derived from an examination of the contemporary sources.

The local distribution of each manufacture has been a subject for elaborate statistical analysis by Professor Day in his latest researches on the "Distribution of Industrial Occupations in England, 1841-1861", the object of which is to determine the extent of the market for industrial products in that period. His conclusions and tabulations have facilitated the present study in locating the chief centers of each manufacture and in determining their relative importance, but an examination of the causes for such localization rests with the present writer. There is a close relationship between localization and industrial organization in each manufacture, for in many instances industrial organization varied with localization. In the woolen manufacture, the West Riding has for centuriesbeen the home of the master craftsman system, while the West Country, including Gloucester and other counties, has preserved to us in 1840 a remnant of the merchant employer system. In the silk manufacture, the Eastern counties such as Norfolk and Essex possessed a system of industrial organization different from that of the North. In the Eastern counties, the merchant employer system prevailed; while in the North it was the factory system rather than the merchant employer system that claimed a larger number of employees[9].

In analyzing the industrial organization of each manufacture two subjects in particular have been chosen for the present study, namely, a differentiation of the industrial systems and an estimate of their extent. In the differentiation of each industrial system, the important elements and effects have been described from a selected viewpoint, - selected because of the vast scope involved and of the previous studies available. I have, for instance, omitted the social effects of the factory system upon the working population of the manufacture. The contemporary (1840) sources, whether official or unofficial, devoted an undue amount of attention to a controversy over that subject. The controversy, even at the present time, seems unsettled with some writers. For my purpose, a summary of the contents of such sources would be an unnecessary duplication.

The approximate extent of the different industrial systems in each manufacture has been examined and determined, and is now summarized in Table 2, in which the numbers 1, 2 and 3 indicate the relative importance of the three industrial systems in each manufacture. These three industrial systems, namely, craftsman, merchant employer and factory, are defined in the following paragraphs. In anticipation of an explanation of the different terms used, it may be pointed out that a new terminology has been adopted in describing the craftsman (or "handicraft") and the merchant employer (or "domestic") systems.

Table 2. Relative Importance of Different Industrial Systems in Selected Manufactures, England, 1840

Manufacture	System		
	Craftsman	Merchant employer	Factory
Textiles			
Cotton	3	2	1
Flax & hemp	3	2	1
Silk		2	1
Worsted		2	1
Woolen	2	3	1
Paper, etc.			
Paper			1
Pottery	2		1
Glass	2		1
Metals			
Iron			1
Engine & machine	2		1
Button, etc.	2		1
Clock & watch	2	1	3
Gun	2	1	3
Cutlery	3	1	2
Nail			
Anchor & chain		1	2
Lock & key	1		
Clothing			
Ribbon		1	2
Hosiery		1	2

TRIUMPH OF FACTORY SYSTEM IN ENGLAND

	System		
Manufacture	**Craftsman**	**Merchant employer**	**Factory**
Hat		1	2
Glove		1	2
Wearing apparel	1	2	3
Lace	1	2	3
Straw plait	1		

The different groups of manufactures, as well as subgroups within each group, have been arranged in order in Table 2, from the most developed to the least developed, developed in the sense of having been affected by the latest industrial system. The prevalence of the factory system in the first two groups of manufactures, its partial triumph in the third, and its relative insignificance in the fourth, are the chief conclusions drawn from an analysis of Table 2. In other words, the factory system in the English manufactures about 1840 under review had become the prevalent system of industrial organization, while the merchant employer system had rapidly declined in importance. The latter, however, existed to a larger extent than the craftsman system.

Many attempts to classify the systems of industrial organization have been made by German scholars[10], of whom Bücher is perhaps the best known. Bücher[11], however, built his theories upon the works of his predecessors, especially Schmoller[12]. In England no systematic formulation of the industrial systems was made until Professor Ashley in 1887[13]. Ashley's classification was based on Thun's, and Thun's classification preceded Bücher's by more than a decade[14].

Bücher's classification of the industrial organization into five systems and periods, namely, housework, wage work, handicraft, commission (Verlagssystem) and factory[15], is well known, and has been frequently quoted by later writers. Its merits also speak for its defects. "Schemes," as Usher has remarked, "are of course incomplete, and their shortcomings are peculiarly evident in this field [economic history]. They can hardly become the basis of sustained historical narrative, but they do serve to bring the larger conceptions of evolution into close touch with history and they vitalize research by emphasizing the problems which are most influential in our thought."[16] Bücher's scheme, therefore, gives a broad interpretation of the evolution of industrial systems, but has its shortcomings in several respects. In the first place, his emphasis that wagework as an industrial system preceded handicraft in chronological order is not supported by the factual investigations into medieval economic history[17]. The parallel existence of the two industrial systems, a phenomenon primarily attributed to the difference in technical and market conditions of the industry, had been overlooked. Secondly, Bücher's assertion that production under the handicraft system was always "custom production" carried on "for a locally limited circle of customers", or that "the customer buys at first hand, the handicraftsman sells to the actual consumer,"[18] proved to be a misstatement of facts even in the Middle Ages[19]. In 1840 several manufactures that were carried on under the handicraft system produced for the market instead of for the consumers. Finally, Bücher's analysis of the commission system disregards the fact that between the merchant employer and his outworkers, the intermediation of the middleman was oftentimes a necessity under different economic circumstances[20].

Other schemes for the classification of industrial systems have been proposed by various German writers, but most of them, not excepting Sombart's[21], tend to confuse rather than to elucidate the understanding of the subject. Sombart's classification of industrial systems into the individual (or small), the transitional (or medium) and the social (or large) "Betriebe"[22] is narrowly based upon one principle, namely, the character and extent of labor co-operation (Arbeitsgemeinschaft). It is consistent as a theory for an abstract thinker like Sombart, but as a guide to the comprehensive interpretation of the different elements in each industrial system, it is highly misleading. Other writers, whose schemes of classification rest upon considerations such as the scale of production[23], the kind of technique[24] and the nature of income derived from the industrial employments[25], have thrown even less light on the subject than Sombart. In England Professor Cunningham's classification of industrial systems into "capitalist" and "domestic"[26] is vague, but it is perhaps the only independent attempt yet made by an English economic historian.

In the present study, Bücher's scheme is accepted as a general background, but is modified in the light of factual investigations into the industrial organization of English manufactures in 1840, in content as well as in terminology. The content of the present scheme differs from that of Bücher's in the inclusion of market production as an element of the "handicraft" system, and of the employment of middleman as an element of the "commission" system. The terminology of the present scheme is a distinct departure from the prevailing terminology today. In our terminology, the industrial systems studied in connection with the English manufactures in 1840 are named after the outstanding feature of each system, namely, craftsman (or "handicraft"), merchant employer (or "commission") and factory. "Craftsman" is to be preferred to "handicraft", as the latter represents the technique of production. The technique of production is not a sufficiently important element to justify the present and past usage to characterize the system itself. Furthermore, it is the same in many cases under the craftsman system, as under the merchant employer system. "Merchant employer" is a term chosen here to replace all others, including "domestic", "commission", "sweating", "putting out" and what not[27]. "Domestic", like "handicraft", is a term that describes one particular feature in the mode of production, namely, the place of work. The place of work - the home - is the same in many cases under the craftsman system as under the merchant employer system. Furthermore, "domestic" has a technical meaning that describes the craftsman system instead of the merchant employer system. It originated in the famous "Report on the State of Woolen Manufacture of England in 1806",[28] and was accepted as such by Professor Cunningham in his pioneer volume on the "Growth of English Industry and Commerce".[29] "Commission" is a translated term for the German"Verlags system", sometimes it is employed to identify the merchant employer system in English industrial organization. The payment of commission to the middleman as employed by the merchant employer was a common practice under the "Verlagssystem" in German industries[30], but not resorted to under the merchant employer system in English industries that we have studied for the year 1840. "Sweating", a term employed in a date as early as 1840 in the London wearing apparel manufacture, was popularized through the appointment of a Parliamentary Committee on the "Sweating System" in 1888. It describes the abuses rather than the elements of the merchant employer system. "Putting out" is perhaps the latest contribution to the already swollen terminology[31], but is unsatisfactory in many respects. In the hosiery manufacture of England in 1840, a "putter out" was either a salaried agent of the merchant employer, or his middleman contractor, whose function consisted in putting out the raw materials to the outworkers to be manufactured into finished products. "Putting out", in other words, represents the mode of distributing the work; its use as a term to characterize the merchant employer system is necessarily of limited application. Finally, for the sake of uniformity, the term "factory" is to be preferred to all other terms in describing the factory system. In the contemporary (1840) terminology in English manhfactures, a "factory"[32] was known under other terms such as "mill", "manufactory", "works", etc.

In the following diagram, the industrial system in the English manufactures in 1840 are classified, in a preliminary fashion, into craftsman, merchant employer and factory systems, with a brief contrast of the prevailing elements of each system such as capital employment, mode of production and method of sale.

	Craftsman	Merchant employer	Factory
Capital employment			
a. Nature	Circulating & fixed	Primarily circulating	Circulating & fixed
b. Extent	Small	Medium	Large
Mode of production			
a. Workplace	Shop, home	Home, shop	Factory
b. Technique	Handicraft	Handicraft	Machinicraft
Method of sale			
a. Market	Local, national	National	National, world-wide
b. Sale	Craftsman to	Merchant employer to	Manufacturer to
	i. consumer	i. consumer	i. consumer
	ii. merchant	ii. merchant	ii. merchant

Craftsman System

Definition and Extent. The craftsman system of England in 1840 may be defined as that system of industrial organization under which the craftsman, with own tools and own (or hired) place of work, purchased the raw materials and worked them up into finished products for sale to the consumers or merchants. Under this definition several types of the craftsman system may be further distinguished. Taking the character of the craftsman as a basis, we may classify the craftsman system into master craftsman and family craftsman systems. In the following diagram the chief differences of these two systems are compared.

Craftsman	Family crafsman	Master crafstman
a. Character	Wife or mother, assisted by own children or not	Master, assisted by journeymen and apprentices or not
b. Activity	By-employment, as worker, foreman and merchant	Full employment, as worker, foreman, employer, merchant (and shopkeeper)
Technique		
a. Skill	Simple; no apprenticeship	Complicated; apprenticeship generally
b. Workplace	Home	Home Shop i. Own ii. Hired
Sale	To merchant	To consumer or merchant

Despite these differences, the master craftsman and the family craftsman under the two systems had a great deal in common. Both retained the ownership of raw materials, tools and generally the workplace; both sold the finished products to consumers or merchants on own account. In the present study, the straw plait, hand lace and shirt button manufactures of England about 1840 were the only examples of the family craftsman system; other manufactures, including cotton, woolen, hemp, pottery, glass, engine and machine, clock and watch, gun, cutlery, lock and key, wearing apparel and machine lace, were partly or wholly organized under the master craftsman system.

Again, taking the economic independence of the craftsman as a basis, the craftsman system may be classified according to three modes of sale, namely, sale to the consumer, sale to the merchant at public markets and sale to the merchant directly. The first mode of sale, namely, sale to the consumer, was the most common and best known type of the craftsman system, and had received undue attention from the theorists to the neglect of the other two types. Under this type the craftsman was most independent. He was not only an industrial artisan, but also a mercantile shopkeeper. In the manufactures under review, hemp spinning, engine and machine, clock and watch, gun, and wearing apparel, furnished good examples of this type of craftsman system. The second mode of sale, namely, sale to the merchants at the public markets, prevailed in the Yorkshire woolen manufacture as well as in the straw plait manufacture of the Eastern counties. In the Yorkshire woolen manufacture, the sale of woolen cloth by the master craftsmen to the merchants at the cloth halls, bore a close resemblance to the sale of straw plait by the family craftsman to the factors at the Luton markets Both the master craftsman and the family craftsman had now lost their direct contact with the consumers, and had relegated the function of distribution to the merchants, - the trading capitalists in possession of the knowledge of market conditions. The craftsman possessed less economic independence under the second mode of sale than under the first; and such economic independence varied in accordance with the manner in which the markets were conducted. The cloth halls in Leeds, Halifax and other woolen cloth centers were organized by the master craftsmen, and therefore retained a certain degree of control or check to the capitalistic influences of the merchants. The Luton markets, on the other hand, were not so organized. As a consequence, the merchants dealt with the family craftsmen on a basis of apparent free competition. In reality, the merchants, being in possession of large capital and sufficient knowledge of the market conditions, struck a better bargain over the family craftsmen than would have been possible should the straw plait markets have also been organized by the family craftsmen. Finally, the craftsman possessed the least amount of economic independence under the third mode of sale, namely, sale to the merchant directly. Under this mode of sale, either the merchant travelled around the different manufacturing districts in order to purchase the finished products from the craftsmen, such as in the manufactures organized under the family craftsman system; or else, the craftsman brought the finished products to the warehouse of the merchant for sale, such as in some of the manufactures organized under the master craftsman system, including lock and key, cutlery, machine lace, etc.

Abuses. Under the craftsman system the abuses were generally attributed to two causes, namely, the economic condition and the industrial organization. Because of the unwholesome economic condition of England in 1840, a result of war, famine, currency inflation, foreign competition and the social and economic effects of the factory system and machinery33, there was an oversupply of labor with consequent unemployment. In several manufactures, such as cutlery, the general economic depression drove the unemployed journeymen to establish themselves as masters, in order to avoid unemployment by manufacturing and trading on own account. The rapid increase of "small masters", which was further stimulated by the small capital required to engage in such manufactures, brought with it an overproduction of goods and a cut-throat competition among the "small masters" in order to win the bids of the merchant. This abuse led directly to other abuses attributed to the industrial organizaion as a primary cause.

Under the craftsman system such features as direct sale to the merchant and shop work gave rise to four abuses under review, namely, discount, truck, irregular employment, and embezzlement. Discount was at first a convenient practice to allow for the frequent fluctuations of price in the lock and key manufacture. A list price would be agreed upon between the master craftsman and the merchant, according to which an order was given by the merchant and executed by the master craftsman. The list price was never changed, but was "discounted" in order to adjust it to the conditions of demand and supply in the market. In the course of time, however, the merchant had exerted such powerful influence upon the master craftsman in the marketing of the product, that the list price was often "discounted" at an unduly high percentage. Truck, the second abuse, was resorted to by the merchants in paying the master craftsman and the family craftsman alike. In the manufacture of hand lace as well as of shirt buttons, the payment of purchase prices in goods was frequently complained of by the family craftsman. In the lock and key manufacture, the presence of the abuse tended greatly to aggravate the wretched economic conditions made possible by other abuses. Irregular employment, the third abuse, was a common feature of shop work. In the lock and key manufacture, the "systematic irregularity" had gone so far that a master craftsman would idle away a large part of the week, and when the time approached for him to turn in the work at the warehouse of the merchant, he would then work continuously from early morning till midnight[34]. Embezzlement was absent under the master craftsman system, but under the family craftsman system, embezzlement only took a special form. In the hand lace manufacture it used to be a practice for the merchant to give out his order to the family craftsman together with the pattern according to which the order was to be executed. As a result, the family craftsman would copy the pattern of the merchant and then sell it to other merchants. Or else, she might work up a larger amount of lace than the order required, and then dispose of it to other merchants at a remunerative price.

Decline and Persistence. The abuses as here described were undoubtedly one factor for the decline of the craftsman system, but other factors, more fundamental than this, have been summed up in comprehensive terms by Bücher[35]. The concentration of demand in large manufacturing and trading centers rendered it uneconomical to supply the demand by means of small-scale production under the craftsman system. The colossal tasks for industry, such as engineering and shipbuilding, were important factors in the rapid disappearance of the small craftsman, for technical as well as economic reasons. The uniformity of output, which became a pre-requisite to manufactures such as tool-making, was only possible under large-scale and factory production, and the change in the character of consumer's demand for ready-made but inexpensive goods hastened the introduction of the factory system.

However, despite the several groups of factors that were working towards the decline of the craftsman system, there were others that accounted for its persistence[36], namely, technique, organization and economic condition. As to technique, many products, of which artistic excellence and personal skill constituted the chief considerations with the consumers, were still produced under the master craftsman system[37]. As to organization, the craftsman system had long been preferred to other systems on account of its over-rated merits. Take but one of the many concrete instances, - the master craftsman system in the Yorkshire woolen cloth manufacture. Under this system, each of the three social classes, the craftsman, the journeymen and apprentices, and the merchant, - possessed certain advantages, including the facility for a young craftsman of steady character to rise in social progression, the industrial solidarity among masters, journeymen and apprentices, and the promptitude with which the large order of a merchant could be executed without considerable outlay of fixed capital. Finally, temporary as well as permanent economic conditions accounted for the persistence of the craftsman system. Certain products, such as those of blacksmiths, tailors, bakers etc., had always a local demand, and could be more economically produced under small scale than under large scale production. The existence of neighborhood stores, in other words, constituted the permanent economic condition. The temporary economic condition that prevailed about 1840 varied with different manufactures. The facility of becoming a small master in cutlery manufacture has been touched upon in a previous paragraph. In the flint glass manufacture about 1840, the existence of the master craftsman system was an outcome of the excise duties levied by the government. In an attempt to evade these taxes, flint glasses, instead of being manufactured in the factories, were made by the small masters in private houses, beyond the observation of the watchful eyes of the excise officials.

Merchant Employer System

Definition and Extent. The merchant employer system of England in 1840 may be defined as that system of industrial organization under which the merchant employer purchased the raw materials, put them out either directly from the warehouse, or indirectly through his own salaried agents, to be worked up into finished or partly finished products by the outworkers, for sale to the consumers or other merchants. The merchant employer, however, might employ the middleman to put out the raw materials, to be worked up by the outworkers into finished or partly finished products, under a work contract. Under the work contract, the middleman is virtually an employee of the merchant employer, but is at freedom to execute the work either by sub-contracting it out to the sub-middleman or outworker, or by employing outworkers in his own shop38. With the employment of middleman and sub-middleman, the merchant employer system may now be called the middleman system. In the following diagram, the merchant employer system and the middleman system, as so defined, are represented.

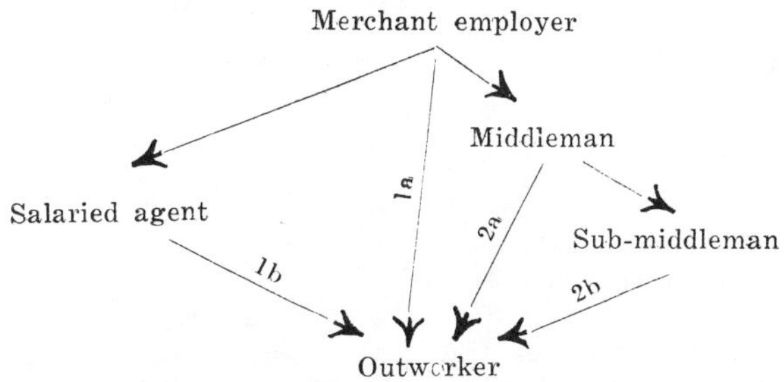

In the above diagram, systems 1a and 1b are variations of the merchant employer system, while systems 2a and 2b are variations of the middleman system. It must be noted, however, that in each manufacture, the merchant employer system might be found independently of the middleman system, while the middleman system always existed parallel with the merchant employer system. In the English manufactures in 1840, the relative importance of the merchant employer system as compared with the craftsman and factory systems is indicated in Table 3 by the numbers 1, 2 and 3; while the relative importance of the merchant employer system and the middleman system is indicated by the small letters a and b[39].

Table 3. Relative Importance of the Merchant Employer and Middleman Systems in English Manufactures, 1840

Manufactures	Merchant employer system	Middleman system
Textiles		
Cotton	2	
Flax (linen)	2	
Worsted	2a	2b
Silk	2a	2b
Woolen	3	
Metals		
Clock and watch	1	
Gun	1	
Cutlery	1b	1a
Nail	1b	1a
Chain	1b	1a
Clothing		
Hat	1	
Glove	1	
Hosiery	1b	1a
Ribbon	1b	1a
Lace	2b	2a
Wearing apparel	2b	2a

Elements. The elements of the merchant employer and the middleman systems consist in the presence of the three social classes and the conditions of outwork. The three social classes include the merchant employer, the middleman and the outworker.

The first class, the merchant employer, may be classified in the following diagram under two heads, namely, capital ownership and activity.

1. Capital ownership
   a. Raw materials
   b. Tools
   c. Shops
2. Activity
   a. As a merchant
      1. Purchase of raw materials
      2. Sale of finished products
   b. As an employer
      1. Puts out raw materials from warehouse.
      2. Puts out raw materials through salaried agents.
      3. Puts out raw materials through middlemen.
   c. As a producer (occasionally)
      1. Finishes partly finished product.
      2. Converts raw materials into intermediate products.

According to the extent of capital ownership a merchant employer may be called large or small. There can be no definite dividing line between a large and a small merchant employer in each manufacture; expediency and conformity to contemporary usage alone must be the guide in an attempt to classify the merchant employers in a summary fashion. In each manufacture, the merchant employer provided the raw materials invariably, the tools frequently, and the shops occasionally. In the handloom weaving of cotton and woolen cloths, the outworker had to provide his own loom. But in other manufactures, such as hosiery knitting, the merchant employer provided the frame, and then let it out to the outworker, by himself or through the middleman, at a rent. In most cases the outworker also provided his own workshop in addition to tools; but there were instances where the merchant employer provided not only raw materials, but also tools and workshop. In the carpet handloom weaving, for example, merchant employer had so completely provided his outworkers with raw materials and looms and congregated them together in large numbers in one or more workshops, that Professor Cunningham was inclined to call such an arrangement a factory [40]. However, in view of the absence of any attempt on the part of the merchant employer to exercise control or supervision over the weavers' work, as well as the presence of features inherent in the merchant employer system, such an arrangement may better be taken to exhibit the most capitalistic development of the merchant employer system. The provision by the merchant employer of raw materials, tools and workshop had a profound effect upon the economic position of the outworker. His dependence upon the merchant employer increased in proportion as his ownership of and control over his means of production is lost, first the raw materials, then the tools, and finally the workshop.

The activity of the merchant employer has been sketched in brief in the above diagram. As a merchant the merchant employer purchased the raw materials to be worked up into the finished or partly finished products by his outworkers, and sold the finished products to the consumers in many cases, but to another class of merchants in a few others. As an employer, the merchant employer put out from the warehouse, through his salaried agents, or through the middlemen, the raw materials to be worked up into finished or partly finished products by the outworkers. A merchant employer, however, might occasionally be a producer under certain economic arrangements. In the hosiery and lace manufactures, for instance, the products, after having been partly finished by the outworkers, must go through a final process of finishing at the warehouse before they were ready in a saleable form. The merchant employer in such cases would employ a varying number of workmen at the warehouse for finishing, under his direction and supervision, or under that of his foremen. The size of such finishing warehouses differed in accordance with the extent of the merchant employer's business. Again, in silk or ribbon weaving around 1840, the merchant employer worked up the raw silk into yarn at his own factory, and then put out the yarn to the outworkers. The assumption of such productive functions by the merchant employer about 1840 was unusual in English manufactures organized under the merchant employer and the middleman systems, and may be taken as an exception.

The second class, the middleman, may be classified in the following diagram under two heads, namely, capital ownership and activity.

1. Capital ownership
    a. No capital
    b. Capital
        1. Tools
        2. Tools and shops
2. Activity
    a. Distributive: As employer (contractor)
    b. Distributive-and-productive: As employer (contractor), foreman and workman
    c. Productive-and-distributive: As employer (contractor); workman and foreman

The capital ownership of a middleman, as a whole, varied with his activity. A purely distributive middleman had no capital, as his function consisted chiefly in putting out the merchant employer's raw materials to be worked up into finished products by the outworkers, under the terms of a work contract. On the other hand, a productive-and-distributive or distributive-and-productive middleman owned a varying amount of capital in accordance with his extent of business, and such capital consisted in tools and workshop, but never in raw materials. Both the productive-and-distributive and the distributive-and-productive middleman differed from the distributive middleman, as the former had taken over the productive in addition to the distributive function. The distinction between a producive-and-distributive and a distributive-and-productive middleman was drawn on the basis of the relative importance placed on the distributive or the productive function. A productive-and-distributive middleman put out the merchant employer's raw materials among the workmen, who worked them up into finished products at his shop with his tools; and made his profits by the difference between the price received from the merchant employer and the wages paid out to the workmen. Such an arrangement was common in the ribbon and cutlery manufactures. On the other hand, a distributive-and-productive middleman had similar functions to perform, and made his profits in a similar way. The difference, however, consisted in his additional function of putting out the merchant employer's raw materials to be worked up into finished products by the outworkers not congregated in his own shop. At his own shop, he was less a workman than a foreman; whereas a productive-and-distributive middleman was more a workman than a foreman. Furthermore, a distributive-and-productive middleman might own tools and shops not worked by the workmen directly under his employ, but under that of other workmen who paid him a certain rent for the privilege of using his tools and shops. Such workmen, in general, received the merchant employer's raw materials to be worked up into finished products from the distributive-and-productive middleman. Both types of middlemen, namely, productive-and-distributive and distributive-and-productive, were found in the hosiery and lace manufactures of England about 1840.

The outworker, finally, may be classified according to his possession of tools and workplace, in the following diagram.

1. Tools
    a. Owns
    b. Hires from merchant employer, middleman or outsider.
2. Workplace
    a. Owns
        1. Home
        2. Shop
    b. Hires from merchant employer, middleman or outsider.

Outwork included home work and shop work, to which we may add such exceptional cases as school work and poor house work. Home work, however, was a commoner form of outwork than shop work. Both home work and shop work under the merchant employer system bore a close resemblance to the same forms under the craftsman system. In case of home work, the common features, namely, family labor and by-employment, were prevalent under the merchant employer as under the craftsman system. Family labor, including child and woman labor, varied in accordance with the amount of skill and strength required, as to its extent. Where both skill and strength were required in order to carry on the employment, family labor was seldom resorted to. On the other hand, where they were not required, family labor was almost universal. Woolen handloom weaving, in this respect, offered a strong contrast to hosiery knitting. In the former the absence of child and woman labor was a blessing in maintaining the level of wages paid to the men; in the latter, its presence constituted an advantage on account of the family earnings derived from the by-employment, but a disadvantage in view of the deplorable fact that the wages for men were frequently brought down to the same level as those for children and women.

Abuses. There were several abuses inherent in the merchant employer and the middleman systems. Some of them, such as embezzlement and sweating, were good indexes upon which to determine the presence of these two systems in each manufacture. Embezzlement was possible only when the work was done beyond the supervision of the merchant employer. It was practiced not only by the outworkers, but also by the middlemen in the hosiery and ribbon manufactures. Its extent varied with the possibility of detection by the merchant employer and the economic conditions of the outworker. In some manufactures, such as Lancashire cotton handloom weaving and the woolen handloom weaving of the West Country, embezzlement of raw materials had been carried on to such an extent as to make the purchase of embezzled materials a regular trade. Sweating was a second abuse attributed to the presence of middleman and sub-middleman. The latter, better known as sweaters in the London wearing apparel manufacture in the 'forties, exploited the outworkers by various means, such as non-delivery of tickets, deduction of shop charges, etc. Truck, a third abuse under the merchant employer system, prevailed in practically every manufacture so organized. In Yorkshire, Lancashire and Staffordshire, the system had been found in extensive operation in woolen, worsted and linen handloom weaving, and in nail and chain manufactures by hand process. Wherever the system prevailed, the outworker was forced either by the merchant employer or by his middleman to accept goods in part payment of the wages, or to purchase them from the shops owned and managed by the merchant employer or the middleman. In some manufactures, such as nail and chain, the profits of a merchant employer derived from the truck shops were so large that he was able to monopolize the market by underselling the non-truck paying merchant employers. Cutthroat competition, a fourth abuse, was thereby introduced. It was not uncommon that as a result, the non-truck paying merchant employers were forced either to substitute truck payment of wages for cash, or to lower the wages of their outworkers in order to retain the customer's demand. Wage abatement, a fifth abuse, was at first intended to compensate the merchant employer or the middleman for the poor work done by the outworker, but once the practice was introduced, its abuses soon became highly oppressive in character. For little or no reason, the outworker's wages were reduced because of abatements, and should any dissatisfaction or complaint be expressed by the outworker, he would in a few days be thrown out of employment. Irregular employment, a sixth and last abuse, was pushed to a farther extent under the merchant employer than under the craftsman system. Under the merchant employer system, its existence was attributable partly to the merchant employer and partly to the outworker. This abuse, in many cases, had been a primary consideration with the merchant employer in retaining his primitive form of industrial organization despite the supreme economy of the factory system. The same may be said of the outworker in refusing to join the factory with better conditions of work and better wages.

Decline and Persistence. The decline of the merchant employer system in the English manufactures in 1840 was an accomplished fact; it was concretely proven by the prevalence of the factory system according to our analysis. In the eyes of the contemporary witness, it seems that the sole cause must have been the supreme advantages of machinery. "Where," the friend of the old merchant employer system complained, in 1833, "is the combing and spinning, cotton-weaving, silk-filling, cruel-filling, hemp-dressing, sack-weaving, paper - making, hand - sawing, hot-pressing, and printing; - where the employment general till the last few years? all absorbed by machinery. Where is the distaff and spindle, - where the employment in knitting, in the manufacture of hemp and flax, - where the employment for women and children, formerly carrying comfort and independence to the *home* of every cottager?- all absorbed by machinery, or sacrificed to the cry of 'cheap'."41

Machinery, however, did not accomplish so much as its opponents believed it to have accomplished. In the English manufactures in 1840, the merchant employer system still had to be reckoned with as the second prevalent system of industrial organization. In some manufactures it had lost its ground completely, but in others it had retained partial or complete domination for obvious raison d'être. In this connection, many reasons given in a previous paragraph on the persistence of the craftsman system applied equally well to the persistence of the merchant employer system. Technique, in the first place, was one factor. In the textile and clothing manufactures, machinery had been applied with varying degree of success. Handloom weaving, the existence of which was chiefly responsible for the maintenance of the merchant employer system, was still a necessity in 1840 for the making of fancy and other cloths. Secondly, the organization itself offered several advantages to the merchant employer and the outworkers alike, of which irregular employment has been briefly touched upon in a previous paragraph. Irregular employment was an abuse in one sense, but in another sense it constituted an advantage. From the viewpoint of the merchant employer, the chief advantages of the system consisted in the possibility of adaptation to the economic conditions of demand. In a great number of English manufactures, including textiles, clothing, and metals, the frequent and harrassing changes arising from business depression or prosperity and other factors such as fashion, climate, etc., required a system of industrial organization such as the merchant employer to protect the merchant employers from heavy financial losses. In all these changes, the important consideration with the merchant employer had always been the attempt to avoid overhead cost indispensable to the investment of fixed capital under the factory system. Under the merchant employer system the absence of fixed capital and therefore of overhead cost was probably the most influential factor for its persistence. From the viewpoint of the outworker, the merchant employer system had several advantaages over the factory system. Home work, with its characteristic features of freedom from discipline, extra earnings from the by-employment afforded to the family members, and social features of home life, were too precious gains to be readily given up.

Factory System

Definition and Extent. The factory system of England in 1840 may be defined as that system of industrial organization under which the manufacturer was the possessor and manager of a whole or a part of the building or several buildings where the wage-earning workers, with the aid of the manufacturer's machinery or fixed instruments of production, were concentrated under the supervision of the manufacturer or his foremen, to manufacture raw materials of the manufacturer under a closely related series of sub-divided processes. Under this definition the essence of a factory consisted in the labor concentration and supervision by the manufacturer and the employment of machinery or other fixed instrument of production42. The former, however, is a more important element than the latter, and a factory may accordingly be called "mature" or "immature" because of the presence or absence of machinery or other fixed instrument of production, in addition to the labor concentration and supervision by the manufacturer. A "mature" factory, furthermore, may be called small, average, or "first-rate", in proportion to the extent to which the two elements are present in a factory. In our analysis of the textile manufactures, a factory is called "first-rate" when it is at least twice the size of a median average factory with regard to the extent of labor concentration. For other manufactures, such as metal, paper, pottery, glass and clothing, the statistics yet available do not permit us to attempt such a clear-cut distinction, and the classification of such factories into the above categories must remain primarily a subject for approximate estimation.

Elements of a Factory. The first of the two elements of a factory is labor concentration. In the English manufactures in 1840 the concentration of labor was always accompanied by supervision except in two cases. First, in the carpet handloom weaving in the county of Worcester, the merchant employer provided his outworkers with raw materials, looms and shops. According to one contemporary, these shops contained from two to thirty-six looms each. Supervision, however, was never exercised, and each outworker might work or quit as he pleased. The outworker, with his assistant or draw-boy, was paid by the piece, and had no relation whatsoever with his fellow outworkers, in an economic sense at least. Second, in the Sheffield cultery manufacture, the so-called "tenement factory" had a large number of outworkers concentrated in it, but each of them worked on his own account or on that of his master—the productive-and-distributive middleman. The element of supervision was therefore absent as a whole, but inside each team including the middleman and his workmen, it was exercised. The "tenement factory", in other words, was composed of a series of shops.

The extent of labor concentration in a factory varied not only in each manufacture, but also among different manufactures. Technical requirement constituted one factor for such variation, while capital investment supplied another[43]. For an accurate determination of the extent of labor concentration in a factory, abundant statistics have been provided by the factory inspectors for the textile factories since 1835, but unfortunately these statistics are defective in many respects. In the first place, a factory as defined by the factory inspectors in the Factory Returns differed from what is meant by a factory in our definition. A factory, in our definition, might occupy a part of the building, a building, or several buildings; while in the definition of the Factory Returns, afactory always meant one building. Perhaps the one-building factory was the commonest type of factories in the textile manufactures of England around 1840, but even so it is quite possible that the several-building factories might have outnumbered the part-building factories. In Inspector Horner's District in 1838, which comprised the leading counties of cotton, woolen, worsted, silk, flax and hemp factories such as Lancashire and Yorkshire, there were 1,798 factories (i.e. buildings) that belonged to 1,541 occupiers[44]. Furthermore, these factories—one-building, part-building, or several-building—were factories in the sense that each of them employed power machinery, with no reference to the extent of labor concentration[45]. As a result, "the occupier of a building, containing a single carding machine, impelled by water, is amenable to the law, although the persons employed in the building, including adults as well as children, do not amount to half a dozen"[46] The occupier was "amenable to the law", and on that account his returns were included in the Factory Returns. In the second place, the Factory Returns of 1835 and subsequent years gave only the total number of factories, employees and motive power, for each county, as a consequence of which the mean size of factories alone, instead of the individual sizes, could be obtained. A mean, however, leaves a good deal untold, and is in general much above rather than below the typical size of textile factories. With these two qualifications in mind, we may now proceed to summarize the results of our computation on the mean size of a cotton, woolen, worsted, silk, flax or hemp factory in England in 1838, to be supplemented by median sizes calculated from the same sources with a different method, or from other sources of a different date.

Table 4. Mean Number of Persons Employed Per Cotton, Woolen, Worsted, Silk, Flax or Hemp Factory, England, 1838

Kind of factory	Total number of factory	Total number of employees	Number of employees per factory
Cotton	1,598	218,177	137
Woolen	1,029	47,175	46
Worsted	416	31.632	76
Silk	263	33,565	128
Flax or hemp	179	16,573	93

According to the above table, the mean size of a factory as measured by the number of persons employed is the largest in the cotton manufacture, namely, 137 persons per factory; while that in the other manufactures is, in order of size, 128 per silk factory, 93 per flax or hemp factory, 76 per worsted factory, and 46 per woolen factory. The mean, however, is far above the median, on account of the presence of extremely large factories. The median, as compiled from statistics available for the years 1833, 1838 and 1841, is presented in Table 5, and compared with the mean. The statistics for 1833 are taken from the Returns by the individual factory occupiers to the Yorkshire Form of Inquiry; while those for 1838 are taken from 1838 Factory Returns, including all factories located in parishes having one or two factories per parish; and those for 1841, from the enumeration of individual factory sizes for the District of Inspector Horner in his Half-yearly Report ending December, 1841.

Table 5. Comparison between Mean and Median Number of Persons Employed Per Cotton, Woolen, Worsted, Silk, Flax or Hemp Factory, England, 1833-41

Kind of factory	Mean	Median		
	1838	1833	1838	1841
Cotton	137	—	92	92
Woolen	46	—	19	15
Worsted	76	56	57	—
Silk	128	—	68	—
Flax or hemp	93	—	50	46

In the above table, the significant features are that the median is invariably lower than the mean; and that the medians computed for different years agree closely with each other. For these two reasons we may venture to conclude that the median is truer and more representative than the mean, and may be preferred to the mean as an index of average factory size. In Table 6 the growth of mean size of a cotton, woolen, worsted, silk, flax or hemp factory from 1838 to 1856 is presented. The mean is computed on account of the lack of statistics necessary for the calculation of the median.

Table 6. Growth of Mean Factory Size as Measured by Number of Persons Employed Per Cotton, Woolen, Worsted, Silk, Flax or Hemp Factory, England, 1838-56

Factory	1838	1850	1856
Cotton	137	166	167
Woolen	46	55	59
Worsted	76	160	170
Silk	128	153	120
Flax or hemp	93	141	143

With the exception of the silk factory, the other factories all increased in size, as measured by the number of persons employed per factory from 1838 to 1856, the increase being due to the growth of large scale production. The decrease in the number of persons employed per silk factory from 153 in 1850 to 120 in 1856 may be explained by the decline of the English silk industry with the coming of foreign competition, especially from the French manufacturers.

The above statistics relate only to the size of an average factory. We may next inquire into the size of a first-rate factory. In the textile manufactures, a first-rate factory has been tentatively defined as a factory whose size is at least twice the median size of an average factory. According to this definition we may derive the median size of a first-rate factory from the same source of statistics from which we obtain the median size of an average factory. In Table 7, the median size of a first-rate factory so derived is presented in comparison with the median size of an average factory.

Table 7. Comparison between Median Size of *Average* Factory and Median Size of *First-rate* Factory in Cotton, Woolen, Worsted, Silk, Flax or Hemp Manufacture, England, 1833-41

Kind of factory	Average factory	First-rate factory
Cotton	92	397
Woolen	17	62
Worsted	57	207
Silk	68	251
Flax or hemp	48	177

Statistics on the factory size in other manufactures, whenever available, are not so comprehensive in nature as those for the textile manufactures. In the iron manufacture, the only statistics applicable to all ironworks in England consist in the amount of pig iron output per ironworks. In the glass manufacture, similar statistics consist in the amount of excise duties paid by each glass factory to the government. In the engine-and-machine and pottery manufactures, statistics on the average size were furnished by contemporary witnesses for the important districts such as the Staffordshire Potteries and Lancashire. In the machine lace manufacture, such statistics were merged together with those for the cotton, woolen, worsted, silk, flax and hemp factories in the Factory Returns, and there has been no way of sifting the one from the other. In other manufactures, only scattered references to the sizes of different factories were found, and in view of the limited space available, a detailed review of them must be left to the various sections devoted to each of these manufactures in this study.

The employment of machinery or other fixed instrument of production in a factory in English manufactures in 1840 was always accompanied by a closely correlated series of sub-divided processes. Machinery was driven by hand or mechanical power, mechanical power consisting of water or steam power. In the textile manufactures as well as in other manufactures, mechanical power was the chief source of motive power employed in the factories. The hand power factories, however, were occasionally found. In the textile manufactures, for instance, handloom or "shop loom" factories were insignificant as to their extent, but furnished interesting examples that illustrated the attempt of the textile manufacturers to adapt their mode of production in the transition from the merchant employer to the factory system. In some factories, such as pottery and glass, fixed instrument of production rather than machinery was employed. Machinery, it may be noted, did not affect the two industries until after the 'forties.

Statistics on the employment of machinery in a factory were scattered in nature for the manufactures such as machine lace, engine-and-machine, and paper; those for the textile factories alone were comprehensive. These statistics, like those on the number of persons employed, were given in the Factory Returns of 1835 and subsequent years. Omitting qualifications already made in connection with the number of persons employed in the textile factories, the employment of power machinery in the textile factories may be summed up under motive power, spindles and power looms.

The motive power employed in the textile factories consisted in steam or water. Water power, however, had already declined to a large extent around 1840, and the triumph of steam power had been achieved with varying degree of success in the cotton, woolen, worsted, silk, flax and hemp factories. In Table 8 the increase in the proportion of steam power in cotton, woolen, worsted, silk, flax or hemp factories from 1838 to 1856 is presented.

Table 8. Increase in the Proportion of Steam Motive Power in Cotton, Woolen, Worsted, Silk. Flax or Hemp Factory, England, 1838-56

(in Percentages)

Factory	1838	1850	1856
Cotton	81	89	92
Woolen	61	67	74
Worsted	82	87	91
Silk	71	76	84
Flax and hemp	74	81	78

The increase of motive power per factory from 1838 to 1856 is presented in Table 9.

Table 9. Increase in Amount of Motive Power Per Cotton, Woolen, Worsted, Silk, Flax or Hemp Factory, England, 1838-56

(in Horse Power)

Factory	1838	1850	1856
Cotton	31	41	42
Woolen	17	16	19
Worsted	17	23	28
Silk	12	13	11
Flax and hemp	24	33	33

The amount of motive power employed per factory, being an arithmetic mean, is subject to the same criticism as the number of persons employed per factory. Moreover, as an index for comparison among the factory sizes in different textile manufactures, it is not so representative as the number of persons employed per factory, the number of spindles per factory, or the number of power looms per factory, for the fact that the spindles and power looms employed in the various textile manufactures required different amount of motive power, depending upon the nature of the raw materials or intermediate products to be worked on.

Two more tables, Tables 10 and 11, are presented below to show the change in mean factory size in the textile manufactures as measured by the number of spindles and power looms employed. In this connection, statistics for 1838 were not available, while those for 1850 and 1856 were classified according to the two types of factories, namely, spinning and spinning-and-weaving[47].

Table 10. Change in Number of Spindles Per Cotton, Woolen, Worsted, Silk, Flax or Hemp Factory, England, 1850-56

Kind of factory	Spinning		Spinning-and-weaving	
	1850	1856	1850	1856
Cotton	11,398	16,765	18,621	16,947
Woolen	1,050	1,119	2,764	2,500
Worsted	1,881	2,861	2,885	4,460
Silk	4,617	3,252	10,083	7,050
Flax or hemp	1,976	3,813	5,532	4,455

Table 11. Change in Number of Power Looms Per Cotton, Woolen, Worsted, Silk, Flax or Hemp Factory, England, 1850-56

Kind of factory	Weaving		Spinning-and-weaving	
	1850	1856	1850	1856
Cotton	160	157	342	337
Woolen	27	25	43	43
Worsted	132	109	128	171
Silk	53	40	132	98
Flax or hemp	78	76	25	31

Causes and Effects. In the English manufactures in 1840 the factory system had become the prevalent system of industrial organization in textiles such as cotton, woolen, worsted, silk, flax and hemp; in metals such as iron, engine-and-machine, button, screw, steel pen, needle, pin; and in paper, pottery and glass. In other manufactures, in metals such as nail, anchor and chain, clock-and-watch, cutlery and gun; in clothing products such as lace, ribbon, glove, hat and wearing apparel, the system was making rapid headway, taking a second or third place as compared with other industrial systems. In short, in the English manufactures that have been analyzed in this study, including about one half of the manufacturing population of England in 1840, the factory system had become the predominent system of industrial organization at the end of the first phase of the Industrial Revolution. In view of the fact that the system was not introduced on a significant scale until the last quarter of the eighteenth century in English industries, a survey of the causes for its rapid rise would undoubtedly throw a great deal of light. "The factory system," in the words of Cooke-Taylor in 1844, "originated in no preconceived plan - it sprang from no sudden exercise of human wisdom, like Minerva from the brain of Jupiter,- but it was formed and shaped by the irresistible force of circumstances, fortunately aided and guided by men who, like the late Sir Robert Peel, were able to profit by circumstances."[48] What, we may ask, constituted "the irresistible force of circumstances"? Briefly, it was the ideas and efforts of man rather than the benevolence of nature that had brought about this revolutionary change. Inventions in textiles and metals had an immediate effect, but without the enterprising spirit of the English manufacturers and the social philosophy of laissez-faire that nourished its growth and maturity, the factory system might have remained a fiction even today. The economic conditions, meantime, had also changed. The extension of market at home and abroad resulted in the attempts of men to increase production by means of better and newer technique; the acquisition of capital by the early English traders supplied the necessary resources for production on a large scale. The development of modern means of transport had just begun; its conquest of continents and of oceans widened the areas of demand indispensable to the English manufactures. Indeed, the establishment of the factory system "required all the forces .... physical, mental, philosophical, commercial, and philanthropical, working in separate yet convergent lines to lay the foundation of an entirely new system of industry, and these forces, coming into existence during the twenty years following the success of efforts of Hargreaves and Arkwright, and extending in their wonderful influences over the earth wherever civilization has a foothold, constituted that period one of the most remarkable since the Christian era. In fact, no generation since then has so completely stamped itself upon the affairs of the world."[49]

The establishment of the factory system in the English manufactures produced profound effects[50] upon the different social classes. From the viewpoint of the manufacturers, production on a large scale resulted in an economy of savings not only from the purchase of raw materials and the sale of finished products, but also from an effective use of machinery and buildings and a close supervision of labor and production. The quicker turnover of capital and the larger output of product brought him on the whole a greater return in terms of monetary profits, although there were disadvantages resulting from the employment of fixed capital and the recurrence of business cycles. Social legislation, which was almost unknown under the merchant employer system, became an additional charge upon the business expenses of the manufacturer, although in the long run, better conditions of work were essential factors for efficient production. From the viewpoint of the worker, the factory system was undoubtedly a benefit in giving him shorter hours of work, higher wages, more regular employment, and stronger organization than would have been possible under the merchant employer system. The introduction of unskilled labor, such as children and women, was perhaps exaggerated in its effects upon the unemployment of men, as in many cases, unskilled labor had created a new market all of its own. Social legislation under the factory system became a new right to labor. Despite its slow introduction, it had afforded adequate protection to an increasing extent once its enforcement was rendered effective by the passage of the 1833 Act. On the other hand, the loss of economic independence, the monotony of machine and factory production, the increase of accidents as created by the introduction of new technique in dangerous industries such as mining,—all these put the life of a worker in an unfavorable light. From the viewpoint of the consumer, lower price afforded the satisfaction of newer wants. The product, however, now became standardized, and for some if not for all, the decline of handicraft and the absence of personal and artistic touch in factory products constituted a permanent loss. On the whole, from the viewpoint of all parties concerned, the manufacturer, the worker and the consumer, the introduction and establishment of the factory system had undoubtedly been a social gain. In conclusion, we may quote a passage from the contemporary champion of the factory system, Cooke-Taylor, in 1844.

> "Every branch of industry in England, and to some extent throughout Europe and America, is daily assuming more and more definitely the aspect of factory organization. This is one of those great facts over which human foresight and legislation can exercise as little control as over the direction of the winds or the motion of the tides, and hence the proper sphere for the exercise of legislative wisdom is, to perfect the system, but not to check its progress."[51]

Chapter II. Cotton

Cotton Manufacture

Cotton manufacture about 1840 was defined in several different ways, and in accepting one definition or the other, its scope or extent became considerably narrowed or widened. With regard to the number of persons employed in the manufacture, Baines included not only those in lace-making and embroidery, hosiery, and calico printing, in addition to spinning and weaving, but also the outworkers in some of these branches who carried on their occupations as by-employments.[1] For the present purpose, the definition as given by the Census of 1841[2] is preferable, the acceptance of which makes available to us a wealth of statistics concerning the manufacture in the Occupation Census of 1841, and gives us, with reasonable assurance of uniformity, a basis for comparison with the Factory Returns of 1838.[3] Under the Census definition, the cotton manufacture of England in 1841 included all persons fully employed in preparing and spinning the raw cotton into yarn, and weaving the yarn into cloth. It employed a total of 263,154 persons,[4] or 52% of all persons employed in the textile manufactures of England, including cotton, woolen, worsted, silk, flax and hemp. The rapid rise of the cotton manufacture was no less remarkable than a revolution in the industrial history of England, for in a date as late as 1783, cotton manufacture ranked as a poor third, with woolen leading the list and linen occupying the middle position.[5] Among the causes that explained this rapid rise, the unbroken series of mechanical inventions leading to the economy of large scale production under the factory system, and the great extension of market at home and abroad, must be given the first importance. These in turn fostered the localization of the manufacture in those counties of England which were most favorably situated for its profitable development. In Table 12 it may be noted that in Lancaster alone, about 70% of the persons employed in the cotton manufacture were concentrated in 1841.[6]

Table 12. County distribution of persons employed in cotton manufacture, England, 1841

County	Number	Percentage
Lancaster	183,945	69.9
Chester	37,368	14.2
York, W. R.	19,210	7.3
Derby	10,263	3.9
Other	12,368	4.7
England	263,154	100.0

The localization of the cotton manufacture in Lancaster and vicinity was an outcome of many factors. The natural advantages that Lancaster possessed over other regions, the excellent harborage that its proximity to Liverpool offered for the import of raw cotton and export of finished yarn or cloth, the cheap power arising from the presence of abundant coal and water, the atmospheric quality so highly valued in spinning and weaving of cotton, and the pure water essential in bleaching and calico printing, were leading instances. Specialization of businesses—of which we may well remember the usefulness of such marketing facilities as the Manchester Exchange[7]—and the proximity of subsidiary industries such as machine making, were distinct advantages in the economies of centralization. In addition, the kind reception extended to the foreign refugees in early times, the existence of a laboring population in connection with the woolen industry already there, and the freedom from the burdensome restrictions upon industrial development that Manchester, not being a Corporation, enjoyed, fostered the growth of cotton manufacture in a rather small region.[8]

The Factory System.

This brief sketch of cotton manufacture as regards its extent and localization brings us to a better understanding of its industrial organization—the crucial issue of our present analysis. In 1840 the industrial organization in the cotton manufacture could be classified into three systems, namely, factory, merchant employer, and master craftsman systems. Through process of elimination, we may conclude that as the Census returns for persons employed in the cotton manufacture of England in 1841 was 263,154, as compared with the factory returns of 218,177 persons for England in 1838,9 the number of persons employed in the cotton manufacture under the other systems was 44,977. Table 13 gives the number of persons employed under the factory and other systems in absolute as well as in relative terms.

Table 13. Persons employed in cotton manufacture under factory and other systems, England, 1838-1841.

System	Number	Percentage
Factory	218,177	83
Other	44,977	17
Total	263,154	100

The return for the persons employed under "other" systems was an underestimate, for according to the leading authorities of the period 1820 to 1840, the handloom cotton weavers in Great Britain were variously estimated at from 240,000 to 250,000,10 of whom Baines assigned 200,000 to England. It is true, however, that the scope of enumeration for the Census officials differed from that as defined by the above authorities. The inclusion by the former, and the exclusion by the latter, of the part-time out-workers such as women and children, resulted in a material difference. Again, the Census authorities themselves admitted that "the numbers employed in and about the manufacture of cotton have been estimated much beyond, and as far below the truth."11

Returning now to the analysis of industrial organization, we may begin with the factory system as being by far the predominant one in the cotton manufacture. The system had its origin much earlier than the appearance of mechanical inventions in spinning or weaving,12 but it was such inventions that brought about its development on a significant scale. Hargreave's jenny, Arkwright's water frame, and Crompton's mule, with the assistance of Watt's steam engine, removed spinning from the cottage to the factory during the last quarter of the eighteenth century; while weaving remained untouched by the new system of organization until fifteen years after the invention of Dr. Cartwright's power loom, viz. 1800. The transfer from one system of organization to the other may thus be divided into two stages. In the first stage, only spinning was done in the factory; in the second weaving followed suit, and was generally added to the spinning branch of the factory—that is, spinning and weaving were now re-united under one roof.13 This change was vividly described by the Children's Employment Commission in 1834:

> "Prior to the invention of power loom, handloom weavers were the original purchasers of cotton yarn; but when the power loom was introduced, the manufacturing of cloth, by its assistance, did not form a separate business, as in the handloom trade, but was carried on under the same roof with the spinning of cotton, the manufacturers of yarn having added power looms to their spinning machinery. The handloom weavers, of course, could not work against the competition of the steam looms, or if they did continue to earn a scanty subsistence from their business, they could only do so by giving a far lower price for the thread they made use of

> than before; the consequence was, that there were hardly any buyers of yarn in the market, and the price was reduced so low that the profits of the makers of it were almost annihilated; many of them have been brought to bankrupcy; and of late years those cotton spinners only have derived a profitable trade, who have added power looms to their spinning establishments."14

Under the factory system, the chief element of a factory is the concentration and supervision of labor in the employer's establishment; while the employment of power-driven machinery is but a secondary one. In the following paragraphs, each of the two elements as found in the cotton factory about 1840 is analyzed in detail.

Labor Concentration.

Concentration of labor in factory brought with it a revolutionary change of habit on the part of the cotton spinners and weavers who had long been accustomed to the liberty and freedom of irregular work under the old systems of industrial organization, namely, the master craftsman and the merchant employer systems. As late as 1800, about thirty years after the establishment of the first spinning factory by Arkwright at Crompton, "the spinners were not held to any regular hours of work; they frequently spent two or three days in the week in idleness and drinking, letting the children in their service linger for them at the public houses till they were disposed to go to their work; to which, when they did return, they would sometimes work desperately, night and day, to clear off their tavern score, and get more money to spend in dissipation."15 This habit of irregularity, synonymous with the dislike for discipline, was the chief cause of labor's hostility to enter the factory. In 1840 the Committee on Mills and Factories was told by one witness that during a recent turnout of the workpeople at a cotton factory in Bolton, "the handloom weavers would not walk across the street to get 10s. per week by work (sic) in a mill, when they complained to Parliament that they were not getting above 6s. or 7s."16 Another cause, that of home attachment, may also be observed. One Manchester manufacturer, Mr. Watkinson, complained that

> "the handloom weavers who live in the outdistrict seem to expect the trade to go to them. I have urged upon people repeatedly, and offered to find them a house if they would follow me to Manchester, and that I would undertake to find them subsistence, and their children and wives work in the factories, but they prefer staying eight or nine miles off in the country district where they can scarcely exist, because they are prejudiced in their original district—their paternal home, I may almost call it, if, on the contrary, they had accepted of that offer which I have urged upon them, that is, to follow the trade, they then would never have suffered the vast number of thousands and thousands of Irish persons to come, and at this moment [1843] to be earning 10s., 15s., and 30s. a week, whereas these poor men are living six miles from Manchester earning 7s., 6s. and 5s. to maintain a family upon."17

A good index of the extent of labor concentration is the number of persons employed per factory. Statistics of this nature, like those on motive power, spindles and power looms, are far below the modern standard. The factory returns, which record average size alone, must be used with caution. In using such average for our purpose, we must first of all notice the great variation of size in the cotton factories. A few factories whose number of persons employed per factory exceeded a thousand might easily raise the sizes of those factories whose number of persons employed per factory did not frequently amount to more than half a dozen.18 Secondly, the definition of a factory as an "occupancy,"—a building, a part of it, or a group of buildings—was adhered to by the factory inspectors in enforcing the factory acts; but such definition was not observed in computing the factory returns. The factory inspectors of England for 1838, including Messrs. Horner, Howell, and Saunders, were unanimous in their practice of reporting a building as a factory. Inspector Howell, for instance, made a remark to this effect in transmitting the Factory Returns of 1838:

> "It is right to remark that a considerable number of mills are returned as single mills [i.e. factories] each of which is nevertheless occupied by several different tenants, unconnected with each other in trade, and who, for the purposes of the Factory Act, are treated as the occupiers of so many distinct factories, although they are not counted as such for the purposes of the present returns."[19]

Inspector Horner, concurring with Inspector Howell, observed that

> "in many instances, one firm has more than one building (I have them with as many as five), and in other cases, several firms are separate tenants in one building. Both mills and firms must be taken into account in estimating the work to be done in inspecting mills, and in communicating with mill occupiers. In this way, there are, at present, 1,798 mills and firms in my visiting books,"

but there were only 1,541 separate firms or concerns.[20] Finally,

> "estimating the number of mills in Saunders' district 'as near as possible', it is 1,342, what we have termed mills, though there may be two or more in a building. They having each a separate set of books, and everything of that sort to inspect, therefore we have called them mills, though the proper term would be 'occupancies.' Calculating each building as a mill would considerably reduce the number in some parts. About Huddersfield there are some mills that have 10 or 12 occupancies." [21]

With these statements made directly by the authorities who compiled the factory returns, it is safe to conclude that in these returns each factory was meant to be a separate building, not a part of the building, nor several buildings. Such definition of a factory facilitated the inspectors' work in computing the returns, but it surely distorted the facts. It is true that a textile factory in the majority of cases was a separate building in itself, but not so an iron factory, a pottery factory, or a glass factory. In the latter class of factories, each factory in general had several buildings of varying size scattered over an extensive area of land.[22] On the other hand, as the inspectors had already pointed out, one building in some parts of the country, such as Huddersfield, contained the factories of several tenants or occupiers.[23] The defect of defining a factory as a separate building is therefore obvious. Nevertheless, we are forced to accept such a definition in using the Factory Returns, the only set of data available for an analysis of the average factory size.[24]

In Table 14 the average number of persons employed per factory for England and important counties is given.[25]

Table 14. Number of persons employed per cotton factory, England, 1838.

County	Number of factories	Number of employees	Employees per factory
Lancaster	1,125	152,150	135.2
Chester	154	36,352	236.1
York	169	12,436	73.6
Derby	85	10,462	123.1
Other	65	6,777	104.3
England	1,598	218,177	136.5

According to the above table the average number of persons employed per cotton factory for England in 1838 was 136.5. Chester, with an average size of 236.1 persons employed per factory, took the lead; while Lancaster, with that of 135.2 persons, qualified poorly as a second. On the whole, the variation of average factory size among the counties was great, ranging from the highest of 236.1 persons employed per factory in Chester to the lowest of 73.6 in York. In Lancaster, the same variation as that among the counties was found among the parishes.

Table 15. Number of persons employed per cotton factory, in important parishes of Lancaster and Chester, 1838.

County	Parish	Number of factories	Number of employees	Employees per factory
Lancaster:	Manchester	182	39,363	216.3
	Oldham	213	16,645	78.1
	Bury	114	13,652	119.8
	Ashton	82	12,143	148.1
	Rochdale	117	10,513	89.9
	Blackburn	44	10,460	237.7
	Whalley	113	9,960	88.1
	Bolton	69	9,918	143.7
	Preston	35	7,161	204.6
	Wigan	37	6,137	165.9
Chester:	Stockport	86	23,772	276.4
	Mottram	37	7,138	192.9
	Prestbury	16	3,084	192.8

In the above table, Stockport in Chester had the largest average factory employing 276.4 persons; while Oldham in Lancaster had the smallest employing 78.1.[26]

In cotton spinning a common explanation for the variation of factory sizes lies in the difference in technical conditions. Jenny and mule spinning factories for high numbers were generally small in size, for here the skill of the operatives was a matter of fundamental importance, and the careful selection and supervision of the hands was essential to success. But in the case of twist-spinning conditions were different; the economies of the factory on a large scale, in respect of power, building, management, marketing and the division of labor, were so evident that twist mills soon assumed considerable proportions.[27]

Coming back now to the average size of a cotton factory for England in 1838, that given by the Factory Returns was 136.5 persons per factory. This average is hereafter referred to as average number 1, and being an arithmetic mean which by its nature is subject to the influence of extremes, is probably an overestimate. Another average, computed with a view to eliminate the influence of extremes, shows a lower sized factory, namely, 92. It is a *median* average obtained after an enumeration of all cotton factories which in 1838 were located in parishes having one or two factories per parish. The method employed is subject to criticism, inasmuch as it lays disproportionate stress upon the isolated rather than typical factories; but lacking statistics that may promise more satisfactory results, such method is perhaps justified, partly if not wholly.

Table 16. 95 cotton factories located in parishes of England having one or two factories per parish, 1838.

Persons per factory	No. factories
10 or below	2
11 to 20	5
21 to 30	3
31 to 40	8
41 to 50	10 (a)
51 to 60	9
61 to 70	4
71 to 80	2
81 to 90	4
91 to 100	4 (b)
101 to 110	3
111 to 120	6
121 to 130	3
131 to 140	2
141 to 150	1
151 to 160	5
161 to 170	3 (c)
171 to 180	0
181 to 190	3
191 to 200	2
200 or above	16 (d)
Total	95 (e)

(a) 42, 42, 44, 46, 49, 49, 49, 49, 50, 50.

(b) 92, 92, 95, 98.

(c) 161, 170, 170.

(d) 7 factories employed 201 to 300 employees.
5 „ „ 301 to 400 „
2 „ „ 401 to 500 „
2 „ „ 801 to 900 „

(e) These 95 factories employed a total of 12,331 persons, or an average of 129.8 persons per factory.

According to the above table, the median is 92, the lower quartile 49, and the upper quartile 161. The quartile coefficient of deviation, therefore, is 0.53. This average is hereafter referred to as average number 2.

Still another average, also a median, gives a size of 92 persons per factory, which is the same as the average number 2. This average is the most significant of all, as it is compiled from an enumeration of the actual sizes of 846 cotton factories in Lancaster, 32 in York, and 8 in Cumberland. The statistics from which this average is compiled—the Factory Returns of 1841 for Inspector Horner's District—are unique, as they constitute the only set of data of such nature available for our purpose. 28

Table 17. 886 cotton factories in the county of Lancaster and vicinity, 1841.

Persons per factory	No. factories
10 or below	76
11 to 20	67
21 to 30	67
31 to 40	50
41 to 50	42
51 to 60	39
61 to 70	46
71 to 80	21
81 to 90	29
91 to 100	30
101 to 110	39
111 to 120	11
121 to 130	17
131 to 140	21
141 to 150	28
151 to 160	23
161 to 170	14
171 to 180	28
181 to 190	11
191 to 200	11
201 to 300	88
301 to 400	43
401 to 500	30
501 to 600	17
601 to 700	9
701 to 800	7
801 to 900	2
901 to 1000	8
Above 1000	12
Total	886 (a)

(a) These 886 factories employed a total of 142,063 persons, or an average of 160.3 persons per factory.

According to the above table, the median is 92, the lower quartile 32, and the upper quartile 196. The quartile coefficient of deviation, therefore, is 0.72. This average is hereafter referred to as average number 3. In the following table, the two averages, namely, average number 2 and average number 3, are compared with respect to their medians, quartiles and quartile coefficients of deviation.

Average	Lower quartile	Median	Upper quartile	Quartile coefficient of deviation
no. 2	49	92	161	.53
no. 3	32	92	196	.72

In other words, the dispersion is greater for average number 3 than for average number 2, which is chiefly due to the difference in the nature of the two sets of data employed in computing the two averages. The first set of data, from which average number 2 is computed, is more homogenous, as it represents all cotton factories located in parishes having one or two factories each. The second set of data, from which average number 3 is computed, is more typical, as it represents all cotton factories located in Lancaster and neighborhood, Lancaster being the most important county in cotton manufacture, with 70 percent of the total employees. On the whole, the high degree of dispersion among the actual sizes of cotton factories in Lancaster and neighborhood confirms our earlier analysis on the variation of average facory sizes among the parishes as well as among the counties.

In Table 18, the increase of the average factory size from 1838 to 1856 is presented. The averages for 1850 and 1856, being arithmetic, are subject to the same criticism as the average for 1838.

Table 18. Persons employed per cotton factory, in England and two important counties, 1838-56.

Year	England	Lancaster	Chester
1838	136.5	135.2	236.0
1850	166.4	174.8	246.7
1856	166.7	175.0	247.3

The following table contains two returns for the years 1816 and 1833 respectively, showing the size of the cotton factories that may be called "first rate".[29] The 1816 return for Manchester and Neighborhood was compiled for the Committee on Employment of Children in Manufactories by a reliable witness,[30] while the 1833 return was an abstract from the Returns to the Lancashire Forms of Inquiry by Mr. S. Stanway of the Children's Employment Commission.[31]

Table 19. "First-rate" cotton factories in 1816 and 1833.

No. employees per factory	1816 (Manchester and Neighborhood)	1833 (Lancashire)
184 to 200	3	7
201 to 300	6 (a)	20 (e)
301 to 400	7 (b)	16 (f)
401 to 500	2	10
501 to 600	2 (c)	7
601 to 700	1	3 (g)
701 to 800	1	7
801 to 900	0	4
901 to 1,000	1	2
1,000 or above	2	7
Total	25 (d)	83 (h)

(a) 202, 210, 220, 238, 252, 288.
(b) 306, 309, 310, 320, 334, 342, 376.
(c) 549,550.
(d) These 25 factories employed a total of 10,800 persons, or an average of 432 persons per factory.
(e) 202, 203, 203, 204, 209, 210, 211, 240, 241, 245, 251, 251, 265, 269, 273, 276, 278, 284, 289, 298.
(f) 304, 312, 319, 319, 342, 343, 346, 348, 349, 350, 382, 387, 388, 391, 397, 400.
(g) 628, 632, 648.
(h) These 83 factories employed a total of 32,287 persons, or an average of 389 persons per factory.

According to the above table the medians, quartiles, and quartile coefficients of deviation for the two years are as follows.

	Lower quartile	Median	Upper quartile	Quartile co-efficient of deviation
1816	238	320	550	0.40
1833	269	397	648	0.41

The median first-rate factory, therefore, employed 320 persons in 1816, but 397 persons in 1833. The close correspondence between the quartile coefficients for the two years showed the stability in the distribution of sizes among the first-rate factories. In 1816 Messrs. George and Adam Murray was the largest factory, employing 1,215, and another one, employing 43, was the smallest; in 1833 Messrs. Birley and Kirk was the largest factory employing 1,692, and George Froggart, employing 19, was the smallest. The following abstract, from the 1833 Returns, gives details on the seven largest cotton factories for that year including all factories whose number of persons employed exceeded a thousand.[32]

Table 20. Statistics on seven largest cotton factories in Lancashire, 1833.

Name & location of factory	No. employees		Monthly wages paid	Average counts spun
1a. Birley & Kirk, Manchester	1502			25
b. ,, ,, Duckenfield	190	1692	3471	30
2. Ormrod & Hardcastle, Bolton		1576	2877	100
3. M'Connell & Co., Manchester		1545	3374	170
4a. E. & W. Bolling, Bolton	728			50
b. ,, ,, ,,	141			43
c. ,, ,, ,,	214			57
d. ,, ,, ,,	273	1356	2447	85
5 T. Houldsworth, Manchester		1201	2457	180
6. Joseph Horsefield, Hyde		1183	2521	39
7. Thomas Ashton, Hyde		1149	2888	19

Note : Factories la, 2, 6, 7 were spinning-and-weaving factories, the rest being spinning factories.

Of these seven huge factories the factory of Messrs. Birley and Kirk was, in the language of Mr. Tufnell of the Children's Employment Commission, "the largest in Manchester" in 1833. In this factory in 1841 "1,600 are employed daily, whose weekly wages amount to £900." The establishment had "thousands of power looms (of which one room alone contains upwards of 600)" and "80,000 spindles which daily twirl in these mills, under a moving power equivalent to 397 horses, and which convert annually 4,000,000 pounds of raw material into cotton thread".[33] Messrs. M'Connel and Kennedy, first established in 1791, had a remarkable growth in the course of 42 years, employing the following number of persons in different years.[34]

Table 21. Growth of the cotton factory of Messrs. M'Connel and Kennedy, 1802-1833.

Year	Number of persons employed
1802	312
1816	1,020
1819	1,150
1833	1,545

Mr. Thomas Ashton's factory at Hyde, in its extensive scheme of welfare work for his employees, rivalled Robert Owen's at New Lanark.[35] In addition to these seven huge factories, others of equal and even greater magnitude may also be mentioned. The earliest and largest factory on record was probably that of the famous statesman and reformer, Sir Robert Peel. He erected in 1773 his first factory in partnership with Mr. Yates at Bury, whose seventeen-year old daughter he married at the age of thirty-three. Other plants were later on started elsewhere, especially at Tamworth, until in 1803 it was said that he had no less than 15,000 persons in his employ, and in a single year he paid more than £40,000 to the Excise Office by way of duty on his printed calicoes.[36] In 1844, Mr. Faucher visited in Manchester the factory of Greg & Co. This factory, like that of Strutt, had given a member to Parliament and held the first rank among the manufacturers. It consumed annually nearly 4,000,000 pounds weight of cotton, possessed five plants with 4,000 power looms, and employed more than 2,000 persons at Bury, Borrington, Caton, Lancaster and Wilmslow.[37] The welfare work carried on here could be compared with that of Mr. Ashton's at Hyde, and to quote again from Faucher, "Mr. Greg has commenced the science, which in my humble opinion, I should call moral economy of manufactures."[38] Mr. Whittaker of Oldham, who commenced the cotton business with the possession of a small carding factory moved by horse power before 1808, had in 1840 a factory employing about 1,000 hands. In 1847, the same factory employed about 1,500 hands, produced 40,000 pounds of yarn and 8,000 pieces of cloth weekly, and paid a monthly wage of £2,500.[39]

Employment of Power Machinery.

The employment of power machinery is the secondary feature of a factory as compared with the concentration of labor. In the following sketch, one section is devoted to each of the three elements that constitute the power machinery in cotton factories, namely, motive power, power looms, and spindles.

Motive Power. While steam power was first applied to the cotton manufacture in 1785 for weaving, water was used in the early days, commencing with the invention of Arkwright's water frame in 1769. In those days, it was not uncommon for a cotton manufacturer to establish his factory in rural districts where water power was available, and to build, in addition to the factory, houses for the workpeople to live in and stay to work permanently.[40] The disadvantages of such an arrangement were obvious. Water power, in the first place, was irregular; it varied not only every hour in the day, but every day in the month and every month in the year. "When the stream was full, work was brisk (we should have called it excessive); when it was dry the factory hands were employed on the lands, in hay-making, or other like occupations. Thus the operatives were both farm laborers as well as factory workers."[41] In the second place, water power was local, and as noted above, it necessitated, if situated in rural districts, the provision of means necessary to the daily life of a laboring population in a secluded region; it also increased the cost for inland transportation of raw materials and finished products. Finally, water power was not expansible, it was inelastic. Even when the manufacturer offset this disadvantage by the additional instalment of a steam engine, he was still not fully protected against unforeseen eventualities of natural as well as seasonal fluctuations. "In four years out of five, perhaps, if he wanted 30 horse power he might have enough; but a period might arise when he wanted 50 horse power, and he is then liable to great inconvenience from the engine and water power together being unable to propel his mill; if he was cut off from working for lost time, both himself and his hands would suffer thereby."[42] With these disadvantages, the attempt to substitute water power by steam grew in the course of a few decades to a very large extent. This substitution tended to destroy whatever advantages there were in the factory villages along the streams, and rendered possible the gradual concentration of population in factory towns. In Table 22, the steady decrease of water power as compared with steam is shown; its eventual disappearance is foretold in the small percentage for England in 1856, viz. 7.6 percent.

Table 22. Motive power in cotton factories, England, 1838-56 (in horse power)

Year	Steam power		Water power	
	Amount	Percentage	Amount	Percentage
1838	40,589	81.1	9,477.5	18.9
1850	62,940	88.5	8,182.0	11.5
1856	79,806	92.4	6,521.0	7.6

Another table is given below to show the increase in the amount of motive power per cotton factory from 1838 to 1856.

Table 23. Motive power per cotton factory, England, 1838-56 (in horse power)

Year	Total amount of motive power	Total number of factories	Motive power per factory
1838	50,066.5	1,598	31.3
1850	71,122.0	1,753	40.6
1856	86,327.0	2,045	42.2

Spindles. The growth in the number of spindles is a matter of great interest, but in this as in other problems of similar nature, statistics are deplorably wanting. We may share our common regret with Ure over the failure of the factory inspectors to report the number of spindles in the 1835 and 1838 Returns, for "it would exhibit the condition of the business more accurately than any other document."[43] Table 24, although compiled from various sources, is worth our attention in view of its consistency in showing the increase in the number of spindles from 1781 to 1856.[44]

Table. 24. Number of spindles employed in cotton maunfacture, England and Great Britain, 1781-1856.

Year	Number of spindles	
	Great Britain	England
1781	50,000 (a)	
1787	1,951,000 (b)	
1812	5,100,000 (c)	
1832	9,333,000	8,000,000 (d)
1835	12,000,000 (e)	
1850	20,977,017	19,173,969 (f)
1856	28,010,000	25,797,000 (f)

The distribution of the number of spindles employed in the cotton factories among the counties in England in 1850 and 1856 is shown in the following table. Lancaster, again, had the largest proportion of the total number, 72.2 percent in 1850 and 78.4 percent in 1856.

Table 25. County distribution of spindles in Cotton factories, England, 1850-56

County	1850	1856
Lancaster	13,955,497	20,227,197
Chester	2,294,703	2,602,376
York	1,943,897	1,813,706
Derby	616,199	737,502
Other	363,673	416,219
England	19,173,969	25,797,000

In a study of the number of spindles employed per cotton factory, the interesting feature has been the larger size of the spinning-and-weaving factory rather than the spinning factory in the first half of the 19th century; while in the second half, the spinning factory began to gain size over the spinning-and-weaving factory, and a movement for the separation of spinning and weaving was started again to reap the economy resulting from specialization. In Table 26, this feature is clearly shown in the changing number of spindles employed per factory from 1831-2 to 1856. For 1850 and 1856 we have a complete enumeration of all spindles in the cotton factories of England, while for 1831-2 the only statistics available are taken from 61 spinning (and spinning-and-weaving) factories at Stockport. [45]

Table 26. Number of spindles per cotton spinning (or spinning-and-weaving) factory in England, 1831-56

Type of factory	No. factories	No. spindles	Spindles per factory
Spinning			
1831-2 (Stockport)	29	141,757	4,888
1850	762	8,685,392	11,398
1856	909	15,239,201	16,765
Spinning-and-weaving			
1831-2 (Stockport)	32	447,464	13,983
1850	540	10,055,410	18,621
1856	623	10,557,799	16,947

The change in size as shown in the above table can be brought out more clearly in the following arrangement.

Year	Spindles per spinning factory	Spindles per spinning-and-weaving factory
1831-2	4,888	13,983
1850	11,398	18,621
1856	16,765	16,947

The increase in the number of spindles per factory was accompanied by a similar increase in the number of spindles per mule. The mule used to be a small machine containing but 30 spindles in 1786; in 1887, it contained over a thousand. In the course of a century, therefore, the number of spindles per mule increased thirty-six-fold.[46]

Table 27. Increase in number of spindles per mule, England, 1786-1887

Year	Spindles per mule
1786	30
1800	200
1822	220
1833	350
1856	480
1873	750
1887	1,080

Power looms. Although power loom was invented early in 1785 by Dr. Cartwright, yet its introduction into general use was strikingly slow.[47] In 1790-1, the inventor established in agreement with Messrs. Grimshaw a factory at Manchester to put his new machine into operation, but opposition from the handloom weavers was aroused instantly. The factory had 24 looms made, 12 of which were set to work at weaving check; it worked exceedingly well, and wrought the cloth for half of the wages which were then paid the handloom weavers. The latter, in consequence, threatened to destroy the factory, and sent letters to the following effect:

> "Sirs, we have sworn together to destroy your factory, if we dye for it, and we have your lifes for ruining our trade; and, if you go on, you know the certainty, which comes unknown to my companions."

The factory was set on fire, it was contended, by the handloom weavers. Because of such threats and acts of destruction, the invention was not used from 1792 to 1800.[48] Another cause for the slow introduction of power loom into general use was its unprofitableness. Best hands were hard to secure; those that were secured generally came from a lower stock of labor, and being unaccustomed to the use of machinery, were inefficient. New machinery was oftentimes imperfect.[49] Before the invention of the "dressing machine" by Messrs. Radcliffe and Ross of Stockport in 1803 and 1804, by which the warp was prepared previously to its being placed in the loom, there existed the continual necessity for stopping the loom, in order to "dress" the warp as it unrolled from the beam.[50] On the other hand, we may well remember a leading manufacturer's remark on the advantage of power loom. "The advantage of the power loom," said Mr. Watkin, "is in being able, and that is a very great advantage, to produce a certain quantity of cloth in a certain time, so that you may with confidence make your contract complete, and also that you keep a control over the manufacturing material; those are the two great considerations which have built up the power loom."[51]

In Table 28, the increase in the number of cotton power looms from 1813 to 1856 is shown.[52]

Table 28. Increase in number of cotton power looms, England, 1813-1856

Year	No. cotton power looms
1813	2,064 (a)
1820	12,150
1829	45,500
1833	85,000
1835	90,679
1850	223,626
1856	275,590

In 1835, the position of the power loom weaving was fully established in cotton, not yet sensibly felt in linen, and in the experimental stage in woolen, worsted and silk manufactures. In the same year, out of 97,560 power looms in England, 90,679, or 93 percent of the total, were employed in cotton factories.[53]

Table 29. Number of power looms in textile factories, England, 1835.

Kind of factory	No. Looms
Cotton	90,679
Woolen	2,245
Worsted	2,856
Silk	1,714
Linen	41
Mixed	25
Total	97,560

Table 30 shows the distribution of power looms in cotton factories among the counties in England, from 1835 to 1856.

Table 30. County distribution of cotton power looms, England, 1835-56

County	1835	1850	1856
Lancaster	61,176	176,947	224,353
Chester	22,491	29,198	31,438
Other	7,012	17,481	19,799
England	90,679	223,626	275,590

Regarding the number of power looms per cotton factory, statistics for 1835 are not complete. Out of 90,679 cotton power looms in England, returns for 187 cotton power loom factories in Lancaster employing 43,021 power looms were made to the factory inspectors; some of these factories might have been for weaving only, while others for weaving and spinning together. The arithmetic average, then, would be 230 power looms per factory; the median average, 144. The distribution of the 187 factories of varying sizes is shown in Table 31.[54]

Table 31. Sizes of 187 cotton power loom factories in Lancaster, 1835.

Power looms per factory	No. factories
50 or below	27
51 to 100	35
101 to 200	55
201 to 300	21
301 to 400	9
401 to 500	18
501 to 600	7
601 to 700	4
701 to 800	7
801 to 900	2
901 to 1,000	2
Total	187

The number of power looms per cotton factory for 1850 and 1856—whether spinning or spinning-and-weaving—is given in the following table.

Table 32. Number of power looms per cotton weaving (or spinning-and-weaving) factory, England, 1850-56

Type of factory	No. factories	No. power looms	Power looms per factory
Weaving			
1850	229	36,544	159.6
1856	419	65,880	157.2
Spinning-and-weaving			
1850	540	184,816	342.3
1856	623	209,609	336.5

The change in size as shown in the above table can be brought out more clearly in the following arrangement.

Year	Power looms per weaving factory	Power looms per spinning-and-weaving factory
1850	159.6	342.3
1856	157.2	336.5

The number of power looms per weaving or spinning-and-weaving factory was about the same in 1850 as that in 1856. The striking change, however, lay in the rapid increase in the number of weaving as well as spinning-and-weaving factories from 1850 to 1856.

So far we have dealt with the three elements in the employment of power driven machinery, viz. motive power, spindles and power looms. We may now briefly summarize a few effects of it, viz., displacement of labor, presence of overhead cost, and increase of productivity.

The displacement of labor, which Bücher called the latest phase of the division of labor, was more keenly felt in cotton than in any other textile manufacture. Here it is true that "the mechanism undertakes only separate movements that until then have fallen to the human hand; and in the business installing the machine the sole initial change is generally the transfer of the workman, who formerly performed those muscular motions, to attendance upon the machine, which demands from him other muscular motions."[55] A new class of labor, primarily unskilled and requiring little strength, were called into existence to perform the new type of work. A burning issue of the epoch has arisen—the coming of the child and woman labor.

In assigning the displacement of labor as the chief cause for the introduction of child and woman labor into the cotton factories, we are not overlooking other causes, of which the indisposition of adult male labor to enter factories has been previously discussed. To the employer, the inexpensiveness in terms of money wages was a powerful motive for the employment of child and woman labor, despite the fact that cheap labor might not always be synonymous with efficient labor. Table 33, which is compiled from the statistics furnished by the fartories in Lancaster to the Children's Employment Commission in 1833, supports this conclusion beyond all uncertainty.[56]

Table 33. Wages in cotton factories, Lancaster, 1833.

	Males			Females		
Age	No. employed	Average weekly wage		No. employed	Average weekly wage	
		s.	d.		s.	d.
Below 11	246	2	3½	155	2	4¾
11 to 16	1,169	4	1¾	1,123	4	3
16 to 21	736	10	2½	1,240	7	3½
21 to 26	612	17	2½	780	8	5
26 to 31	355	20	4½	295	8	7¾
31 to 36	215	22	8½	100	8	9½
36 to 41	168	21	7¼	81	9	8¼
41 to 46	98	20	3½	38	9	3½
46 to 51	88	16	7¼	23	8	10
51 to 56	41	16	4	4	8	4½
56 to 61	28	13	6½	3	6	4
61 to 66	8	13	7	1	6	0
66 to 71	4	10	10	1	6	0
71 to 76	1	18	0	–	–	–
76 to 81	1	8	8	–	–	–
Total	3,770			3,844		

From 1835 to 1895 the proportion of the adult male labor in the cotton factories in England was never lower than 24.1%, nor higher than 28.8%.[57] The change in the proportion of child and woman labor in England is shown in Tables 34 and 35 compiled directly from the factory returns for the respective years.[58]

Table 34. Child labor in cotton factories, England and Lancaster, 1838-47 (in percentage)

Year	England	Lancaster
1838	42.08	42.96
1847	35.84	35.73

Table 35. Woman labor in cotton factories, England and Lancaster, 1838-56 (in percentage)

Year	England	Lancaster
1838	54.23	54.06
1847	55.49	55.47
1850	54.88	55.08
1856	56.52	56.64

The decrease in the proportion of child labor since 1838 was a phenomenon due to the passage and enforcement of the Factory Acts, and the changed social and economic conditions that discouraged the employment of child labor. The proportion of woman labor from 1838 to 1856, however, remained fairly constant.

The employment of power-driven machinery in cotton factories brought to the front for the first time the problem of overhead cost,[59] which made it vastly more profitable to continue "the last hour of each day's work."[60] According to one calculation of the time, "in the case of a manufacturer who invests £100,000 to £80,000 in his mill and machinery, and £20,000 in raw materials and wages, if (prices remaining the same) the factory could be kept at work 13 hours instead of 11½, by an addition of about £2,600 to the circulating capital, the net profit would be more than doubled."[61] According to another, a factory containing 52,000 spindles, which with the machinery had cost £40,000 and which required a floating capital of £12,000, incurred fixed expenses equal to £121 16s. per week, or £6,334 per year. A factory of 52,000 spindles produced 12,000 lbs. of yarn per week, which with the fixed expenses of £121 16s. gave a total of £413 16s., and which sum was equal to about 8¼d. per lb. for the yarn produced; but when the demand slackened, and the manufacturer was obliged to work short time, say three days a week, the weekly expenses were raised to £267 16s. for 6,000 lbs. of yarn, and this advanced the average outlay each pound of yarn, from 8¼d. to 10¾d.; the loss consequent upon working short time would thus be £60 18s. per week, or £3,167 8s. per annum.[62] The significance of overhead cost in the private business economy of a factory is clear enough from a study of these two calculations; it explains in a logical way the advantages as well as the disadvantages of the factory system as compared with the merchant employer system. Under the former, it is a cardinal principle to keep up the operations of a factory so long as there is a demand for the product at a paying price. In fact, production was often increased, when the selling price declined. If the manufacturer found that the temporay loss would be less by producing to the utmost of his power, than by working short time, he would of course adopt that plan, which was generally the case with the Lancaster manufacturers. Inspector Horner said that

> "a factory may be working full times and yet the occupier of it may be losing. It may be a struggle to keep himself afloat. By the custom of the trade, he can buy his cotton with a credit of three or even four months, and he can sell his yarn at 14 days' credit. Now as the cotton may be converted into marketable yarn in a period varying from 14 days to a month, the means of meeting his acceptances for the raw material will be provided. How long a necessitous man might carry on this losing trade would depend upon the strength of his credit, and his own hope of better times."[63]

That the employment of power-driven machinery brings with it an increase in productivity is an established fact. Farey, in his "Treatise on the Steam Engine" published in 1827, calculated that in one cotton factory where 50,000 spindles with 100 horse power were employed, 750 people were enough to attend the operations that might take 200,000 in a factory without machinery, in order to produce 62,500 miles of thread in every day of twelve hours.[64] Baines remarked in 1835 that "a very good hand weaver twenty-five or thirty years of age will weave two pieces of shirting (of specified quality and dimension) per week. In 1823 a steam loom weaver about fifteen years of age attending two looms could weave nine similar pieces in a week. In 1826 a steam loom weaver about fifteen attending to four looms could weave twelve similar pieces in a week, some fifteen. In 1833 a steam loom weaver from fifteen to twenty, assisted by a girl of twelve, attending to four looms, could weave eighteen pieces."[65] Tufnell of the Children's Employment Commission, in his computation on the weekly production of yarn per spindle for 1812 and 1830, which Schulze-Gavernitz extended up to 1890, showed a two-fold to four-fold increase for yarns of different counts.[66]

Table 36. Increase in weekly production of yarn per spindle, in cotton manufacture, England, 1812-90.

Year	Weekly working hours	Weekly production per spindle (in hanks)		
		40's Twist	60's Twist	200's Weft
1812	74	12	9 (a)	4.5
1830	69-70	16.5	15	5.4
1890	56 1/2	28-30	23.5	17-18

This three-fold effect of the employment of power driven machinery—displacement of labor, presence of overhead cost, and increase in productivity—by no means exhaust the list. We may add, for instance, its bearing on the social welfare of the community which has produced a vast amount of literature of a controversial nature, official as well as unofficial.

Capital.

The question of capital investment in cotton factories before 1840 has attracted the attention of leading authorities including Kennedy, Burns, M'Culloch, Stanway, and Baines. Their estimates, however, were mainly intelligent approximations. Granting that they were true to the facts, they dealt in toto with the cotton manufacture which, in their definition, included lace-making and embroidery, hosiery, and calico printing in addition to spinning and weaving. They dealt, furthermore, with the cotton manufacture as a whole, no attempt being made to separate the factory from the other branch.[67] In view of these limitations, it would be hardly possible for us to derive deductions from their conclusions. However, on authority, a manufacturing firm, gave an independent estimate for the information of the Children's Employment Commission in 1833. This manufacturing firm was the famous Messrs. Greg & Co., which we had already touched upon in a previous paragraph.[68] These gentlemen stated that they were the owners of five plants for spinning and weaving in all branches; that they employed 450 horse power in steam and water wheels, and 2,000 workpeople, and consumed 3,586,000 pounds of cotton yearly, or 1.29% of the total consumption of Great Britain. They spun into cloth 3,260,000 lbs., weaving by power loom 2,273,000 lbs. of that quantity. With regard to four of these plants, they made the following estimate:—

Plant number	Sunk capital £	Floating capital £	Wages paid £	Yarn spun (lbs.)	Yarn woven into cloth (lbs.)	Value of produce £
1	10,000	7,000	2,600	363,000		16,000
2	26,000	18,000	6,800	877,000		38,000
3	25,000	20,000	14,000	733,000	733,000	48,000
4	37,000	25,000	10,660	740,000	740,000	55,000
Total	98,000	70,000	34,060	2,713,000	1,473,000	157,000

Taking this as their basis, Messrs. Greg calculated that in 1832 there were employed in the cotton spinning and weaving factories of Great Britain:—69

Sunk capital	Floating capital	Wages paid	Persons employed
£10,000,000	£5,000,000	£4,000,000	160,000

Taking the above estimate as our basis, we may proceed to calculate for the 218,177 persons employed in the cotton spinning and weaving factories in England in 1838 as Messss. Greg did for the 160,000 persons in Great Britain in 1832, although we are well aware of the fact that with the progress of time, the amount of money expended per employee under some of the above headings, such as "sunk capital," might have decreased, while that under others such as "wages paid" might have increased on account of the greater efficiency resulting from the improved machinery and other economic changes.

Sunk capital	Floating capital	Wages paid	Persons employed
£14,454,000	£6,818,000	£5,454,000	218,177

If we divide the different amounts under above headings among the 1,598 cotton spinning and weaving factories in England in 1838, we get data of the same per factory.

Sunk capital	Floating capital	Wages paid	Persons employed
£9,045	£4,267	£3,413	136.5

This composite average per cotton factory is the only estimate yet available. In the following paragraphs, we quote from the contemporary authorities calculations of capital investment for several first-rate cotton factories of varying size.

Spinning factory: Mr. Symons, a member of the Assistant Handloom Weavers' Commission, published valuable statistics on capital investment as furnished by a Glasgow manufacturer in 1838.[70] He gave minute figures for two spinning factories with slightly varying amount of sunk capital and produce of cotton yarn.

Factory A:—"Probable expense of erecting a spinning cotton factory, with hand mules, calculated to produce no. 40 of a fair quality, would be £23,000. If patent self actors, £2,000 additional. The produce annually, taking the present prices of cotton, and the rates at which the yarn could be sold, would be about £25,000."

Factory B:—"Cost of a cotton spinning factory of 10,000 throstles, calculated to produce a fair quality of no. 24, would be almost £20,000. Taking present value of produce, the amount annually would be £23,000."

The cost of the annual produce for both factories was further subdivided as follows:

	Factory A	Factory B
Interest, depreciation charges	£	£
say 10%	2,300	2,000
Raw cotton	14,000	13,300
Steam power, upkeep & repair of machinery	1,800	2,500
Wages	5,400	3,800
Profits	1,500	1,400
Total	25,000	23,000

Factory A, spinning higher numbers, paid higher interest and depreciation charges, higher price for raw cotton and higher wages for labor. Factory B, using throstles rather than mules, spent more on power. Each factory represented a distinct type, one for mules and the other for throstles, each having the same amount of weekly produce at 10,000 pounds. Another factory, given by Holland Hoole and quoted by Ure, seems to represent a combination of the above two types of factories, using both mules and throstles, and spinning the same numbers of yarns as those spun by Symons' two factories. This factory was built on fire-proof principle of 24 windows long (exclusive of engine house), 9 feet bays, 42 feet wide, and 7 stories high, with a steam engine of 100 horse power, to turn 24,000 mule spindles spinning no. 40 and 12,000 throstle spindles spinning no. 30; with all necessary preparations for the same. Its sunk capital amounted to £47,134; its weekly produce of yarn, 19,800 pounds.[71] In comparison with the two spinning factories as cited by Symons, Hoole's factory was twice as big in sunk capital and weekly produce.

	Sunk capital	Weekly produce
Factory A	£23,000	10,000 lbs.
Factory B	£20,000	10,000 lbs.
Hoole's factory	£47,134	19,800 lbs.

Weaving factory: We quote here from Symons a Glasgow weaving factory, which did not spin itself but purchased the yarn from the market[72] Thus, "expense of erecting a power loom factory of 500 looms, calculated to weave a good fabric of calico or shirting, such as is generally made in Glasgow, would be about £18,000. Annual produce, say 150,000 pieces of 24 yards at 6/," would be £45,000, which cost:

Interest, & depreciation charges, say 10%	£1,800
Steam power, upkeep & repair of machinery	2,000
Yarns and flax	32,000
Wages	7,500
Profits	1,700
Total	£45,000

Spining-and-weaving factory. Ure gave in 1836 an analytical estimate of the equipment and cost of two of the most recent and complete cotton spinning-and-weaving factories in England, a condensed statement of which is given in the following table.[73]

	Factory A	Factory B
Amount of motive power (H.P.)	250	220
Number of spindles	45,860	40,000
Number of power looms	1,100	1,280
Amount of sunk capital	£85,000	£88,000

If factory B increased its number of power looms to 1,480, the sunk capital would then be raised to £100,000.

Ure's two factories were undoubtedly first-rate, employing a sunk capital of £85,000 and £88,000 respectively. Holland Hoole, in his "Letter to Lord Althorp, in defense of Cotton Factories" (p.8), said that "a cotton factory, upon the fire proof principle, adapted for the employment of 1,000 persons, cannot be built, filled with machinery, and furnished with steam engines, gas works, for a less sum than £100,-000."[74] Dodd gave the same opinion in observing that "a first-rate cotton factory, with its machinery, costs very little short of £100,000."[75]

Statistical Summary

Table 37. Factory statistics on cotton manufacture in England, 1838-56[76]

Year	No. factories	No. employees	Amt. motive power (in horse power)	No. spindles	No. power looms
1838	1,598	218,177	50,066.5	....	......
1847	.....	276,797	....	....	......
1850	1,753	291,662	71,122.0	19,173,969	223,626
1856	2,045	340,944	86,327.0	25,797,000	275,590

If we take 1838 statistics as a base of 100 percent, we may then express the above table in relative terms below.

Year	No. factories	No. employees	Amt. motive power (in horse power)
1838	100.0	100.0	100.0
1847	....	126.9	....
1850	109.7	133.7	142.1
1856	128.0	156.3	172.4

The more rapid increase since 1838 in the percentage of employees over that of factories indicated the expansion in the average factory size as measured by employees; the same may be said of the more rapid increase in the percentage of motive power over that of factories. We may conclude, therefore, that the average cotton factory in 1856 was a great deal larger than that in 1838 with regard to the number of persons employed, or the amount of motive power. Unfortunately, the Factory Returns for 1838 gave no statistics for spindles and power looms, and thus left us without basis of comparison with the Returns of later years.

Handloom Factories.

Before we conclude our study on the power factories, hand factories in the weaving branch of the cotton manufacture may be discussed in brief. No statistics as to their extent were available. As far as we can judge, such factories were exceptions, and could not have been established for the weaving of plain rather than fancy cloth. For plain weaving, the advantage of power loom over handloom factories was an established fact in 1840. Were there any handloom factory for plain weaving still in existence, it would have been a makeshift arrangement. In such case the manufacturer had perhaps owned too many handlooms which he could not easily dispose of in the market without incurring considerable loss from depreciation, and as a compromise measure, he would rather choose to work them temporarily in the factory at a smaller loss. As these handlooms began to depreciate and wear out, he would most likely install power looms and expand his factory under a new arrangement.

Handloom factories, however, preceded power loom factories by many centuries. With regard to the woolen manufacture, Cunningham observed that "instances of the organization of hand weaving in factories had occurred as early as the fifteeth century, and it is hopeless to try and obtain information as to the gradual extension of that system."[77] But the practice began to spread around 1840. In the latter year, Mr. Hickson of the Handloom Weavers' Commission observed that

> "the cotton weavers, in most cases, work at home; but the practice is beginning to extend itself of assembling them in factories. There are handloom factories, as well as power loom factories. In large manufacturing towns, a saving of time is regarded as a saving of money. One thousand pounds capital, if it can be returned four times in the year, is equal to a capital of £4,000 returned once; and the interest on £3,000 is the saving effected. Hence the anxiety of every good man of business to despatch his orders quick, and hence the urgency of merchants, when writing to his manufacturer, to ship without delay. In fact, promptitude of execution is often a more important consideration than price. A merchant, not limited by his foreign correspondent, but left to his own discretion, will give his orders to the manufacturers who, on a given day and month, will engage to have his goods on board a ship in the export docks, and will disregard the offer of another manufacturer less punctual, and more dilatory in the conduct of his business, although cheaper, perhaps, by 5 percent. On this account factory labor is much more advantageous to the manufacturer than domestic labor. The domestic weaver is apt to be irregular in his habits, because he does not work under the eye of a master. At any moment the domestic weaver can throw down his shuttle, and convert the rest of the day into a holiday, or busy himself with some more profitable task; but the factory weaver works under superintendance; if absent a day, without sufficient cause, he is dismissed, and his place supplied by one of greater power of application. One hundred webs, therefore, in a factory of handloom weavers, would be finished, even in Manchester, in the time in which 50 would not be finished by an equal number of domestic weavers. But in Ireland the disparity is much more striking. I was assured by Mr. M'Cauley of Belfast, that it would be necessary for him to employ 400 country weavers to get him through the same quantity of work in a given time which he could produce from 100 handlooms employed in his factory, under his immediate superintendence."[78]

Besides, the handloom factory system

> "is also favorable to a large manufacturer, in protecting him, to some extent, against the embezzlement of yarn. His property is safe in his own possession, and he runs no risk of the work being taken out of the loom to be sold or pawned by a dishonest weaver. The subjects of wrangling and dispute between his foreman and the men are also less numerous upon the factory than upon the outdoor system. The men have not to lose hours and days in dancing attendance upon the foreman's leisure; and the daily inspection of the master enables him to see that his directions are understood and followed by all parties."[79]

The above account was a first-hand summary of the advantages of the handloom factory system over the merchant employer system as it was written by an authority who had the opportunity of making his observations only after an extensive field survey of the conditions of the handloom weavers throughout the United Kingdom Quick turnover of capital, prompt execution of the customer's order, great regularity of work, protection against embezzlement, and absence of dispute between foremen and workmen—these were the distinct advantages of handloom factory system. In addition, we may mention the saving of middleman's expense and the uniformity in the quality of cloth woven.

Handloom Weaving: Extent.

The number of handloom weavers in England around 1840 had been variously estimated at about 200,000, including all those who were either fully or partly employed in the occupation.[80] Some of these handloom weavers were scattered throughout the whole of England, while a large majority of them were concentrated in the counties of Lancaster and Chester. "Manchester and its neighborhood," reported the Assistant Handloom Weavers' Commission in 1838, "contains the greatest number of handloom weavers . . . in any part of the district. . . . Their members were variously estimated by intelligent employer and others at from 8,000 to 10,000. They are employed mainly in the weaving of the cotton, silk, and mixed goods."[81] In 1834 there were at Bolton about 7,000 to 8,000 handloom weavers;[82] while other estimate, referring also to Bolton but probably including the neighborhood, gave a return of 23,000.[83] At Preston in the same year, there were about 3,000 handloom weavers in the town and a great many in the country,—about 10,000.[84]

Master Craftsman System.

These handloom weavers, numbering about 200,000 in England around 1840, were organized chiefly under the merchant employer system, and rarely under the master craftsman system. The master craftsman system was a growth of the existing economic condition, which, in 1834, was described by a Bolton manufacturer, Mr. Halliwell. "There is a facility to become a master in our business, which has a direct tendency to produce this pernicious effect [i.e. cut-throat competition]. If a man can purchase a winding machine and a warping mill, and can get credit for a skip of yarn, he can get into motion as a master; and with respect to embezzlement, perhaps he may weft his work with embezzled stuff."[85] These embezzled articles, according to Needham, a Bolton weaver, "have, in many instances, caused a reduction of wages upon the whole of the stock in manufactured goods. There are many persons that manufacture on their own looms; instead of working for a warehouse, they have three or four weavers. These persons purchase their materials considerably lower, say 6d. a pound lower, than a regular master; and the consequence is, that the 60 reed, which weighs 5 pounds, would be half a-crown cheaper by the embezzled material than the regular manufacturers. Those men have come to Bolton to sell those pieces; they wanted to have their men back with them, and when they could not sell their pieces before evening, they would then sacrifice them at 2s. less."[86]

Merchant Employer System.

Social classes. The merchant employer system in the handloom weaving branch of the cotton manufacture could be traced back to the year 1740,[87] immediately after the invention of the warping mill.[88] Under this system,

> "the most prominent functionary ... and a pivot of the system was the Manchester merchant. He warehoused goods received from the weavers, and distributed them for export or consumption in the country. ... The other relationship of the merchant to be explained is that to the weavers. From some of these he simply bought cloth, the weavers having provided themselves with warps and cotton. To others he gave out warps and cotton, paying merely for workmanship. At first the weavers prepared the warp for the loom by the system of peg-warping ... but after the invention of the warping mill the merchants as a rule gave out warps ready prepared for insertion in the looms. As the weavers were scattered throughout the country many Manchester merchants put out their work through local agents."[89]

This account of the merchant employer system for the last part of the eighteenth century applied, with slight changes, to the conditions around 1840, at which date it was said that generally "the weaver finds the loom and the shuttle, the manufacturer finds the warp, weft, heels and the reed through which the web is woven."[90] The manufacturer, i.e. the merchant employer, purchased the yarn from the market or spun it at his own factory, and then gave it out to the weavers. His business, according to George Smith, a large Manchester merchant employer in handloom calico weaving, did not require much capital, probably from £500 to £1,000, and yet a large number of weavers could be kept under his employ. Smith himself had 1,300 to 1,400 weavers,[91] while Dixon, of the firm of Messrs. Peter Dixon and Sons of Carlisle, had an even larger number scattered throughout the United Kingdom. In 1840 Dixon was employing 3,571 handloom weavers, namely, 2,389 in England, 599 in Scotland, and 583 in Ireland, of whom 11/18 were men, 2/18 women, 3/18 boys, and 2/18 girls.[92]

Under the merchant employer system, there were two classes of weavers, viz. those residing in the town and those living away in the rural districts. "It was upon the latter," the Committee on Emigration (1826-7) reported, "that the distresses of the times fell with peculiar severity, though both classes were reduced to deplorable poverty. They were engaged chiefly on the rough work, and they were the first to suffer when work was slack."[93] They were, to borrow a simile from the classical economics, on the margin of production; when prices fell and profits declined, they were dropped first from the list of employment.

Outwork. Outwork included home work or shop work. In whichever case, the working conditions for the handloom weavers, especially those residing in the towns, were miserable beyond description in 1840, and no greater instances of human sufferings could surpass those of the handloom cotton weavers. Commissioner Hickson reported that

> "the great majority of handloom cotton weavers work in cellars, sufficiently light to enable them to throw the shuttle, but cheerless because seldom visited by the sun. The reason cellars are chosen is, that cotton requires to be woven damp. The air, therefore, must be cool and moist, instead of warm and dry. Unhappily, the medium which might be preserved without injury to the constitution, and which is preserved in the best

> power-loom factories, the impoverished handloom cotton weavers are obliged often to disregard. I have seen them working in cellars dug out of an undrained swamp; the streets formed by their houses without sewers, and flooded with rain; the water therefore running down the bare walls of the cellars, and rendering them unfit for the abode of dogs and rats. The descent to these cellars is usually by a broken step ladder. The floor is but seldom boarded or paved; a proper place for coals and ashes, but less fitted for a workshop than even an Irish hovel, because underground."94

Under such conditions, Kohl, a keen German observer, wondered

> "how any of these can continue to earn a penny, in the unequal competition with the immense machinery and capital employed in the great factories. . . . These work from morning till night, in close places, with scanty nourishment and clothing, and suffering grievously from privation and want. The earnings of the poor hand weavers are probably, by this time, reduced to their minimum, and they will soon be forced to quit the field. I went through a number of their cellars, and I found many in which the loom was already at rest. Before one such, sat in unwilling idleness, the very image of silent despair, a poor half-starved weaver, who had not a penny to buy cotton, and who told me that he had been in vain seeking employment for eight days. On his table was a small plate of cold watery potatoes, saved from the scanty sustenance of the preceeding day, to satisfy the hunger of the next."95

Abuses.

The abuses of the merchant employer system in cotton handloom weaving in 1840 included wage abatement and embezzlement.

Wage abatement. In 1840 wage abatements were quite commonly practiced by the merchant employers upon their outworking weavers in cotton and silk manufactures in the county of Lancaster and vicinity. These abatements were originally designed to compensate the merchant employers for the poor work done by the weavers. but in the majority of cases they were arbitrarily exacted. Thomas Brogan, a weaver of ginghams and plain cotton goods, complained that "the abatements are unfair. Occasionally, of course, a weaver may take home bad work, and deserve abatement. There is no remedy, at least the remedy is worse than the disease."96 Other witnesses confirmed Brogan's evidence, and generally in such cases, the weavers, despite the Arbitration Act, 5 George IV, c.96, were afraid of resorting to redress. Commissioner Muggeridge observed that

> "the influences, inseparable from the relative position of master and servant, which give to the former the means and power of inflicting oftentimes irreparable injury on the refractory or discontented workman, by depriving him of further employment, will always be found to exercise, in effect, prevention from annoyance of this kind, except under circumstances of peculiar aggravation; or in cases where the latter has the clearest conviction of right on his side . . . . The subject matter of dispute is almost invariably but one, two, or at most a few shillings, the attempted recovery of which usually involves the loss of more of the workman's time than is adequate to the sum disputed. Even when successful, he is so, in 90 cases out of 100, with the moral ceratinty of his success involving with it the loss of future employment. He has also to forego, pending the time preceding the hearing and adjudicating on his case, the receipt even of the undisputed portion of his wages and earnings, and it is as difficult to depict, as it is for those more happily circumstanced to feel, the ad-

ditional suffering inflicted on an expectant and halfstarving family, when a poor man so mulct of a portion of his scanty and inadequate earnings, if insubmissive, returns home penniless."97

Embezzlement. The distress of the handloom cotton weavers which had prevailed so extensively around 1840 intensified the evils of embezzlment, although embezzlement, according to several Bolton merchant employers, had been in existence ever since the end of the eighteenth century.98 About 1840 witnesses from Manchester, Bolton, Preston and elsewhere reported the prevalence of the evil,99 of whom Mr. Makin from Bolton rendered perhaps the best account of it. In enumerating the causes for the drastic reduction of wages among the handloom cotton weavers, Mr. Makin remarked that

> "another cause is the introduction of embezzled materials, which has gone on progressing to my knowledge for forty years, from the testimony of men on whom I can rely, and my own observation, that embezzled materials are openly purchased, and the purchasers of them are openly known, and whose cloth is bought by the most respectable merchants and manufacturers under the belief that they are made partially out of embezzled materials, and which production of embezzled materials, as they cost less than the honest yarn can be bought for, enables these manufacturers to bring in their goods at a less cost than the honest manufacturers can do .... I know individuals who have been servants to manufacturers, who occupy situations that would call respectable and their employment has been in sorting embezzled weft, which has been bought at a price from 1/3 to 1/2 its value .... As a proof of the prevalence of embezzlement, I have a piece here which I will prove to the satisfaction of any gentleman conversant with the manufacture, is made out of twenty different kinds of yarn; I have also a letter from Mr. Thomas, the Deputy Constable of Manchestor, stating his knowledge that the practice of embezzlement is beyond all belief or calculation; I have also hand bills that have been published in Manchester, stating that very large quantities of this sort of materials were seized, and advertising for the owners; and notwithstanding the fact of this, the state of the embezzlement laws is such that the conviction of the party cannot be obtained."100

Yet in face of such facts, the same witness had to admit that to prevent embezzlement would not be desirable, for it would reduce the weavers' wages.

> "The poor man only lives from hand to mouth; when he has not a farthing in his pocket nor a mouthful of bread in his cupboard, what will he not do to satisfy the cries of his children; what does he care about the great question of political economy, or honesty, or religion, or anything else of importance!"101

This practice of embezzlement under the merchant employer system was not confined to the weavers alone; it extended also to the agents who gave out work to the weavers. One witness even went so far as to declare that

> "it is not the weaver that embezzles; it is the manufacturer's agent that is sent to put out work ten or twenty miles from his own place. This agent of his will weigh the weaver perhaps two ounces of weft or one ounce of weft less than he knows will weave his cut, and then the man has not sufficient to work with; and supposing he has 100 weavers, and he weighs every weaver two ounces less than he ought to have, that is two hundred ounces in a day, and when that amounts to a skip, that fellow sells that skip to some person he is connected with." 102

Persistence.

In giving reasons for the persistence of the merchant employer system, we may first deal with the advantages that the system itself entails. From the viewpoint of the merchant employer, Commissioner Muggeridge remarked that

> "the employer of the handloom weaver may at once suspend or limit production by withholding work, without any loss to himself. The amount of his fixed capital is considerably less than that of the power loom manufacturer. He has not, as in the case of the latter, a large amount vested in building and machinery, and a superintending establishment, the expenses of which are going on, whether employed or idle; and he has besides comparatively little difficulty in calling together, at any time he may want them, any number of handloom weavers, whereas it is well known that factory hands, when dispersed, are not again collected in a state of efficiency without considerable time, trouble, and loss."103

From the viewpoint of the weaver, handloom weaving as carried on under the merchant employer system was preferred to factory work for obvious reasons. To quote again from Commissioner Muggeridge:

> "It gratifies that innate love of independence which all more or less feel, by leaving the workman entirely to the master of his own time, and the sole guide of his actions. He can play or idle, as feeling or inclination leads him; rise early or late; apply himself assiduously or carelessly, as he pleases; and work up at any time, by increased exertion, hours previously sacrificed to indulgence or recreation. Beyond the necessity imposed upon him of yielding a given quantity of labor to produce a given amount of earnings, he has little, if any, control. In the proportion he is willing to sacrifice the one, he can dispense with the other, and idleness carries with it no punishment, beyond the restrictions of enjoyment which arise from its being unremunerated. There is scarcely another condition of any portion of our working population thus free from external control. . . . Another cause of the attachment of the weaver to his precarious occupation arises, I am persuaded, in many cases, from better, and more commendable feelings. It concentrates the family under one roof; gives to each member of it a common interest; leaves the children under the watchful eye of the parent; and, the fate of one being the fate of all, it is borne, be it harsh or otherwise, without repining. '*Home is home*, be it ever so homely,' is a trite but an apposite proverb, the truth of which cannot be more practically evinced than it has been by hundreds of families of handloom weavers. A third cause, and an influentially operating one, is the remembrance of bygone days. Few weavers but recount the 'good old times' when their labors were fourfold remunerated compared with their present rate of earnings; and it is hard to cast off altogether the cherished, tho it may be but delusive, hope, that such days are past never to return. Many, who, if interrogated on the subject, would perhaps deny being actuated by such feelings, would nevertheless admit that an influence is exercised over them by such reminiscences; and that some latent prospect of an ulterior change for the better induces them to continue in an employment at present so ill requited. 'It is a long lane that has no turning' has been more than once offered me as a significant apology for the weaver's continuance at his loom."104

The apparent advantages that the merchant employer system offered to the merchant employer as well as the weavers, therefore, constituted the first cause for the persistence of the merchant employer system. The second cause for such persistence was technical rather than industrial, namely, the necessity for handloom weaving. Power loom weaving was distinctly a new branch of cotton manufacture; it created for itself a new market, "market almost sufficient to carry off its own productions, leaving the demand for handloom cotton cloth nearly as great as before."[105] It was, in other words, the plain cloth that has been manufactured by the power looms. The kind of cloth that could only be woven by handloom, even in 1804, was the fancy goods commonly called, such as "fine jaconets, fine twills, and in fact, all our fancy goods that required to be wrought in damp situations." [106] These goods had a limited and uncertain demand, "for which it would never pay to erect complicated and costly machinery." [107] Also, "the experiment of new fabrics will always be made first on the handloom, until public favor creates for them a demand sufficient to make it worth the while of a large manufacturer to apply power."[108]

Chapter III. Woolen and Worsted

A. Woolen

The woolen manufacture of England in 1841 may be defined as including all persons who were fully employed in preparing (scribbling, carding, slubbing) and spinning the woolen fibre into yarn, weaving the woolen yarn into cloth, and finishing (fulling and dressing) the woolen cloth for sale at the market. This definition agrees with that given by the Census of 1841, and excludes all persons employed in the manufacture of worsteds, stuffs, carpets, and fancy goods. The worsteds, stuffs, and carpets are treated in the following section entitled "Worsted". The omission of the fancy goods is due to the fact that the manufacture employs as raw materials not only wool, but also cotton and silk.

Under our definition, the woolen manufacture of England in 1841 employed 106,966 persons,[1] and as compared with the cotton manufacture, it took a second place in the textile industry. The decline of its relative importance was significant when we realize that, previous to the nineteenth century, the woolen manufacture was pre-eminently the staple manufacture of the country. It was England's premier industry from the twelfth to the nineteenth century; it employed the raw materials whose fortunes were interwoven with those of the English agriculture. It illustrated the different forms of English industrial organization in its gradual evolution from the master craftsman to the factory system; and, as a favorite child of legislature, it was the first branch of English industry to be subjected to national control and a uniform system of regulation.[2]

The distribution of the woolen manufacture of England in 1841 was another significant feature in English industrial history. It used to be the most widespread of all English industries, and "there was probably not a town, village, or hamlet throughout the length and breadth of the country which was not connected at some time or other with the manufacture of cloth."[3] But in 1841 about 77 percent of the persons employed in the manufacture were concentrated in a relatively small region, namely, the Northern counties of West Riding of York, and Lancaster.

Table 38. County distribution of persons employed in woolen manufacture, England, 1841

County	Number	Percentage
York, W. R.	67,175	62.8
Lancaster	14,868	13.9
Gloucester	6,739	6.3
Wilts	4,386	4.1
Somerset	2,567	2.4
Devon	2,246	2.1
Other	8,985	8.4
England	106,966	100.0

The Western counties, generally known as the "West Country", were once the leading centres of woolen manufacture. But in 1841, Gloucester, Wilts, Somerset, and Devon had a combined total of but 15 percent in the number of persons employed. The extent of the migration of the industry from the Western counties to West Riding is seen in Table 39, showing the growth of population in the chief manufacturing areas in terms of percentages.[4]

Table 39. Population growth in West Riding, Gloucester, and Wilts, 1700-1851

Year	York, W.R.	Gloucester	Wilts
1700	100	100	100
1801	236	159	126
1851	547	292	167

Note: 1700 = base year (100)

The migration of the industry from the Western counties to the North of England was an outcome of many factors. In the first place, West Riding in its early years had already a promising start in the manufacture. The number of pieces of broad cloth manufactured in West Riding grew (in round figures) from 27,000 in 1726 to 60,000 in 1750, and to 173,000 in 1790; while the output of narrow cloth increased in fifty years (1740-90) from 59,000 pieces to 140,000.[5] Secondly, the advent of the Industrial Revolution had the most remarkable effects upon the localization of the manufacture in West Riding. The introduction of machinery in the North was an easy triumph as compared with that in the Western counties. The flying shuttle and factory yarn were used in West Riding about 1800, and wages were "comparatively high," while all machinery appeared to have been tabooed in the Western counties.[6] Finally, the natural advantages that West Riding possessed,—the abundance of water power, the presence of coal and iron mines, the facility of transport by canals and railroads, and the vicinity to the eastern port of Hull and the western port of Liverpool—all these tended to promote the growth of the manufacture.[7]

The Factory System

In the woolen manufacture the systems of industrial organization varied according to the nature of the technical processes employed. In fulling,[8] the economy resulting from the application of machinery was realized long ago; and "so early as the year 1316 a fulling mill stood near the town [Bradford], and it may be inferred from the considerable annual rent which it brought to the Lord of the Manor, that at it large quantities of cloth were fulled."[9] In 1806 the Committee on Woolen Manufacture were in a position to report that "various processes [fulling and dressing], the chief of which were formerly done by hand, under the manufacturer's own roof, are now performed by machinery, in public mills, as they are called, which work for hire. There are several such mills near every manufacturing village, so that the manufacturer, with little inconvenience or loss of time, carries thither his goods, and fetches them back again when the process is completed."[10] In 1840, according to Mr. Chapman of the Assistant Handloom Weavers' Commission, "all the processes to which the wool is submitted previous to that of spinning, are carried on in factories, called 'scribbling and slubbing mills'. The spinning of the yarn, and the weaving of the cloth, though sometimes carried on in factories (especially in the case of the superfine cloths made in the town of Leeds), are generally performed in the clothmaker's family; whilst all the finishing processes subsequent to weaving, such as scouring, fulling, raising, cutting and pressing, are completed in factories."[11]

For our purpose, we may classify the industrial organization of the woolen manufacture into three systems, viz. master craftsman, merchant employer, and factory systems. The factory system, like that in the cotton manufacture, is again subdivided according to the nature of motive power employed,—human or mechanical. The factories employing human power—generally known as the "shop loom" factories in weaving—were negligible in extent as compared with those employing mechanical power—water or steam.

The factory system, hereafter taken to include only the factories employing mechanical power, claimed 44.1 percent of the total employees in the woolen manufacture. As compared with the master craftsman and merchant employer systems, it was the predominant form of industrial organization. The master craftsman system ranked second in importance, while the merchant employer system bottomed the list.

Table 40. Persons employed in woolen manufacture under factory and other systems, England, 1840

System	Number	Percentage
Factory	47,175	44.1
Other	59,791	55.9
Total	106,966	100.0

The factory employees were chiefly centered in the two important counties of the North—West Riding of York and Lancaster; while a small minority resided in the "West Country"—Gloucester, Wilts, Somerset and Devon.

Table 41. County distribution of factory employees in woolen manufacture, England, 1838

County	Number	Percentage
York	27,548	58.4
Lancaster	4,947	10.5
Gloucester	5,515	11.7
Wilts	3,218	6.8
Somerset	2,133	4.5
Devon	1,810	3.8
Other	2,004	4.3
England	47,175	100.0

The growth of factories in woolen manufacture can be best seen in Table 42, showing the number of factories erected or occupied from 1779 to 1833 in every decennial period in Yorkshire, which is compiled from the Returns of 129 factory occupiers to the Yorkshire Form of Inquiry.12

Table 42. Number of woolen factories erected or occupied in Yorkshire from 1779 to 1833

Decennial period	No. factories erected or occupied
1780 or earlier	1 (a)
1781 to 1790	14
1791 to 1800	25
1801 to 1810	18
1811 to 1820	19
1821 to 1833	46
Date unknown	6
Total	129

(a) The earliest woolen factory was erected in 1779 and occupied by William Stancliffe, Hopton, Mirfield; it was a scribbling, carding and slubbing factory employing in 1833 six steam horse power, and 19 persons. "The first power employed was a horse; afterwards water, from a small brook; at present [1833], steam." (Factory no. 212) Another factory occupied in 1833 by Messrs. Thomas and John Brook, near Milnsbridge, Longwood—fulling and scribbling coarse woolen goods—was erected "time immemorial." It employed 32 water horse power and 20 persons in 1833.

A further classification of these 129 factories is based on the nature of technical processes in which these factories were employed in 1833, and is shown in Table 43. This classification is significant in revealing the fact that the woolen factories of that year were generally employed in preparing and finishing processes,[13] which alone could be more readily and profitably carried on along capitalistic lines, while processes such as spinning and weaving were still confined chiefly to the small clothiers' (i.e. master craftsmen's) homes.

Table 43. Classification of 129 woolen factories in Yorkshire in 1833, according to technical processes employed

Nature of processes	No. factories employed in these processes
Preparing & finishing	40
Preparing only	19
Preparing, spinning, weaving & finishing	12
Finishing	11
Preparing & spinning	7
Preparing, spinning & finishing	3
Spinning & finishing	1
Unclassified	36
Total factories	129

Labor Concentration

The chief element of a factory, as we have already defined in connection with the cotton manufacture, is the concentration of labor under the supervision of a capitalist employer. An index of such concentration is found in the number of persons employed per woolen factory in the different counties of England in 1838.[14]

Table 44. Number of persons employed per woolen factory in important counties of England, 1838

County	No. factories	No. employees	Employees per factory
York	606	27,548	45.5
Lancaster	101	4,947	49.0
Gloucester	96	5,515	57.4
Wilts	53	3,218	60.7
Somerset	30	2,133	71.1
Devon	39	1,810	46.4
Other	104	2,004	19.3
England	1,029	47,175	45.8

The average number of persons employed per woolen factory in England in 1838, according to the above table, was 45.8. This average is hereafter referred to as average number 1. Excepting the unimportant counties included under the heading "other," Somerset, with an average size of 71.1 persons per factory, took the lead; while York, with that of 45.5, bottomed the list. On the whole, the variation of average factory size in woolen manufacture was smaller than that in cotton manufacture. In cotton manufacture it ranged from 236.1 persons per factory in Chester to 73.6 in York. Among the parishes, however, the variation of average factory size in woolen manufacture was much greater than that among the counties. It extended from the smallest average woolen factory of 29 persons in Huddersfield, York, to the largest average woolen factory of 93 persons in Stroud, Gloucester.

Table 45. Number of persons employed per woolen factory in important parishes of York, Lancaster, Gloucester and Wilts in England, 1838

County	Parish	No. factories	No. employees	Employees per factory
York:	Leeds	106	9,738	91.9
	Almondbury	67	3,158	47.1
	Huddersfield	100	2,881	28.8
	Halifax	63	2,074	32.9
	Batley	21	1,332	63.4
	Birstall	33	1,332	40.4
	Calverly	20	1,309	65.5
	Dewsbury	25	1,179	47.2
Lancaster:	Rochdale	56	2,546	45.5
	Whalley	21	1,317	62.7
Gloucester:	Stroud	14	1,299	92.8
Wilts:	Trowbridge	19	1,278	67.3

An average, however, is always misleading, especially an arithmetic average such as we are now confronted with in our analysis of the woolen factory size. That such an average cannot be readily accepted as a foregone conclusion has been pointed out in our treatment of the cotton factories. Baines, commenting on the Factory Returns of 1838 ad 1856, gave us the same impression. According to Baines, "the woolen factories differ so much from each other, that the most careful returns from some of them do not afford safe grounds of calculation for the rest. In some of them there are power looms or hand looms, but in two thirds of the whole there is no weaving carried on. In some the cloth is finished, but in a much greater number it is not finished; whilst about one seventh of the woolen factories in the Returns are finishing establishments exclusively."[15]

Following our practice in dealing with the cotton manufacture, two averages for the woolen factories are now computed on exactly the same principles and from the same sources as those for the cotton factories, in the years 1838 and 1841 respectively. The first average is a median compiled after an enumeration of all woolen factories located in parishes having one or two woolen factories per parish. Omitting all details, the full enumeration is given in Table 46.

Table 46. 136 woolen factories located in parishes having one or two factories per parish, England, 1838

Persons per factory	No. factories
(a) 10 or below	39
(b) 11 to 20	35
21 to 30	16
(c) 31 to 40	12
41 to 50	7
51 to 60	6
61 to 70	5
71 to 80	3
81 to 90	1
91 to 100	4
101 to 200	4
201 to 300	1
301 to 400	3
(d) Total	136

(a) 5 factories employing 10 persons each.
(b) 5 „ „ 19 „ „
2 „ „ 20 „ „
(c) 3 „ „ 40 „ „
(d) These 136 factories employed a total of 5,136 persons, or an average of 37.8 persons per factory.

According to the above table, the median is 19, the lower quartile, 9, and the upper quartile, 40. The quartile coefficient of deviation is, therefore, 0.61. This median average, 19, is hereafter referred to as average number 2.

The second average is compiled after an enumeration of the actual sizes of 205 woolen factories that were at work in Inspector Horner's districts in 1841-122 factories in Lancaster, 55 factories in York, 28 factories in Westmoreland, Cumberland and Durham. This enumeration is presented in Table 47.[16]

Table 47. 205 woolen factories at work in the counties of Lancaster, York, etc., England, 1841

Persons per factory	No. factories
10 or below	71 (a)
11 to 20	53
21 to 30	22
31 to 40	5
41 to 50	12
51 to 60	13
61 to 70	3
71 to 80	3
81 to 90	—
91 to 100	2
101 to 200	7
201 to 300	10
301 to 400	2
401 to 500	2
Total	205 (b)

(a) 29 factories employing 11 to 14 persons each.
4 „ „ 15 persons each.
(b) These 205 factories employed a total of 8,635 persons, or an average of 42.1 persons per factory.

According to the above table, the median is 15, the lower quartile 7, and the upper quartile, 50. The quartile coefficient of deviation, therefore, is 0.75. This average is hereafter referred to as average number 3. In the following table, the two averages, viz. average no. 2 and average no. 3, are compared with respect to their medians, quartiles and quartile coefficients of deviation.

Average number	Lower quartile	Median	Upper quartile	Quartile coefficient of deviation
2	9	19	40	0.61
3	7	15	50	0.75

The dispersion of woolen factory sizes is greater for average no. 3 than for average no. 2, as is also the case with the dispersion of cotton factory sizes. The striking fact, however, is the close correspondence between the two averages, viz. 15 and 19.

This great variation in the sizes of woolen factories as well as the small size of a woolen factory demand an explanation.

The variation of sizes among woolen factories I am inclined to attribute to the persistence of the master craftsman system (historically known as the "domestic" system in Yorkshire woolen cloth manufacture) on the one hand, and the rise of merchant or merchant employer to the position of manufacturer on the other. Both features were common in 1841, and therefore accounted for the simultaneous appearance of factories whose size among 129 woolen factories in York in 1833 varied from 6 persons per factory (Factory no. 193) to 687 (Factory no. 247).[17] The master craftsman—the small clothier—might save enough to purchase a wheel or engine to be driven by water or steam power, and thereby changed the character of his domestic workshop into a factory in a legal sense. He might, in other cases, combine with his fellow clothiers to set up small commission factories in which those processes alone that had been taken over by machinery could be performed.[18] On the other hand, the merchant who had control of trading capital, instead of buying cloth in the half state from the small clothiers and then having it finished in his own workshop, might add to his old functions the whole of those of the clothier; while a merchant employer, whose position was not so prominent in Yorkshire as in West Country, might cease putting out woolen yarns to be woven into cloth by his workmen, and in place of the old practice, established a factory where they could be concentrated and worked under his immediate supervision. In either case, the merchant or merchant employer lost his original character as a trading capitalist, and became from thence an industrial capitalist—a manufacturer.[19]

The small size of the woolen factory in 1840 was an outcome of many factors, of which we have briefly touched upon the persistence of the master craftsman system, under which the small clothiers of Yorkshire maintained their stand against competition from the large manufacturers employing power driven machinery. Another factor is less industrial than technical, namely, the absence of specialization in woolen manufacture. Spinning and weaving were generally processes carried on under the roof of a single factory; [20] so were the preparing processes which were short and relatively simple. Under this arrangement, no large factory could be expected to carry on its operation with success, and "there is some difficulty, in the case of what Professor Marshall would call the representative firm, in securing that every one of the numerous processes of the woolen manufacture shall be carried on with that continuous skill which alone would lead to steady growth."[21]

In Table 48, the increase in the average (mean) factory size as measured by persons employed is shown for the period 1838 to 1856.

Table 48. Persons employed per woolen factory, in England and three important counties, 1838-56

Year	England	York	Lancaster	Gloucester
1838	45.8	45.5	48.9	57.5
1850	54.7	46.1	339.1	75.5
1856	59.2	53.3	95.0	83.9

The increase of average factory size in woolen manufacture since 1838 is a normal development resulting from the growth of the industry and the economy of large-scale production. In the following paragraph, we may say a few words regarding the size of the "first-rate" woolen factories.

The following return, compiled from reliable data given before the Children's Employment Commission in 1833, shows the average size among the "first-rate" woolen factories. The omission of those factories employing 34 or less gives the return a character comparable with that of the "first-rate" factories.[22]

Table 49. 90 "first-rate" woolen factories in Yorkshire, 1833

Persons per factory	No. factories
34 to 40	14
41 to 50	16 (a)
51 to 60	13
61 to 70	10 (b)
71 to 80	6
81 to 90	7
91 to 100	1
101 to 200	15
201 to 300	6 (c)
301 to 400	—
401 to 500	1
501 to 600	2
601 to 700	1
Total	92 (d)

(a) 8 factories employing 41-46 persons each.
1 factory ,, 47 ,,
(b) 3 factories ,, 61, 62, 62 persons respectively.
7 factories ,, 63-70 persons each.
(c) 202, 208, 215, 231, 261, 273.
(d) These 92 factories employed a total of 9,659 persons, or an average of 105 persons per factory.

According to the above table, the median is 62, the lower quartile, 47, and the upper quartile, 215. The quartile coefficient of deviation is, therefore, 0.55. In Table 50 a few details about the four largest "first-rate" woolen factories from the 1833 Yorkshire Returns are given.

Table 50. Statistics on "first-rate" woolen factories in York, 1833

Number & name of factory	Motive power (In H.P.)	Employees
247. Taylor, Son & Gibson	54	687
138. Pim, Nevins & Son	61	552
134. Starkey Bros.	84	521
191. Williams & Sons	70	473

Besides these four "first-rate" factories in woolen manufacture, several still larger ones could be found. In Frome, Somerset, Messrs. Sheppard had 2,000 persons in his employ, of whom 500 worked in the factory employing a total motive power of 126 H. P.[23] In Bath of the same county, Messrs. Charles Wilkins, an establishment for the manufacture of broad cloth and kerseymere and said to be the most considerable in the West of England, were preparing to give dinners and various other refreshmets to about a thousand of his workmen in the employ of his factory—children and grown-up people, at different times of the day, with music and other amusements, as Commissioners Horner and Woolriche passed by the proprietor's residence near his factory.[24] Mr. John Brooke, one of the large manufacturers of woolen cloth at Huddersfield, Yorkshire, employed upwards of 1,000 persons at his factory, where all branches of woolen manufacture were carried on. He told the Committee on Manufactures, Commerce and Shipping that several of his own class employed respectively from 300 to 500 persons.[25]

Employment of Power Machinery

The employment of power machinery in woolen manufacture took place first in the preparing and finishing processes, whereas the spinning and weaving processes retained primarily their handicraft character until the second quarter of the nineteenth century. In following our previous practice in dealing with the cotton manufacture, we may divide up our analysis on the employment of power machinery in woolen manufacture into three sections, viz. motive power, spindles and power looms.

Motive power. Motive power consisted of water and steam. In the early days of woolen manufacture, the presence of water was essential not only for its power, but also for the necessary service it rendered to the process of fulling. As late as 1833, Stroud, the great clothing district of Gloucester for the manufacture of superfine broad cloth and fine kerseymere, contained nearly a hundred factories relying primarily on water for these two purposes.[26] However, the decline of water as a source of power since that date and probably much earlier was rapid. In the same county of Gloucester, Commissioner Miles in 1838 reported that the number of water power factories "has decreased, but this does not prove a decrease in trade, because, since the introduction of steam in order to generate a moving power, these mills are dispensed with, and consequently fall in value."[27] The result was that out of 137 water power factories in Gloucester, only 79 were at work during the same year, 4 factories having been taken down, 17 used for other purposes and 37 void.[28] Table 51 showns the steady decrease of water power as compared with steam from 38.7 percent. in 1838 to 25.9 perecnt. in 1856. In woolen manufacture the decrease of water power is not so rapid as that in cotton manufacture, for water, in addition to being a source of motive power, is essential for fulling purposes. In the latter, only 7.6 percent. of the total motive power employed in 1856 was water, as compared with 18.9 percent. in 1838.

Table 51. Motive power in woolen factories, England, 1838-56
(in horse power)

Year	Steam power		Water power	
	Amount	Percentage	Amount	Percentage
1838	10,847	61.3	6,844	38.7
1850	12,530	66.7	6,261	33.3
1856	16,229	74.1	5,686	25.9

Another table is given below to show the increase in the amount of motive power per woolen factory from 1838 to 1856. The decrease from 17.2 horse power per woolen factory in 1838 to 16.3 horse power in 1850 is probably due to the rise of small joint stock factories where only the preparing and finishing processes were performed.

Table 52. Increase in amount of motive power per woolen factory, England, 1838-56 (in horse power)

Year	Total amount of motive power	Total number of factories	Motive power per factory
1838	17,691	1,029	17.2
1850	18,791	1,151	16.3
1856	21,915	1,138	19.3

Spindles. In woolen manufacture the introduction of mule spinning was almost as gradual as that of power weaving. It took place for the most part between 1820 and 1860, but was not completed, even in Yorkshire, in the latter year.[29] The distribution of the number of spindles employed in the woolen factories among English counties in 1850 and 1856 is shown in Table 53.[30] The North of England, including the counties of York and Lancaster, again, had the largest proportion of the total number of spindles, 87.4 percent. in 1850 and 84.5 percent. in 1856.

Table 53. County distribution of spindles in woolen factories, England, 1850-56

County	1850	1856
York	925,449	992,897
Lancaster	238,492	266,079
Gloucestor	61,896	63,256
Wilts	33,804	34,473
Somerset	22,604	22,879
Devon	13,250	18,605
Other	36,093	67,485
England	1,331,588	1,465,674

The absence of specialization in spinning and weaving has been touched upon in a previous paragraph. It is interesting to note that what little specialization there was in 1850, was considerably reduced six years later. The number of woolen spinning factories decreased in number from 724 in 1850 to 637 in 1856; whereas that of woolen spinning-and-weaving factories increased in number from 206 in 1850 to 301 in 1856.

Table 54. Number of spindles per woolen spinning (or spinning-and-weaving) factory, England, 1850-56

Type of factory	No. factories	No. spindles	Spindles per factory
Spinning			
1850	724	760,531	1,050
1856	637	713,257	1,119
Spinning-and-weaving			
1850	206	569,347	2,764
1856	301	752,417	2,500

The change in size as shown in the above table can be brought out more clearly in the following arrangement.

Year	Spindles per spinning factory	Spindles per spinning-and-weaving factory
1850	1,050	2,764
1856	1,119	2,500

Power looms. The introduction of power weaving into woolen manufacture took place more slowly than that of power spinning; many of the reasons given for the same change in the cotton manufacture were also applicable here, only to a greater extent. The hostility of weavers, for instance, was a strong factor. We may cite the famous incident that happened at the Rawford's woolen factory occupied by the ill-fated inventor, Dr. Cartwright. During the Luddite outbreak throughout the country in 1826, Cartwright, foreseeing a possible attack upon his factory, turned it into a fortress, trained and armed four of his loyal workmen, and procured five soldiers to assist him in the defense. On April 11 the attack took place, and a sanguinary fight, "ending in the complete beating off of the besiegers, who left two of their number lying dead outside the walls of the mill and carried off a considerable percentage of wounded", ensued. As a testimonial to his bravery, the manufacturers of the neighborhood presented to Cartwright a sum of £3,000.[31] Secondly, the technical obstacle had to be overcome before the power loom could be successfully applied to the woolen manufacture. The friction which resulted from frequent and harassing change of fashion was an impediment to continuous growth. Both woolen and worsted manufactures suffered here, but the impediment had hindered the former most. Such friction had almost killed what was known as the broadcloth trade, for instance.[32] Again, in the cloth trade, where the power loom was employed in 1833, it was confined only to the coarser description of goods. "It appears to be an object with the manufacturer that the yarn should be spun as loose as possible to facilitate the interlacing of the fibres in the process of fulling, which transforms the web into a solid piece of cloth. The yarn being so soft is more liable to break, and hence will not admit the rapid and uninterrupted shoot which alone gives the power loom its advantage in webs of harder material."[33] Finally, the willingness of the handloom weavers to submit to repeated reductions of wages unnaturally prolonged the misery and sufferings consequent upon their unequal competition with the power loom—the better tool. "Mr. Peter Playne, a manufacturer [in Gloucester], says his weavers know his opinion, that it will not be to the interest of the manufacturer [i.e. merchant employer] to use the power looms if the men will work, as he says at reasonable wages; he considers that the weaver is racing against steam, and that the slightest rise for wages would cause their general adoption of power throughout the country."[34]

The introduction of power looms into woolen manufacture was therefore slow; but it is not correct to say that "power looms were not used at all in the woolen industry until after 1839."[35] In 1835, there were in England alone 1,927 "woolen" power looms and 318 "woolen and worsted."[36]

Table 55. County distribution of woolen power looms, England, 1835-56

County	1835	1850	1856
York	914	3,849	6,275
Lancaster	1,142	4,839	5,733
Gloucester	4	224	421
Wilts	–	170	252
Somerset	74	27	251
Devon	–	16	93
Other	111	43	438
England	2,245	9,168	13,463

The concentration of power looms was greater than that of spindles in the North of England, including the counties of York and Lancaster; it is seen in Table 56 computed directly from the statistics given in the above table.

Table 56. Proportion of spindles and power looms in Northern England, 1835-56

(in percentages)

Year	Spindles	Power looms
1835	—	91.5
1850	87.4	94.8
1856	84.5	89.2

The number of power looms per woolen factory for 1850 and 1856—whether spinning or spinning-and-weaving—is given in Table 57.

Table 57. Number of power looms per woolen weaving (or spinning-and-weaving) factory, England, 1850-56

Type of factory	No. factories	No. power looms	Power looms per factory
Weaving			
1850	9	245	27.2
1856	11	276	25.1
Spinning-and-weaving			
1850	206	8,923	43.3
1856	301	13,049	43.4

The change in size as shown in the above table can be brought out more clearly in the following arrangement.

Year	Power looms per weaving factory	Power looms per weaving-and-spinning factory
1850	27.2	43.3
1856	25.1	43.4

There was apparently little change in size with regard to the number of power looms employed for either weaving or spinning-and-weaving factory. The common feature in both years, however, is the relative insignificance of the weaving factory as compared with the spinning-and-weaving factory. In 1850 the weaving factory employed only 2.7 percent. of the total number of woolen power looms; in 1856 it employed 2.1 percent.

So far we have given an analysis of the important elements that constitute the employment of power machinery in woolen manufacture. The effects of it, including displacement of labor and increase in productivity, may be dealt with briefly. For further details, we refer again to the preceding chapter on the "Cotton Manufacture."

In the woolen manufacture " the labor of women and children is not so available as in others. The loom, required for weaving blankets or broad cloth, is of great width and corresponding strength; the labor, therefore, is much more severe than in weaving light and narrow fabrics of silk or cotton, and the physical powers of a woman would be unequal to the muscular exertion required."[37] The small percentage of woman and child labor employed in the woolen manufacture is shown in Tables 58 and 59. It is also shown that child labor decreased since 1838, while woman labor increased slightly in 1847, dropped off in 1850, and increased again in 1856.

Table 58. Child labor in woolen factories, England and Northern Counties, 1838-47 (in percentage)

Year	England	York	Lancaster
1838	47.8	49.9	50.0
1847	42.5	45.3	42.4

Table 59. Woman labor in woolen factories, England and Northern counties, 1838-56 (in percentage)

Year	England	York	Lancaster
1838	38.1	32.0	37.6
1847	40.6	36.5	42.3
1850	38.9	34.5	42.1
1856	41.9	36.5	46.7

Excellent statistics on the increase of productivity as a result of the employment of power driven machinery were furnished by Mr. Austin of the Assistant Handloom Weavers' Commission.38 On account of the limited space available, we can only quote the conclusions without details.

Table 60. Quantity of labor required to make a piece of superfine broad cloth; the expense of labor for the same, 1781-1828

Period	Total quantity of labor			Total expense for labor		
	Days	Hours	Minutes	£.	s.	d.
1781-1796	262	8	24	8	18	2¾
1796-1805	85	—	35	6	3	5
1805-1820	80	6	5	5	15	5
1820-1827	77	10	5			
1828	64	9	57	4	7	3

In terms of percentages, we may quote the above table in the following fashion.

	1781		1828		1781:1828
	Hours	Minutes	Hours	Minutes	
Man's labor	1,002	22	310	26	100: 31
Woman's ,,	797	2	144	2	100: 18
Child's ,,	1,353	—	322	29	100: 24
Total Labor	3,152	24	776	57	100: 25
Expense for labor	£8. 18s. 2¾d.		£4. 7s. 3d.		100: 50

The total quantity of labor required in making a piece of superfine broadcloth was reduced to 1/4 of its original amount in the course of half a century (1781-1828), while the expense for the same quantity of labor was halved. An increase in productivity therefore resulted in a saving of time as well as of expense, together with a simultaneous increase of wages.

Statistical Summary

Table 61. Factory statistics in woolen manufacture, England, 1838-56

Year	No. factories	No. employees	Amt. motive power	No. Spindles	No. power looms
1838	1,029	47,175	17,691	....	......
1847	....	62,057	......	....	......
1850	1,151	62,972	18,791	1,331,588	9,168
1856	1,138	67,385	21,915	1,465,674	13,463

If we take 1838 as a base of 100 percent., we may express the statistics in the above table in relative terms below.

Year	No. factories	No. employees	Amt. motive power
1838	100.0	100.0	100.0
1847	....	131.5	....
1850	111.9	133.5	106.2
1856	110.6	142.8	123.9

The more rapid increase in the percentage of employees over that of factories indicated the expansion in the average size of factories as measured by employees; the same may be said of the more rapid increase in the percentage of motive power over that of factories since 1850. We may conclude, therefore, that the average woolen factory in 1856 was a great deal larger than that in 1838 with regard to the number of persons or the amount of motive power employed per factory. Unfortunately, the factory returns of 1838 give no statistics on spindles and power looms, and leave us without basis of comparison with those of later years.

Handloom Factories

As in the cotton manufacture, handloom (more technically known as "shop loom") factories were found in the woolen manufacture. In fact, woolen manufacture as the first staple industry of England, supplied the earliest instance of such factories. The factory as erected by Thomas Blankett, afterwards bailiff of Bristol, and his fellow burgesses in 1339, would have been a success but for the presence of factors that hastened its downfall. At that time and as late as the eighteenth century, social antipathy, government legislation, and economic structure were strong obstacles to the establishment of factory system.[39] In West Riding in 1805, when the factory output of narrow and broad cloths did not amount to more than two percent. of the total, "the most serious apprehensions" were already expressed by the small clothiers who gave evidence before the Committee on Woolen Manufacture of England, "lest the factory system should gradually root out the domestic [i. e. master craftsman system in our terminology] and lest the independent little master manufacturer, who works on his own account, should sink into a journeyman working for hire".[40] In 1840 even, the "shop loom" factories were negligible in extent; most of them centered in West Country—particularly in the county of Gloucester.[41] Here these factories were located at Stroud, Wootton, Darsley and other places. All the firms of large merchant employers had factories attached; and when trade was brisk, they gave chains to the outdoor weavers at their own cottages.[42]

From the returns supplied by the merchant employers, there were in Gloucester in 1838 911 "shop loom" weavers in the factories, as compared with 1,755 outdoor weavers.[43] Of the 911 "shop loom" weavers, a large number must have been also "factory" weavers, legally speaking at least; for many of the "shop loom" factories had power looms as well as other sorts of machinery driven by power.

Among the 43 merchant employers who gave returns of varying degree of completeness, 29 had handlooms in their factories.[44] The total number of shop looms was 1,054, of which 824 were employed at the time of the returns, and 230 were unemployed. The employed looms afforded work to 694 men, 184 women, and 116 children—a total of 994 weavers of all sorts. The last figure, namely, 994 weavers, did not agree with that given by the same Commissioner at another occasion and quoted here above (viz. 911 factory weavers). There being no way of checking up the two sets of figures, we may venture to conclude that the inconsistency was another index of the unsatisfactory character of early industrial statistics.

Table 62. Number of shop looms employed by 29 woolen manufacturers in Gloucester, 1838

No. looms per manufacturer	No. manufacturers	Total no. looms
1 to 10	6	32
11 to 20	7	114
21 to 30	4	101
31 to 40	1	35
41 to 50	1	50
51 to 60	4	233
61 to 70	1	69
71 to 80	2	142
81 to 90	2	178
91 to 100	1	100
Total	29	1,054

Note : Nine of the 29 manufactures also employed power looms, totalling 96. These power looms were introduced during the years 1836-8.

The average number of shop looms per factory is, therefore, 36; while the average number of weavers employed, 34. It stands in fair comparison with the average number of power looms per power weaving, or power weaving-and-spinning, factory.

Kind of factory	Average looms per factory	
Shop loom	36.0	(1840)
Power weaving	27.2	(1850)
Power weaving-and-spinning	43.3	(1850)

The above averages, however, are arithmetic means, and are therefore subject to the influence of extremes. The statistics available do not permit us to calculate the median, instead of the arithmetic average, except in the case of handloom factories. For the handloom factories, the median average, as derived from the above table, is 26 handlooms per factory.

A critical summary of the handloom factories is at its best incomplete because of the limited amount of information now available. It is, furthermore, hardly possible without a just appreciation of the significance which the transition from one system of industrial organization to another may fully imply. Many of the arguments against the system appear to be of little moment today if we fail to bear in mind the hard fought battle of the outwork weavers in resisting the adoption of the factory system. The following quotation from a contemporary witness well acquainted with the clothing districts of Gloucester and the habits of the weavers, points clearly towards the direction of conduct and thought on the part of the weavers that we have here broadly indicated.

> "Although there appears throughout this [i.e. weaving] department of the factories great regularity as regards work, every hand being, in full employ, chain after chain following in quick succession, yet few factory weavers will acknowledge, and these but faintly, that the shop system is a benefit to them; but with a strange inconsistency, state that they would rather take the chance of uncertain work at the master weaver's as they used to do. These weavers at the same time confess that they earn more money in their present situation as factory weavers: that they get more food and constant work; yet they are not contented as when they were under the acknowledged *arbitrary* master weavers, with uncertain employ, and consequent low wages. The objections to shop factories as stated by them are:—
>
> 1. The unjust assumption of the weavers' trade by a manufacturer;
> 2. The prison-like situation of the weaver;
> 3. The non-employment of their wives and children, who formerly could act as quillers, and occasionally, as regards the wife, relieve the man from the loom while he worked in his garden, or earned a trifle another way—to which I may add, or went to the beer shop and 'lost' a trifle another way;
> 4. The deprivation of their little messes, such as a cup of tea, and hot vegetables at dinner time, which the master weaver used to supply them with at a moderate charge, which in the shop factory is impractable, owing to the distance many of them live from the shop;
> 5. The distance that many of them have to walk to their work in all weathers, without a change of clothes;
> 6. The confinment to a certain number of hours per day, at the option of manufacturer, whereby the weaver is prevented from earning what is usually called 'overtime money';

> 7. The constant tendency to reduce wages: A manufacturer knows that it is difficult for a weaver to get work if he is once discharged for a stand against wages; and thus the manufacturer so constantly trying to find out a clue to lower the price, taking, perhaps, as an unfair standard, a man who can probably weave a chain in 10 or 11 days.
>
> These are the objections to shop factories, made by the weavers; but when it is considered that a weaver, whether in prosperity or adversity, is the same restless being in reference to his master with whom he considers he has no interest in common, allowance must be made for his prejudices. A weaver is no economiser of time; when it is at his own disposal, too frequently neglects his work for even the most trivial causes, and too often attributes the scantiness of his earnings to the low wages of the manufacturer, when it might be more fairly chargeable to his own want of energy and industry."45

Throughout this whole quotation, it may be noted that the features of the shop loom factory to which objection was made were chiefly those which differentiated the factory system from its predecessor, the merchant employer system. The dislike for discipline, the loss of by-employment for family members, and the increased economic dependence were pertinent points at issue.

On the other hand, from the viewpoint of the employer, the "shop loom" factory system offered a number of distinct advantages. The concentration of labor gave him immediate supervision over his workmen and the output. "The master knows the nature of the work he has on his looms, and he knows how many hours the men work. and the quantity they ought to do; he regulates his wages accordingly; the cloth is better made." He could, furthermore, check the evils of embezzlement. Many advantages. which inured chiefly to the benefits of the workmen, indirectly helped the employer in the form of better work done. "The workmen being a part of the establishment, his master," for instance, "takes an interest in his moral and pecuniary welfare, makes him put aside money for sickness, when he is earning good wages, sees to the education of his children etc., and would only, in case of absolute necessity, lower those wages below the point which will enable a man to live by his earnings."46 Constancy of employment and regularity of habits tended to build up the character of the workman, which under the merchant employer system would have been impossible. Again, a handloom factory was unlike a power loom factory; as it did not require any large capital investment in power driven machnery and buildings. The overhead cost was extremely low, as both the handlooms and the shops necessary for their accomodation could be cheaply found. In a period of commerical depression, a handloom factory manufacturer could as easily stop the operation of his factory as a merchant employer, without having to incur considerable loss incumbent upon a power loom manufacturer during such occasions.

In conclusion, we may quote a passage from the Report of the Committee on Woolen Manufacture in 1806. In this passage, the economic advantages of the factory over the 'domestic' systems were summarized in other respects than those mentioned above.

> "The factories, to a certain extent at least, and in the present day, seem absolutely necessary to the well-being of the domestic system; supplying those very particulars wherein the domestic system must be acknowledged to be inherently defective: for it is obvious, that the little manufacturers cannot afford, like the man who possesses considerable capital, to try the experiments which are requisite, and incur the risks, and even losses, which almost always occur, in inventing and perfecting new articles of manufacture, or in carrying to a state of greater perfection artic-

les already established. He cannot learn, by personal inspection, the wants and habits, the arts, manufactures, and improvements of foreign countries; diligence, economy, and prudence are the requisites of his character, not invention, taste, and enterprize; nor would he be warranted in hazarding the loss of any part of his small capital: He walks in a sure road as long as he treads in the beaten track; but he must not deviate into the paths of speculation. The owner of a factory, on the contrary, being commonly possessed of a large capital, and having all his workmen employed under his own immediate superintendence, may make experiments, hazard speculation, invent shorter or better modes of performing old processes, may introduce new articles, and improve and perfect old ones, thus giving the range to his taste and fancy, and, thereby alone, enabling our manufacturers to stand the competition with their commercial rivals in other countries. Meanwhile, as is well worthy of remark (and experience abundantly warrants the assertion), many of these new fabrics and inventions, when their success is once established, become general among the whole body of manufacturers; the domestic manufacturers themselves thus benefiting, in the end, from those very factories which had been at first the objects of their jealousy. The history of almost all our other manufactures, in which improvements have been made of late years, in some cases at an immense expense, and after numbers of unsuccessful experiments, strikingly illustrates and enforces the above remarks. It is besides an acknowledged fact, that the owners of factories are often amongst the most extensive purchasers at the halls, where they buy from the domestic clothier the established articles of manufacture, or are able at once to answer a great and sudden order; while, at home, and under their own superintendence, they make their fancy goods, and any articles of a newer, more costly, or more delicate quality, to which they are enabled by the domestic system to apply a much larger proportion of their capital. Thus, the two systems, instead of rivalling, are mutual aids to each other, each supplying the other's defects, and promoting the other's prosperity."[47]

Master Craftsman System

In the Yorkshire woolen manufacture in 1840 three systems of industrial organization were in simultaneous operation to a varying extent, viz. master craftsman, merchant employer, and factory systems.[48] Probably in the whole range of English industrial organization, no industrial system has received more numerous but confused terms than the master craftsman system, in the Yorkshire woolen manufacture. According to Professor Ashley, this system was called "domestic" for the first time by the Committee on the State of Woolen Manufacture of England in 1806,[49] "domestic" primarily because the work was carried on at home. In the local terminology, however, it was known as the "small clothier" system, as contrasted with the "master clothier" system of "West Country," i.e. the merchant employer system, a "small clothier" being in our terminology a master craftsman, or a "master manufacturer" in that of the Parliamentary Committee of 1806.[50] Professor Cunningham, to take another instance, was inclined to identify this system with the "gild" system, although he accepted the term "domestic" in following the tradition laid down by the Parliamentary Committee of 1806.[51] In our opinion, all these terms were inadequate, and another term, "master craftsman," would better express the essence of the system and meantime agree with the whole scheme of our classification. Our reasons for the adoption of this new term "master craftsman" have been sketched elsewhere in the first chapter, and need not be repeated here.

Under the master craftsman system, as described by the Committee on Woolen Manufacture in 1806,

> "the manufacture is conducted by a multitude of master manufacturers, generally possessing a very small, and scarcely ever any great extent of capital. They buy the wool of the dealer, and, in their own houses, assisted by their wives and children, and from two or three to six or seven journeymen, they dye it (when dyeing is necessary), and through all the different stages work it up into undressed cloth. .... When it has attained to the state of undressed cloth, he [the manufacturer, or, master craftsman in our terminology] carries it on the market day to a public hall or market, where the merchants repair to purchase. Several thousands of these small master manufacturers attend the markets of Leeds, where there are three halls for the exposure and sale of their cloths; and there are other similar halls, where the same system of selling in public market prevails, at Bradford, Halifax, and Huddersfield. .... The greater part of the domestic clothiers live in villages and detached houses, covering the whole face of a district from twenty to thirty miles in length, and from twelve to fifteen in breadth. ...... A great proportion of the manufacturers occupy a little land, from three to twelve or fifteen acres. They often like-wise keep a horse, to carry their cloth to the fulling mill and the market."[52]

The master craftsman system as above described applied equally well in 1840 as in 1806, for the changes, if any, were relatively unimportant.[53] In the following paragraphs, the chief elements of the master craftsman system in Yorkshire woolen manufacture are sketched in brief, viz. social classes, home work and cloth halls.[54]

Social classes. In the Yorkshire woolen manufacture under the master craftsman system, there were two social classes, namely, industrial and commercial.[55] The industrial class was composed of the master craftsman and his family members, the journeymen, and the apprentices. The commercial class was composed of the merchants, such as the travelling merchants, the merchants who supplied the Blackwell Hall of London, and the merchant exporters who had commercial agents of their own scattered all over the world.[56] The local merchants, however, played the major role.[57]

Home work. Home work carried with it two interesting features, viz. family labor and by-employment. Family labor—the labor of husband, wife and children—was almost universal in Yorkshire; it was excellently described in a colloquial poem "Description of the Manners of the Clothiers," written about 1730.[58] At the evening meal the master of the house gave instructions to the working members of the family, including, beside his own wife and children, perhaps the apprentices and journeymen employed from outside.

> "Lads, work hard I pray,
> Cloth mun be pearked next market day.
> And Tom mun go to-morn to t'spinners,
> And Will mun seek about for t'swingers.
> And Jack, to-morn by time be rising
> And go to t'sizing mill for sizing.
> And get your web and warping done
> That ye may get it into t'loom.
> Jo, go give my horse some corn,
> For I design for t'wolds to-morn.
> So mind and clean my boots and shoon.
> For I'll be up i' t' morn right soon.
> Mary - there's wool - tak thee and dye it."

His wife objected that she had her house work to do: "To bake and swing and blend, and milk, and bairns to school to send," as well as "washing up morn, noon, and neet." But the husband retorted:

"All things mun aside be laid,
When we want help about our trade."

The young folk were then left to themselves, and they sat round the fire telling tales and "merry jokes,"

"Till ten gives warning by the dock,
Then up they start—to bed they run."

At five o'clock the next morning they commenced again the day's round.

Conditions in 1840 had changed only slightly. The small clothier, for instance, no longer put out his raw wool to be worked up into finished yarn by the spinner, as the poem above described. The small clothier could now purchase the yarns more cheaply from the factory manufacturer. In other respects, however, home work retained the same features as were prevalent in the days of yore.[59]

Besides family labor, by-employment was another feature of home work. In some parts of the North in the eighteenth century—for example South Lancaster—weaving was carried on as a by-employment, while farming was the principal job. In other cases, the reverse was generally true.[60] In Yorkshire in 1840, the common practice was for the master craftsman to annex a little farm to his other businesses, to support a cow or horse for carrying cloth to the market or factory for preparing and finishing processes, or to raise vegetables for home consumption.

Cloth halls. Cloth halls were never established in West Country as in the North of England.[61] In the North, they became indispensable media to the master craftsmen—the small clothiers—for the disposal of their cloth.[62] In Leeds,[63] for instance, there were three cloth halls. Two of these, one erected in 1758 ond another in 1775, afforded accomodation to those clothiers who had served a regular apprenticeship; while the third one, called "Tom Paine's Hall" and erected in 1792, admitted all those who were so excluded from the other two cloth halls. The hall last established, however, was soon abandoned, for in 1797 the trustees of the two other cloth halls - the "colored" and the "white" cloth halls - had already adopted new regulations admitting persons having served an apprenticeship or exercised the trade of a clotheir for five years, and a few years later, further extensions were made by admitting all persons without condition whatsoverer.[64]

The "colored" and the "white" cloth halls had almost similar organization and development. Their buildings, as Commissioner Drinkwater reported in 1833, were "the most striking objects of attention which it [Leeds] contains. They are extensive galleries, spaced off into several hundred separate holdings, each of which.. ....is a separate freehold, and each bearing the name of its proprietor. The only furniture of these galleries is a row of tressels to hold the cloth which is brought in and exposed for sale by the weavers and manufacturers of a district extending ten or twelve miles around Leeds."[65] These cloth halls were under the management of a certain number of trustees, who served for three years, being elected from the clothiers at large, "as the most direct and intelligent of their body," to watch over the general interests of the manufacture, and in particular to consider and superintend the execution of the regulations and bye-laws made from time to time for the government of the halls.[66]

The mode of conducting business in these cloth halls bore close resemblance to the modern exchange. According to a description of it by Parsons, the historian of Leeds,

"The market days are on Tuesdays and Saturdays, on which days alone the merchants are permitted to buy or inspect the cloths in the halls. In both the halls the regulations are similar and are admirably calculated to subserve both regularity and expedition. The markets are held in the morning; they are opened by the ringing of a bell; each manufacturer or clothier appears behind his stand and the sales immediately commence. At the end of an hour, a warning bell announces the approaching close of the market; and the sound of the third bell, in a quarter of an hour afterwards, terminates the business of the day. Each merchant now quits the hall on pains of a penalty of 5s. for every five minutes that he continues in it after the last bell has rung; and thus, in an hour and a quarter all the transactions are completed. The mixed [i.e. colored] cloth hall opens at half past eight in Summer; nine in Spring and Autumn; and half past nine in Winter. The white cloth hall opens when the other closes; and strangers passing through the town frequently gratify themselves by visiting the halls during the hours of business to which there is no objection. The cloth is brought to these halls in the unfinished state, and it is dressed under the merchant's direction, either by his own workmen or by persons whose business is to dress and finish woolen goods."[67]

Merits and Persistence

The merits of the master craftsman system in the woolen manufacture of Northern England were excellently summarized in the following quotation from the Committee on Woolen Manufacture, in 1806, which also applied in 1840 with slight changes.[68] Thus, from the viewpoint of the master craftsman,

"a young man of good character can always obtain credit for as much wool as will enable him to set up as a little manufacturer, and the public mills, which are now established in all parts of the clothing district, and which work for hire at an easy rate, enable him to command the use of very expensive and complicated machines, the construction and necessary repairs of which would require a considerable capital. ...... Any sudden stoppage of a foreign market, any failure of a great house, or any other of those adverse shocks to which our foreign trade especially is liable, in its present extended state, has not the effect of throwing a great number of workmen out of employ, as it often does, when the stroke falls on the capital of a single individual. ...... The loss is spread over a large superficies; it affects the whole body of the manufacturers; and though each little master be a sufferer, yet few if any feel the blow so severely as to be altogether ruined. Moreover, it appears in evidence, that, in such cases as these, they seldom turn off any of their standing set of journeymen, but keep them at work in hopes of better times."[69]

On the other hand, the merchant no less than the master craftsman found his interest and convenience promoted by the master craftsman system. The Committee on Woolen Manufacture remarked that,

"while it continues, he [i.e. the merchant] is able to carry on his trade with far less capital than if he were to be the manufacturer of his own cloth. Large sums must then be irrevocably invested in the extensive buildings and costly machinery; and, which perhaps is a consideration of even still more force, he must submit to the constant trouble and solicitude of watching over a numerous body of workmen. He might then often incur the expense of manufacturing articles, which, from some disappointment in the market, must either be kept on hand, or be sold at a loss. As it is, he can agree with his cus-

> tomer, at home or abroad, for any quantity of goods; and whether on a long expected or a sudden demand, he can repair to the market, and most probably purchase to the specific extent of his known want; or, if the market happen not to furnish what he wishes to purchase, he can give out his sample, and have his order executed immediately." 70

With all these merits from the viewpoint of the master craftsman and that of the merchant, there were other reasons that bore more directly upon the persistence of the master craftsman system in the Yorkshire woolen cloth manufacture. Technically speaking, power weaving, and to a smaller extent, power spinning, were not common in woolen manufacture until the second and third quarters of the last century. With regard to power weaving, one authority in 1907 asserted that the handloom "survives not merely as a curiosity in out-of-the-way districts, or as a means of weaving a few elaborate and expensive fabrics, of which only small quantities are required, but also as the indispensable tool of the best textile designers and pattern weavers. New designs can be and are worked out on the power loom, and it may be that in time the handloom will disappear altogether; but this seems improbable, for the older appliance is both economical and convenient for experimental purposes."71 Industrially speaking, the remarkable power of adaptation on the part of the small clothiers to the revolutionary changes of a machine age had a lasting effect in perpetuating the master craftsman system. The establishment of the so-called "joint stock mills" (or "company mills") in the early nineteenth century was rapid, and soon became the saving grace of a declining trade. On this subject, we cannot do better than to quote the excellent brief summary on the history and organization of the joint stock mills by the Committee on Joint Stock Companies in 1844.

> "The history of Joint-Stock Company Woolen Mills exhibits a singular instance of energy amongst the smaller capitalists of the manufacturing districts. In the earlier periods of the woolen manufacture, the wool was scribbled and carded at home, then spun and woven, and afterwards carried to the fulling mills to full, and then returned and sold in the balk state. The next step was the employment of power by asses, horses, and air to scribble the wool by their machinery, which soon, however, merged into water-power, scribbling machines being added to the power which fulled the cloth. These fulling, and afterwards fulling and scribbling, mills were everywhere situated on the banks of streams, as the rivers Aire, Wharf, and Calder, but particularly on the River Aire, the sites of many of which are of very ancient date. Leeds, Bradford, Shipley, and the neighborhood of Dewsbury are all places where these ancient fulling mills were to be found. When steam became a moving power, and was employed in the woolen manufacture, the mills were first erected in the neighborhood of towns, and being a more available power than water generally, and less liable to stoppages, the current of manufacture was directed to towns from the country districts. One can easily imagine this on looking at the rapid strides which were made in Yorkshire in a very short space of time—from about 1810 to 1830—and that the old fulling mills became not only inconvenient on account of the distances at which they were situated from the clothiers, but an absolute loss, owing to the loss of time which was occasioned by the transit of goods to and from the mill. It was customary, for instance, for carts to come as many as twelve miles into the clothing district for wool three times a week, which wool had to be brought first into the district from neighboring towns; then it had to return, scribbled, to be spun and woven; then it had to be resent to the mill to be fulled, and then returned to be prepared for the market, and, lastly, to be sent to the market for sale. The

inconvenience of all this need not be further pointed out, nor the difficulty of getting the work done within the specified time, especially for particular trades, which require goods to be shipped to a day; and hence the clothiers of certain country districts, such as Paisley, Idle, Ecclesfield, Batley, Dewsbury, etc., put their heads together, and subsequently their purses, to overcome these difficulties, and, by erecting mills at home, to scribble their own wool and full their own cloth without the cost of labour of so much carriage and dependence upon variable circumstances. The erection of the steam-engine in towns paved the way for their erection in the country, and as coal was near and plentiful these difficulties were very readily overcome. The first company mill near Leeds was erected at Stanningley, about thirty years ago, by a few persons who ultimately got into difficulties and were dissolved. The second was commenced at Ecclesfield, the third at Paisley, and they have been continued to be erected until there are more company mills now than work can be found for. In the formation of a company mill a number of clothiers (for they must be clothiers to be partners) of small capital meet together and determine to become a company of so many partners, from ten, to fifty in shares generally of 25 each, each person taking as many shares as his capital will enable him .... With this subscribed capital deeds of partnership are drawn, land is bought, a mill erected, and machinery put up. ...... The processes which are carried on in these company mills are scribbling, carding, slubbing, and fulling cloth, which are the preparatory processes of the cloth manufacture, and the remaining processes—viz., spinning, warping, weaving, and burling—are done at home by members of the family or by persons employed for that purpose."[72]

These joint stock mills generally charged a regular price to members and non-members alike, and divided up their profits at the end of the year among the former class. In 1907 it was observed by one authority that "although this type of mill is now practically extinct, mills that accomodate more than one manufacturer and firms that start in a humble way by working for commission for their neighbors are still common enough. The small employer may rent a few rooms or a whole floor in a spinning mill. Looms belonging to two different manufacturers, driven by engines which belong to neither, can often be found running in one weaving shed."[73]

Merchant Employer System

The merchant employer system had one of its earliest examples in the woolen cloth manufacture. Its origin went back to the fourteenth century, while its full development extended from the seventeenth to the eighteenth century.[74] In West Country the system was generally known as the "master clothier" system—a rather misleading term when compared with the "small clothier," which, in the Northern terminology, represented the master craftsman.[75]

Social classes. In the eighteenth century, a merchant employer fell into one of the two classes—the master clothier or the market spinner. The latter had disappeared from the industrial scene in 1840;[76] while the former, in some cases, had taken over the function of a manufacturer in addition to that of a merchant. Whenever the merchant employer retained his original position, his role remained unchanged in 1840 as in 1806. To quote again from the Committee on Woolen Manufacture,

"the master clothier of the West of England buys his wool from the importer, if it be foreign, or in the fleece, or of the wool-stapler, if it be of domestic growth; after which, in all the different processes through

> which it passes, he is under the necessity of employing as many distinct classes of persons; sometimes working in their own houses, sometimes in those of the master clothier, but none of them going out of their proper line. Each class of workmen, however, acquires great skill in performing its particular operation, and hence may have arisen the acknowledged excellence, and, till of late, superiority of the cloths of the West of England."77

In other words, the function of a merchant employer was two-fold in character. As a merchant, he purchased the raw wool or yarn and sold the finished cloth; as an employer, he put out the raw wool or yarn to be spun and woven into cloth by the workman at his own home. Sometimes, the merchant employer purchased the spun yarn from the factory manufacturer; or else he might start a spinning factory to produce it himself. He had, of course, his own warehouse and finishing shop to dress the cloth for sale at the Blackwell Hall of London through his agents, or for export to foreign countries.

Turning now from the merchant employer to the workman, we find him in many respects resembling his Northern comrade except in his loss of economic independence. He became now an employee working for wages while still owning his instrument of production, the handloom. His independence, if any, lay in his freedom to work for several employers at one time.

Home work. In West Country, the conditions of home work were essentially similar to those in the North. Family labor was employed, and by-employment was not uncommon. As regards the latter, the prevalence of the allotment system may deserve a brief mention. Under this system a weaver generallly secured half an acre or one acre of land by paying a reasonable rent, and enhanced earnings by plowing it at spare times or during a period of unemployment. This additional employment was desirable for other reasons; it gave him a change from one type of job to another, and refreshed and invigorated a weaver depressed by the sedentary nature of his work. Besides,

> "it will make him a better husband, a better father, a better subject, and a better man. Let him have as much land as will occupy his waste hours. Every moment engaged in this manner will not only be property to himself, but capital added to the wealth of the country. Idleness is the reservoir of mischief. Opportunity is the source. Look at the recorded facts, - that those who have enjoyed the benefit of the allotment system, have not only not joined the incendiary of his own neighborhood (and those who know how difficult it is for one laboring man to stand apart from his neighbor under such circumstances, can alone appreciate the fact,) but have been amongst the foremost to extinguish the flames."78

Abuses

Among the abuses of the merchant employer system, the most vicious were embezzlement on the part of the employee and payment of wages in truck on the part of the employer. Both abuses were inherent in the system, the existence of which, with a few exceptions, might well serve as an index of the existence of the merchant employer system. On the one hand, the employer had no means to check embezzlement while home work prevailed; on the other, the employees, without any organized resistance against the exactions of the employer, had to submit helplessly to the injurious system of truck payment. Each was in possession of the means of retaliation against the other, but in the long run, the employee was probably the loser. The drastic reduction of real wages among the handloom weavers had amply proven the truth of the statement, partly through the lowering of money wages, and partly through the exorbitant prices charged for goods bought under the truck system. In the remaining paragraph, each of the two abuses is briefly taken up in turn.

Embezzlement. Wherever the merchant employer system prevailed, the evil practice of embezzlement could also be found. In the woolen manufacture, for instance, embezzlement was extensively practiced in West Country; while around the Leeds districts it was largely absent.[79] In Gloucester embezzlement was carried on to such an extent that the employers looked upon it as a certain deduction from profit, though uncertain as to amount; while the weavers looked upon it as their peculiar pre-requisite.[80] Out of this condition, a new class of dealers in the embezzled yarn (technically called "slinge") arose. These "receivers," as they were called, acted with so much cunning and precaution that it was difficult to detect and convict them. It was usual for them to purchase a small lot of wool, merely for the purpose of having an invoice to produce in case their premises should be searched. In other cases, they lived in remote districts, carrying on some ostensible calling, and were well known to the weavers. There would be, for example, slinge shops near the factory under the disguise of beer shops or small cake and apple shops. To these shops, the weavers brought their slinge, and sold them at different prices, depending upon their needs at the time. There were instances where slinge valued at 7s. a pound had been sold for 10d. Oftentimes, the slinges were collected by children and women in the employ of the receivers. One woman and her daughter at Pinfarthings, in the parish of Hampton, had been so employed for twenty-five years. Persons who had neither house, home, nor furniture were also employed in this trade. The extent of slinge dealing could be seen from the fact that at one public auction, 140 bags of seized slinges were offered for sale worth £1,000![81]

Truck. In 1840 payment of wages in truck had spread far and wide among many branches of English manufacture that were carried on under the merchant employer system. The only difference was merely a matter of degree and extent. In woolen manufacture, West Country again offered a fertile field for the exploitation of trucksters. In Chalford, Gloucester, the system had taken such hold of the community as to affect all classes;—"shoemakers, landlords, beershop keepers, and others, all declare they take bacon, cheese, and candles in payment."[82] In the same place, the system had been in operation for seven or eight years, except for one money-paying master.[83] Many instances might be cited, but one would serve our purpose as well as the other. The high prices charged was an oppressive form of wage reduction through indirect means. One witness observed that "supposing the weaver and his family (when at work) could earn together 12s. per week, I should say without fear of contradiction that he would not, all things considered, realize more than 9s. in actual amount." Another bought tea and cheese from the truckster; he believed that he could obtain the same article at 1s. 3d. the quarter pound, for which the truckster charged him 1s. 6d. The cheese was charged 9d. per pound, and the witness declared that he had cheese of the same quality in his own house, for which he only paid 7d. per pound. From the operation of this system of truck, it could be readily seen that serious consequences to all parties ensued, including not only the workman, but also the master and the local grocer. First, the system tended to render the weaver a slave of the employer. Thus, Ross, grocer, draper and druggist at Woodchester, told the Commission that,

> "They [the weavers] are obliged to work, to live, and act in a way of total degradation. They must consent to his [the master's or the truckster's] prices for their work. They must lay out their earnings where

he pleases, which is always at his shop, or at a shop which he countenances (when he refuses to own it as his), but from which he is sure to reap no small profit. They dare not complain of short weight, though they have much cause to do so. They dare not complain of inferiority in the articles supplied. They dare not utter a word, however exorbitant the price charged; if they did, the master would make but little reply; but, in a few days, or at the next settling, some trifling fault is found, or 'not being in want of so many hands', they are turned off, but not a word is said about the shop. The workman is left to draw his own inferences. The workmen know this and therefore are compelled to remain quiet and submit in silence. Second: the system tends to produce poverty in the family of the weaver. Thus, he does not receive the full amount of his earnings by from 15 to 20 percent., the deficiency being left behind him at the truck shop over and above the ordinary profit of the fair dealer. . . . . . Again, the truck master is enabled by his not having to provide cash payments for his workmen,—by his exorbitant profit on his shop goods to undersell very considerably the cash paying manufacturer, so that the latter, to be enabled to compete with the truckster, is compelled, contrary to all the feelings of humanity, to reduce the weavers' wages, and thus grind them down, and that when he himself admits the wages are already too low. And this example, in a few weeks, is followed by the truckster, though he takes the pride to himself (however falsely), that it did not originate with him, while, at the same time, he is the sole cause of it. Thirdly: that the system of truck tends to their degradation in society. It makes them dishonest even against their will; I have had customers who, when they worked for those masters who paid them in money, were punctual, honest men, settling their accounts with me with evident pleasure and satisfaction; but, a scarcity of work ensuing, they have been induced to apply for work to the truck master —they have succeeded; but for that labor they have received nothing but goods; they are called upon to pay the few shillings, or, may be, the few pounds they left unpaid for goods while out of employ, but they are rendered totally incapable to meet the demand; they have no money, neither can they get any; and, moreover, were the tradesman to proceed to extremities and employ an attorney, still the man could not pay but must go to prison. . . . . . Many pounds are lost to the tradesman in this way, and many an honest man thus ranked with rogues."

These, in short, were the consequences of the truck system upon the poor weaver. "Now, look at the tradesman robbed of his customer; I instance myself who have lost at least one-third of my business; I assert it as a positive fact; cheated by my customers, who, compelled by shortness of work to change masters (whereas no money may transpire, or the money paid is expected to be left behind at the shop), are obliged to cheat me, and, moreover, with every prospect of losing the customer that yet remains unless the system is stayed . . Look, again, at the upright, conscientious manufacturer, the man acting from a principle of integrity on the one hand, and a desire for the welfare of the working classes on the other; he really is an object of pity,—he labors under disadvantages that are ruinous to his trade, and which, if not removed will certainly prove so. He provides cash payments for his workman once a fortnight; the truckster, on the contrary, has no such payments to prepare for, his payments are not required under three or four or six months, and hence he is enabled to turn and wind his money all this while, which necessarily obtains for him considerable advantages over and above the other; look, again, at the profits arising from the sale per week of from £200 to £300 worth of goods at an exorbitant profit, and all may be considered as for *ready* money, and what is the result of this? Why this alone enables the truckster to go into the markets and sell the cloths at the *cost of manufacturing*, and yet at the same time realize a living profit. Can the upright manufacturer stand against this? I think not. There is but one way in which he can at all relieve himself, and that is either by reducing the weavers' wages down to the lowest sum possible, or by adopting himself the system he so much abominates, and which from principle he has ever avoided."[86]

This quotation illustrates clearly the effects of the pernicious system of truck. The helplessness of the weaver could have been removed at that time only through organized resistance, which, under the merchant employer system, was impossible because of the extensive dispersion of the weavers in the rural districts. Other causes, such as surplus of labor resulting from the employment of family labor—the impetus that the profitableness of child labor gave to early marriage and high birth-rate—and the decline of the hand weaving trade, were undoubtedly at work.

B. Worsted

The worsted [87] manufacture of England in 1840 may be defined as including all persons who were fully employed in combing[88] and spinning the woolen fibre into yarn and weaving the woolen yarn, independently or together with cotton, or silk, or cotton and silk yarns, into worsted cloth.[89] Under this definition the manufacture employed in 1841 a total of 37,561 persons,[90] including the three branches of worsted, stuff and carpet.[91]

Table 63. Number of persons employed in worsted manufacture, England, 1841

Branch	Number	Percentage
Worsted	26,706	71.1
Stuff	7,437	19.8
Carpet	3,418	9.1
Total	37,561	100.0

These 37,561 persons in the worsted manufacture were distributed in the different counties of England as follows.[92]

Table 64. County distribution of persons employed in worsted manufacture, England, 1841

County	Number	Percentage
West Riding, York	29,823	79.4
Norfolk	3,418	9.1
Worcester	1,653	4.4
Other	2,667	7.1
England	37,561	100.0

West Riding of Yorkshire, employing about 80 percent of the total number of persons, was the undisputed centre in 1841. Its rapid rise since the introduction of the industry in the early part of the 18th century offered a strong contrast to the equally rapid decline of Norfolk, the birthplace and centre until about 1800. There were many reasons that explained the transference of the worsted manufacture from the East to the North, of which the enterprising spirit of the northern manufacturer rather than the abundant supply of natural resources may be given the first importance. It is true that West Riding, as compared with Norfolk, had the advantages of raw materials and climate for worsted spinning, but that alone, without the progressive character of the early industrial leaders to introduce new machinery and to apply capital on a large scale, would not have given the county such an impetus of growth.[93] West Riding, furthermore, had a ready market near at hand, for the worsted yarn was no sooner produced than consumed by the factories in West Riding, Norfolk, and the Midland counties of Leicester, Nottingham and Worcester. This contrast of the relative position between West Riding and Norfolk is more clearly brought out by a comparison of population growth in the leading worsted centres within these two counties.[94]

Table 65. Statistics of population growth in Yorkshire and Norfolk, 1801 and 1851

Place	1801	1851	% increase
Bradford	13,264	103,778	782
Halifax	12,010	33,582	279
Norwich	36,238	68,195	183

The Factory System

In the worsted manufacture in 1840 two systems of industrial organization were in operation, namely, merchant employer and factory systems. Each system, in worsted as in other textile manufacture, predominated in accordance with the technical process employed. Worsted weaving, although partly taken over by the power looms in 1840, still employed a large number of weavers under the merchant employer system. Carpet weaving, for instance, illustrated the complete operation of this system; the same was true of the weaving of bombazine—a mixed fabric of silk and worsted. Worsted spinning, and to a certain extent worsted weaving, were organized under the factory system. In 1784 the first worsted spinning factory, with machinery driven by horse or water power, was erected at Dolphine Holme in Lancashire, 95 while in 1787 another one was erected at Addingham on the Wharfe in Yorkshire.96 The new system of power spinning developed rapidly, and by 1820 hand spinning under the old merchant employer system became "almost extinct in Yorkshire".97 As to worsted weaving, power looms under the factory system were widely adopted from 1825 to 1840. 98 In the following table the number of factories, the total amount of motive power, and the motive power per factory, were given for Bradford, the foremost worsted metropolis, for the different years from 1810 to 1833; they illustrated typically the growth of factory system in the worsted manufacture.99

Table 66. Number of factories, total amount of motive power, and motive power per factory in Bradford, Yorkshire, 1810-33

Year	No. factories	Motive power (in H. P.)	Motive power per factory (in H. P.)
1810	5	120	24
1815	10	250	25
1820	20	538	27
1825	26	706	27
1830	31	862	28
1833	34	1,148	34

This speedy triumph of the factory system was due to several factors. The raw materials were more expensive and needed more capital for their purchase; and the work was less difficult and required less skill on the part of weavers and other operatives. Capital took control and operative skill became subservient.100

Labor Concentration

In worsted as in other manufactures there are two elements of a factory, namely, the concentration of labor and the employment of power machinery. With regard to the concentration of labor an index is found in the number of persons employed per worsted factory in the different counties of England in 1838.101

Table 67. Number of pesons employed per worsted factory in important counties, England, 1838

County	No. factories	No. employees	Employees per factory
York	348	26,603	76.4
Leicester	26	1,734	66.7
Lancaster	12	924	77.0
Worcester	9	696	77.3
Norfolk	3	385	128.3
Other	18	1,290	71.7
England	416	31,632	76.0

According to the above table, the average worsted factory size for England in 1838 is 76. This average is hereafter referred to as average number 1. The close agreement between the average factory size for Yorkshire and that for England as a whole is striking but not convincing, as Yorkshire has more than four-fifths of the total number of factory employees, or 84 percent. An analysis of the average factory size among the different parishes in Yorkshire is necessary in order to find out the variation of average factory sizes.

Table 68. Number of persons employed per worsted factory in important parishes, York, 1838

Parish	No. factories	No. employees	Employees per factory
Bradford	142	10,896	76.7
Halifax	80	5.614	70.2
Leeds	13	2,149	165.3
Keighley	38	2,125	55.9
Bingley	18	1,312	72.9
Wakefield	13	916	70.5
Other	44	3,591	81.6
Yorkshire	348	26,603	76.4

With the exception of Leeds the average factory sizes among the parishes of Yorkshire varied from 55.9 persons per factory in Keighley to 76.7 persons per factory in Bradford (i.e. disregarding the unimportant parishes grouped together under the heading "other"). This variation of average factory sizes among the parishes of Yorkshire in the worsted manufacture is small, and serves to bring out their uniform distribution as a whole.

The average factory size for England, namely, 76 persons per factory, is an arithmetic mean, and is therefore subject to the influence of extremes. Following our practice in dealing with the cotton and woolen manufactures, another average for the worsted factories is now computed on the same principles and from the same sources as that for the cotton or woolen factories, for the year 1838.102 This average, a median, is compiled after an enumeration of all worsted factories located in parishes having one or two woolen factories per parish. Omitting all details, the full enumeration is given in the following table.

Table 69. 34 worsted factories located in parishes having one or two factories per parish, England, 1838

Persons per factory	No. factories
1 to 10	3
11 to 20	3
21 to 30	6 (a)
31 to 40	1
41 to 50	0
51 to 60	4 (b)
61 to 70	1
71 to 80	1
81 to 90	3
91 to 100	0
101 to 110	6 (c)
111 to 120	2
121 to 130	1
Above 130	3 (d)
Total	34 (e)

(a) 22, 23, 25, 25, 26.
(b) 51, 55, 56, 57.
(c) 101, 103, 106, 107, 110, 110.
(d) 180, 214, 215.
(e) These 34 factories employed a total of 2,465 persons, or an average of 72.5 persons per factory.

According to the above table, the median is 57, the lower quartile, 25, and the upper quartile, 107. The quartile coefficient of deviation is, therefore, 0.62. This average is hereafter referred to as average number 2.

A third average is compiled after an enumeration of the actual sizes of 79 factories in Yorkshire in 1833. 103 This enumeration is presented in the following table.

Table 70. 79 worsted factories in Yorkshire, 1833

Persons per factory	No. factories
11 to 20	8
21 to 30	13 (a)
31 to 40	4
41 to 50	12
51 to 60	6 (b)
61 to 70	3
71 to 80	1
81 to 90	4
91 to 100	3
101 to 110	4
111 to 120	0
121 to 130	4 (c)
131 to 140	0
141 to 150	0
151 to 160	2
161 to 170	2
171 to 180	2
181 to 190	0
191 to 200	0
Above 200	11
Total	79 (d)

(a) 21, 21, 22, 22, 24, 25, 26, 27, 28, 29, 29, 30, 30.
(b) 51, 56, 56, 59, 60.
(c) 125, 129, 130, 130.
(d) These 79 factories employed a total of 8,407 persons, or an average of 106.4 persons per factory.

According to the above table, the median is 56, the lower quartile 30, and the upper quartile, 125. The quartile coefficient of deviation is, therefore, 0.61. This average is hereafter referred to as average number 3. In the following table, the two averages, viz. average no. 2 and average no. 3, are compared with respect to their medians, quartiles, and quartile coefficients of deviation.

Average Number	Lower quartile	Median	Upper quartile	Quartile coefficient of deviation
2	25	57	107	0.62
3	30	56	125	0.61

In the above table, the close correspondence between the two medians, viz. 56 and 57, is striking. These medians for the worsted factories, like those already computed for the cotton and woolen factories, give a lower sized and more representative average than the arithmetic mean (76).

In the following table, the increase of the average factory size from 1838 to 1856 is presented. The averages for 1850 and 1856, being arithmetic, are subject to the same criticism as the average for 1838.

Table 71. Persons employed per worsted factory, in England and York, 1838-56

Year	England	York
1838	76.0	76.4
1850	160.1	169.6
1856	169.6	177.5

Referring back again to our compilation of the 79 Yorkshire factories in 1833, there were 21 factories employing more than 113 persons per factory.[104] These 21 factories were in our definition "first-rate" factories, employing a total of 5,597 persons. The arithmetic average of persons employed per factory was therefore 266.5 persons, while the median average, 207 persons. The largest factory employed 610 persons, while the smallest, 125 persons. In the following table a few details regarding the four largest factories are given.

Table 72. Statistics on four largest woolen factories, Yorkshire, 1833

Name of occupier	Location	Year of erection	Amt. motive pwr. (in H.P.)	No. employees
Stansfield, T. W.	Leeds	1822	70	610
Akroyd, James	Halifax	1825	60	550
Wood, John (jr)	Bradford	1812	134	527
Wm. Rouse & Son	Bradford	1815	102	438

Among these four factories that of John Wood, a stuff weaving factory, was well-conducted. The factory was "as clean, as light, and as comfortable as a drawing room, or rather as a series of drawing rooms." The employees, being principally girls, were provided with seats "when they are not actually employed, which state of leisure, from the nature of their occupation, very frequently occurs." There were always a surplus number of children "in order that they may be sent by instalments to a school room on the premises, where they learn to knit and to sew, as well as to read and write," in order to be good wives in the future. "If any one among them [girls] was known to misconduct herself, the rest would immediately apply to him [Mr. Wood, the proprietor] to dismiss her from among them." Furthermore, they were to appear in clean clothes twice a week, and "a medical man is engaged to visit the factory weekly to examine into the general health of the children, besides which he gives more frequent attendance to those who may be ill."[105]

Employment of Power Machinery

Motive power. The first element in the employment of power machinery is the motive power—steam or water. In 1838 the percentage of steam power employed in worsted factories was 81.7, the same as that for the cotton factories. In the following table the increase of steam motive power over water since 1838 is shown.

Table 73. Motive power in worsted factories, England 1838-56 (in horse power)

Year	Steam motive power		Water motive power	
	Amount	Percentage	Amount	Percentage
1838	5,863	81.7	1,313	18.3
1850	9,769	86.7	1,501	13.3
1856	13,182	91.0	1,310	9.0

Another table is given below to show the increase in the amount of motive power per worsted factory from 1838 to 1856.

Table 74. Amount of motive power per worsted factory, England, 1838-56 (in horse power)

Year	No. factories	Amt. motive power	Motive power per factory
1838	416	7,176	17.3
1850	493	11,270	22.9
1856	511	14,492	28.4

Spindles. The second element in the employment of power machinery is the spindles in worsted factories. Specialized spinning factories in the worsted manufacture were in existence long before the general introduction of machinery into the woolen manufacture. "There was a steady demand for their yarns, independent of the local manufacture; and it is the continuance of this external demand in various forms which, more than anything else, has favored specialization at the beginning of the nineteenth century. The Yorkshire 'mill yarn' went in considerable quantities into the old worsted manufacturing districts of East Anglia. As the manufacture in those districts decayed, in face of northern competition, the demands from the hosiery trade and from abroad sprang up—demands which, roughly speaking, have increased ever since."[106] In fact, the factories in worsted manufacture were largely spinning factories during the early years of the nineteenth century. In Yorkshire in 1833, for instance, 66 out of 79 factories were spinning factories according to one compilation.[107]

Table 75. Yorkshire factories in worsted manufacture, 1833, classified according to technical processes employed.

Technical process	No. factories
Spinning	66
Spinning & weaving	7
Weaving	1
Unclassified	5
Total	79

However, as the years advanced, spinning-and-weaving factories were more generally built. In 1850 and 1856 the proportion between spinning factories on the one hand, and spinning-and-weaving factories on the other, had changed; although the spinning factories were still in the majority.

Table 76. Worsted factories in England, 1850-56, classified according to technical processes employed

Technical process	No. factories 1850	1856
Spinning	222	220
Spinning-and-weaving	155	150
Weaving	97	121
Not included in either	19	20
Total	493	511

The distribution of the number of spindles employed in the worsted factories among English counties in 1850 and 1856 is shown in the following table. Yorkshire, with 86.3 % in 1850 and 93.4 % in 1856, had the largest proportion of the total number of spindles.

Table 77. County distribution of spindles in worsted factories, England, 1850-56

County	1850	1856
Yorkshire	746,281	1,212,587
Leicester	27,525	13,015
Lancaster	27,190	20,364
Norfolk	19,216	23,464
Worcester	15,146	18,510
Other	29,516	10,386
England	864,874	1,298,326

In the following table the number of spindles per spinning or spinning-and-weaving factory from 1850 to 1856 is shown. It is to be noted that the spinning-and-weaving factories had a larger number of spindles, collectively as well as per factory than the spinning factories. Evidently large manufacturers in the weaving branch, who used to buy worsted yarns from the spinners, had found it profitable to embark capital in both branches, and to operate his factory on a large scale.

Table 78. Number of spindles per worsted spinning (or spinning-and-weaving) factory, England, 1850-56

Type of factory	No. factories	No. spindles	Spindles per factory
Spinning			
1850	222	417,473	1,881
1856	220	629,345	2,861
Spinning-and-weaving			
1850	155	447,177	2,885
1856	150	668,981	4,460

The change in size as shown in the above table can be brought out more clearly in the following arrangement.

Year	Spindles per spinning factory	Spindles per spinning-and-weaving factory
1850	1,881	2,885
1856	2,861	4,460

Power looms. The third element in the employment of power machinery is the power looms. The introduction of power looms into the worsted manufacture was not effected without opposition from the laboring population. For instance,

"In the year 1822, Mr. Warbrick, a worsted stuff manufacturer [of Leeds], got one of these looms made as secretly as possible, and sent it privately to be set to work in a mill, at the village of Shipley; but it was scarcely put in motion, ere the bellman was sent round to the neighboring villages to give notice of its arrival. Consequently a great number of weavers assembled round the mill, and threatened the whole fabric with destruction, if the obnoxious machine was not instantly removed; it was therefore immediately taken down, and placed in a cart, under the convoy of a body of constables, but it was no sooner seen than it was attacked by the enraged weavers, who routed the civil power, destroyed the loom, and dragged its roller and warp in triumph to Baildon."[108]

Four years later, in 1826, another riot broke out to attack the factory of Messrs. Horsfall in May. Several meetings were held, following which attempts to destroy the machinery and building were made without great success. At last the weavers were so enraged as to start shooting. Two of them were thus killed, while another two were sent to the York Castle for imprisonment. Many others, meantime, were wounded. After 1826 the introduction of power looms met with less opposition.[109] In 1835 there were 3,174 power looms in use, while fifteen years later a ten-fold increase was brought about.[110]

Table 79. Number of power looms in worsted factories, England and Yorkshire, 1835-56

Year	England	Yorkshire	
		County	West Riding[111]
1835	3,174	3,082	2,768
1841	....	....	11,458
1843	....	....	16,870
1845	....	....	19,121
1850	32,617	30,856	29,539
1856	38,809	35,298	......

In the following table the county distribution of the power looms from 1835 to 1856 is given.

Table 80. County distribution of worsted power looms, England, 1835-56

County	1835	1850	1856
Yorkshire	3,082	30,856	35,298
Lancaster	.....	1,112	2,365
Norfolk	.....	428	349
Other	92	221	797
England	3,174	32,617	38,809

In 1835, 2,856 out of 3,082 power looms in Yorkshire were distributed among the following parishes.

Table 81. Parish distribution of 2,856 power looms, Yorkshire, 1835

Parish	No. power looms
Bradford	1,015
Leeds	846
Halifax	596
Keighley	267
Other	132
Total	2,856

These 2,856 power looms in Yorkshire were employed by 37 factories, the size of which varied from 2 to 479 power looms per factory. The arithmetic average of power looms per factory was accordingly 77.2, while the median average amounted to only 35, as shown in the following table.

Table 82. Power looms in 37 worsted factories, Yorkshire, 1835

Power looms per factory	No factories
1 to 10	6
11 to 20	4
21 to 30	6
31 to 40	5
41 to 50	4
51 to 60	3
61 to 70	1
71 to 80	2
81 to 90	1
Above 90	5 (a)
Total	37

(a) 237, 302, 357, 378, 479.

In the following table the number of power looms per weaving or weaving-and-spinning factory from 1850 to 1856 is shown.

Table 83. Number of power looms per worsted weaving (or weaving-and-spinning) factory, England, 1850-56

Type of factory	No. factories	No. power looms	Power looms per factory
Weaving			
1850	97	12,814	132.1
1856	121	13,225	109.3
Weaving-and-spinning			
1850	155	19,797	127.7
1856	150	25,584	170.6

The change in size as shown in the above table can be brought out more clearly in the following arrangement.

Year	Power looms per weaving factory	Power looms per weaving-and-spinning factory
1850	132.1	127.7
1856	109.3	170.6

With regard to the effects arising from the employment of power machinery, the first one is the displacement of adult male by child and woman labor. Although child and woman labor had long been in existence under the old systems of industrial organization, the first employment of power machinery under the factory system greatly extended it. Since 1838, however, both child and woman labor have decreased in proportion. The decrease in the proportion of child labor was a result of the Factory Acts not uncommon in other textile manufactures, while the decrease in the proportion of woman labor was a unique feature in worsted manufacture. In the following table the proportion of child and woman labor in worsted factories for various years is shown.

Table 84. Child labor in worsted factories, England and Yorkshire, 1838-47 (in percentage)

Year	England	Yorkshire
1838	59.07	60.69
1847	43.54	44.42

Table 85. Woman labor in worsted factories, England and Yorkshire, 1838-56

Year	England	Yorkshire
1838	72.51	73.72
1847	70.78	71.01
1850	66.12	66.49
1856	65.68	66.13

Another effect of the employment of power machinery is the increased production. "In 1715, with the old single-spindle it took 10 spinners to keep one stuff-loom at work; one spinner with the jenny will keep two cloth-looms at work; whilst one mule-spinner will keep, perhaps, 10 looms at work; the throstle, on which worsted is spun, will do the same. To do all the spinning now done by machinery would require some millions of people on the single spindlewheel."[112] In fact, "considering the constancy of the employment, and the freedom from all deductions, the young women and girls employed upon the power loom—the better tool—are actually earning more, in some cases considerably more, than the father employed on the hand loom, the inferior tool." [113]

Statistical Summary

Table 86. Factory statistics on worsted manufacture, England, 1838-56

Year	No. factories	No. employees	Amt. motive power (in H.P.)	No. spindles	No. power looms
1838	416	31,632	7,176	....	......
1847	...	51,796	.....	....	......
1850	493	78,915	11,270	864,874	32,617
1856	511	86,690	14,492	1,298,326	38,809

If we take 1838 as a base of 100 percent., we may then express the above table in relative terms below.

Year	No. factories	No. employees	Amt. motive power (in H.P.)
1838	100.0	100.0	100.0
1847	.....	163.7	.....
1850	118.5	249.5	157.1
1856	122.8	274.1	202.0

Comparing the statistics on factories, employees and motive power, we may conclude that the factories in 1856 were considerably larger than those in 1838, with regard to the number of persons employed as well as the amount of motive power per factory.

Merchant Employer System

(IN WORSTED AND STUFF WEAVING)

In Yorkshire in 1840 there were about 14,000 handloom weavers fully or partly employed in worsted and stuff weaving.[114] Here the prevalence of the merchant employer system in the worsted handloom weaving offered a strong contrast to the master craftsman system in the woolen handloom weaving. This difference of industrial system in the woolen and worsted handloom weaving demands a word of explanation. In the first place, the minute specialization in the processes of combing, spinning and weaving in the worsted manufacture was an early factor for the adoption of the merchant employer system. Even before the industry was transferred to Yorkshire from East Anglia the system had been introduced except in Colchester. "Outside of Colchester, the industry appears to have been originally established by capitalists, who had no share in production, and the Colchester bay makers themselves seem to have engaged less and less in production, and in the eighteenth century were merchants pure and simple."[115] In Norwich the transition of worsted weaving from the master craftsman to the merchant employer system was said to have been completed by the year 1662.[116] Secondly, the more capitalistic character of the worsted industry led to the early establishment of the merchant employer system. In Yorkshire there were small master craftsmen in the worsted manufacture who transacted their business in the worsted or stuff halls at Bradford, Halifax and other important centres. But by 1840 these halls had begun to decline. The stalls were then by far too incommodious and unsuitable for the extensive business of the merchant employers (called "manufacturers"), who now required extensive warehouses. In fact, these stalls had been let at a rent, and transformed into private business offices or warehouses.[117]

In 1838 there were in Norwich, Norfolk, a total of 5,075 handloom weavers, of whom only 4,054 were partly or fully employed, and the rest unemployed. About 3,000 of the 4,054 weavers employed worked on the worsted fabrics of various descriptions, while the other 1,054 were engaged in the making of silk, woolen, linen and other cloths of mixed materials.[118] Most of these weavers carried on their occupations at home, under the merchant employer system. A small number, about 650 weavers, however, were congregated together in what was generally known as the handloom factory.[119]

Merchant employer and middlemen. In Yorkshire as well as in Norfolk, the merchant employer in worsted handloom weaving performed very much the same function as the West Country merchant employer in woolen handloom weaving. The chief difference, however, consisted in the existence of a distinct class—the "undertakers"—in the Norwich District. These undertakers, or "carriers", were primarilly distributors of work from the merchant employer to the handloom weavers working at their own houses. They were, in other words, the connecting link between the two classes when distance so separated them as to render their direct dealing an impossibility. In the surrounding villages of Norfolk, such as Wyndham, the usual system was stated to be "that a man, called an 'undertaker', brought the work out from Norwich, and no body knew who were the manufacturers from whom he received it; and got it done at as cheap a rate as he could. The difference between what the manufacturer allowed him, and the price for which he got it done, constituted his profit."[120] These weavers at the outlying districts around Norwich were in worse distress than those in the city and suburbs.[121] "They had no masters to whom to make their complaints, and when they went to the farmers to ask for work, the farmers told them that their own men had a preferable claim."[122]

The number of weavers employed by each merchant employer in Norwich and vicinity varied, and each merchant employer was not always confined to the manufacturing of worsted fabrics alone. In 1838 Mr. J. Francis employed 348 weavers, "consisting of male and female, young and old, good and bad, quick and slow, on the different articles", [123] while Messrs. Shickle, Towler & Co. employed only 32 weavers. [124]

Home work. In Yorkshire ,family labor was commoner in worsted than in woolen handloom weaving. "After weaving, the piece requires *picking*, that is, going over with a pair of steel nipples, and taking off knots, threads and other imperfections left by the process of weaving. If the weaver do this himself, as he often does, it will take about one-third as much time as the weaving; but this is often done by the wife, and perhaps still more frequently by the weaver and his wife together .... In other cases the weaver will have two looms, at one of which he himself weaves, his wife weaving at the other."[125] Besides the wife, the weaver's children were also employed in worsted handloom weaving. "The great facility of acquiring the art of weaving, the encouragement given to parents to have their children taught, have tended to keep the weaving business overstocked with hands."[126] In Norwich, according to one merchant employer, "for every 100 men and boys employed in weaving, 64 women and girls are employed." [127]

Abuses

Truck. The existence of truck among the worsted weavers may be judged from the following statement by Commissioner Chapman for West Riding in 1838.

> "At Idle, where I held a public inquiry for the townships, Shipley, Eccleshill and Bolton, a written statement was handed in, complaining of the existence of the truck system, at a small hamlet called Windhill, in the first named township. It is a fact worthy of notice, that I was begged not to ask questions on the point, as any testimony on the part of a weaver who had suffered from the custom, would have been followed by loss of employment. I was, however, assured by many persons, both employers and weavers, that the allegation was correct. This shows how completely impotent is the law, however stringent it may be, where both employer and employed will consent to violate it. The employed is compelled to submit: the constantly overstocked state of the labor market places him at the mercy of the master, and the same condition prevents him taking a single step to expose the fraud to which he is subjected."[128]

Embezzlement. That the merchant employer system had been well established in the worsted branch of handloom weaving around 1840 may be further substantiated by the frequent complaints of embezzlement among the worsted weavers. In 1838, one weaver in Bradford told the Handloom Weavers' Commission that the temptation to embezzle was "very great; the temptation is this:—Some manufacturers compel the weavers to put in a certain quantity of weft; if two or three hanks be left out, their wages are abated; the consequence is, that, to prevent that deduction, the weaver is tempted to conceal what he may have been unable to weave in." [129] Other weavers gave contradictory evidence on the same subject, while the Commissioner, after an extensive inquiry, concluded that "poverty does certainly occasionally produce the offence [i.e. embezzlement]; in some cases also it may arise from bad moral training—depravity, as it is called; but that embezzlement exists to an extent to be called a practice, I feel myself entitled, from the care I have taken to come at a right decision, to disbelieve."[130]

Merchant Employer System

(IN CARPET WEAVING)

In 1841 the handloom carpet weaving employed a total of 3,409 persons, [131] of whom about one half were found in the county of Worcester alone. The rest were distributed in the various other counties including York, Durham, Salop, Middlesex and Westmoreland. Kidderminster, the carpet metropolis of England, had 2,021 carpet handlooms in 1838.[132] In this city a unique example of the merchant employer system was then found. In fact, the system had been so modified on account of the industrial conditions in carpet handloom weaving that it has been compared with the factory system by leading authorities.[133] The weavers were here concentrated within a building where a number of carpet handlooms were installed. Such concentration, however, lacked the key element necessary to the establishment of a factory system, namely, employer's supervision. The provision of workshop and looms by the employer, together with the raw materials, was necessary because of expensive capital outlay; but such provision did not necessarily carry with it the authority and responsibility of the employer to supervise the weavers in the execution of their work. In this connection we may well quote from a contemporary employer on the operation of the system before proceeding to further analysis. Thus, Henry Brinton, an employer of 120 weavers in 1833, told the Children's Employment Commission that

> "the master manufacturer provides shops, looms, and materials, which are under the care of operative weavers, who provide assistants, generally young persons and children. These young persons and children being under the control of weavers, the masters do not interfere between the weaver and assistant, except in cases of immorality, when the master might require them to be dismissed. The factories contain various numbers of looms, from 2 to 36; and they are generally open at five in the morning ,and closed at nine in the evening. The usual practice is to pay the weaver by piece work, and the manufacturer does not inquire in what way the weaver employs his time."[134]

In the above quotation it is clear that employer's supervision was wholly absent except in unimportant matters such as immorality—unimportant from the viewpoint of industrial organization. The irregularity of work, for instance, would indeed be hard to conceive, and on that account, the mere concentration of labor could not be a significant change towards the factory system. This irregularity was a common experience whether the weaver be industrious or otherwise. To quote from the Children's Employment Commission of 1841,

> "every Saturday and Thursday the work is given out from the warehouse in the shape of bobbins, with which the workman departs to the distant shop, and with his draw-boy, or girl, begins to 'fettle' his frames: if he happens to be an industrious, steady, and sober man he begins at once to weave, and leaves off at a reasonable and proper hour, thereby earning his 30 s. upon the average w eekly, deducting 5s. or 6s. for his drawer; he will, however, have to wait occasionally for yarns undyed, or orders; so that taking the year through he does not work more than 4 1/2 days in the week, or complete a piece in less than 9 days from the beginning. If, on the contrary, he is idle, dissipated, or dissolute, having possession of the key of the shop, he will commence work when he pleases, and leave off when he pleases, always requiring the child to be in waiting for him; the beerhouse is his too constant place of resort, here he wastes his time and money, and impoverishes his family until the moment arrives that he must work in good earnest 'up for the fall'; the children are aroused from their sleep wretched and miserable; at two, three, or four o'clock, and no

matter, they must go to their cold, dark, damp and dreary work-room from such hour until twelve or one at noon without leaving for a meal, their breakfasts on such occasions being brought to them."[135]

On the other hand, the chief elements of a merchant employer system were clearly present. The merchant employer here provided the workshop, looms and raw materials. The raw material, according to one merchant employer, 'is given out from the warehouses when it is wanted, and that is at all hours of the day; it is given out in bobbins, not by weight, we have no criterion to determine whether the same weight is returned in manufactured goods; there are good and bad amongst the people, but we must depend upon their honesty. "[136] In other cases, the merchant employer gave out the yarns "on Thursdays and Saturdays, which they expect will be brought back woven, before 11 o'clock of the morning of one of the above days of the following week, which is designated 'the day of the fall.'' [137] On these 'falling days' the weavers received their wages in return for the work done, and generally spent the rest of day in idleness.[138]

The extent of the merchant employer's business varied, for which a good index is found in the number of looms owned and of weavers employed per merchant employer. In Kidderminster, for instance, there were in 1838 twenty-four merchant employers, owning a total of 2,011 looms. They employed 4,016 persons including men, women and children. In the following table the number of looms owned and of persons employed by each of the twenty-four merchant employers is given.[139]

Table 87. Number of carpet looms owned and of persons employed by each of the 24 merchant employers, Kidderminister, Worcester, 1840

Name	No. looms	No. persons
Hooman & Co	242	360
Lea	209	440
Talbots	157	323
Woodward	123	198
Morton, J.	116	227
Humphries, T. & J.	110	214
Butcher etc.	108	214
Hopkins, T.	93	183
Watson & Son	90	198
Newcomb & Son	76	158
Barber & Cole	75	159
Bowyer & Dixon	75	167
Brinton, H.	67	133
Wright & Crump	67	140
Gough & Sons	64	138
Stoddard & Woodward	60	152
Dobson & Son	55	112
Kiteley & Fawcett	46	105
Cooper, William	42	90
Potter & Duckham	36	79
Hiles, J.	34	79
Bough	24	53
Robinson, J. H.	22	50
Silk, William	20	44
Total	2,011[140]	4,016

Hooman & Co., the largest of the merchant employers in 1838, was reported to have owned 232 carpet looms in 1841, each of which required the service of a weaver and a drawboy. These 232 looms were distributed in six buildings as follows: 30, 30, 30, 34, 42, 66. [141]

Chapter IV. Silk, flax and hemp

A. Silk.

The silk manufacture of England in 1840 may be defined as including all persons who were fully employed in throwing the raw silk or spinning the waste silk into thread or yarn, and weaving[1] the silk thread or yarn into silk goods. Silk goods are most diverse in nature, and no two definitions agree in their scope and extent. It is by process of exclusion rather than of inclusion that an adequate notion of silk goods is gained. In our definition, although ribbons represent one class of silk goods, yet they are excluded and treated in another chapter for two reasons, namely, conformity with the Census definition of 1841 and general acceptance of the separate treatment of ribbon manufacture from silk.[2]

Under our definition, the silk manufacture of England in 1841 employed 66,630 persons,[3] who were distributed in different counties as follows:

Table 88. County distribution of persons employed in silk manufacture, England, 1841

County	Number	Percent.
Chester	18,723	28.1
Lancaster	18,590	27.9
Derby	2,865	4.3
Middlesex	7,263	10.9
Norfolk	2,132	3.2
Essex	1,932	2.9
Other	15,125	22.7
England	66,630	100.0

In an analysis of the local distribution of silk manufacture in England, two features stand out as being unique. In view of the high degree of concentration already attained, the dispersion of the industry since 1860 constitutes a feature of no small interest. The early concentration quickly disappeared in face of the foreign competition after the conclusion of the famous Cobden Treaty, and until today such dispersion in the English silk manufacture became a characteristic absent in other textile manufactures as well as in the silk manufacture of foreign countries like France.[4] The relative decline of the Eastern counties as compared with the Northern is another feature, the causes of which are not far to seek. The rapid growth of the throwing and spinning factories in the North, a result of proximity to other textile manufactures, gave the Northern manufacturer a distinct advantage over the Eastern. It not only permitted him to choose the kind of thrown or spun silk fit for his particular manufacture, but also relieved him from the uncertainty of foreign import. The Northern manufacturer, again, was generally of a more enterprising type than the Eastern; whereas the latter was reluctant to introduce mechanical changes into his business, the former was always ready to adopt latest improvements.

The Factory System

Like other textile manufactures, the silk manufacture was organized under different industrial systems that varied with the technical processes employed. In 1841 silk throwing and spinning were carried on inside the factories; while silk weaving, including winding and warping, was being slowly taken over as a factory process, and the "undertakers" whose former function was partly to wind and warp the thrown or spun silk and give it out to the weavers to be woven into silk goods began to disappear. In short, the silk manufacture in 1840 was carried on under two industrial systems, namely, factory and merchant employer systems. In the following table the extent of each system in silk manufacture is given.

Table 89. Persons employd in silk manufacture under factory and merchant employer systems, England, 1840

System	Number	Percent.
Factory	33,565	50.4
Merchant employer	33,065	49.6
Total	66,630	100.0

The factory employees were chiefly located in the Northern counties of Chester, Lancaster and Derby, while a small minority resided in other counties such as Norfolk, Essex, etc.[5]

Table 90. County distribution of factory employees in silk manufacture, England, 1838

County		Number	Percent.
Northern		21,733	64.7
Chester	11,832		
Lancaster	5,591		
Derby	3,226		
York	1,084		
Eastern		4,964	14.8
Norfolk	2,274		
Essex	1,441		
Hertford	784		
Suffolk	465		
Western, &c.		6,868	20.5
Somerset	2,065		
Other	4,803		
Total		33,565	100.0

Labor Concentration

In following our previous practice in treating the cotton, woolen and worsted manufactures, the index of labor concentration as found in the number of persons employed per silk factory in the important silk counties of England in 1838 is given below.

Table 91. Number of persons employed per silk factory in important counties, England, 1838

County	No. factories	No. employees	Employees per factory
Chester	92	11,832	128.6
Lancaster	31	5,591	180.4
Derby	20	3,226	161.3
Norfolk	4	2,274	568.5
Somerset	24	2,065	86.0
Stafford	12	1,494	124.5
Essex	7	1,441	205.9
Other	73	5,642	77.3
England	263	33,565	127.6

The average number of persons employed per silk factory in England in 1838, according to the above table, was 127.6. Lancashire, with an average size of 180.4 persons per factory, took the lead,[6] while Somerset (excepting the unimportant counties included under the heading "other"), with that of 86.0, bottomed the list. Similar variation of the average factory sizes among the counties was found among the parishes, as may be seen in the following table.

Table 92. Number of persons employed per silk factory in important parishes of Chester, Lancaster and Derby, England, 1838

County	Parish	No. factories	No. employees	Employees per factory
Chester	Prestbury (incl. Macclesfield)	48	7,779	162.1
	Astbury (incl. Congleton)	35	3,276	93.6
Lancaster	Manchester	20	4,245	212.3
Derby	Derby	17	3,004	176.7

The average factory size in silk manufacture of England in 1838 was 127.6 persons per factory. This average, an arithmetic mean, is hereafter referred to as average number 1. In following our previous practice in treating the cotton, woolen and worsted factories, another average, a median, is compiled after an enumeration of all factories located in parishes having one or two factories per parish. The full enumeration, omitting details, is presented in the following table.

Table 93. 82 silk factories located in parishes having one or two factories per parish, England, 1838

Persons per factory	No. factories
11 to 20	4
21 to 30	8
31 to 40	9 (a)
41 to 50	4
51 to 60	7
61 to 70	9 (b)
71 to 80	5
81 to 90	2
91 to 100	6
101 to 110	4
111 to 120	3
121 to 130	1 (c)
131 to 140	0
141 to 150	2
151 to 160	2
161 to 170	0
171 to 180	2
181 to 190	2
191 to 200	2
201 to 300	2
301 to 400	6
Above 400	2
Total	82 (d)

(a) 31, 31, 32, 35, 37, 38, 39, 39, 40.
(b) 63, 63, 64, 64, 64, 64, 65, 66, 68.
(c) 125.
(d) These 82 factories employed a total of 9,436 persons, or an average of 115.1 persons per factory.

According to the above table, the median is 68, the lower quartile, 40, and the upper quartile, 125. The quartile coefficient of deviation is, therefore, 0.52. This median average of 68 persons per silk factory is much below the arithmetic average of 127.6 persons per silk factory. The arithmetic average for the silk factory, like that for the cotton, woolen, or worsted factory, is undoubtedly too high, and is distorted by extremes. In view of our analysis of the cotton, woolen, or worsted factory, we may accept the median average as being more representative of the silk factory size than the arithmetic average.

In the following table, the increase of average (mean) factory size in England since 1838 is shown.

Table 94. Persons employed per silk factory, England and two important counties, 1838-56

Year	England	Chester	Lancaster
1838	127.6	128.6	180.3
1850	153.3	127.7	283.0
1856	120.2	105.6	239.7

The increase of average factory size in silk manufacture from 1838 to 1850 is primarily due to the establishment of large throwing or spinning factories to make use of the mechanical improvements in textile machinery on the one hand, and to supply the greater demand of the manufacturers for home than for foreign thrown or spun silk on the other. Its decrease since 1850 is chiefly a result of foreign competition in throwing and spinning, which rendered it unprofitable to operate throwing or spinning factories on a large scale. Such decrease in average factory size is true not only of the number of persons employed, but also of the amount of motive power and the number of spindles and power looms employed.

The following return, compiled from the replies to the Inquiry of the Children's Employment Commission in 1833, shows the average size of the "first-rate" silk factories as measured by the number of persons employed per factory.[7]

Table 95. 11 "first-rate" silk factories in Manchester, Lancashire, 1833

Persons per factory	No. factories
136 to 200	3
201 to 300	3 (a)
301 to 400	1
401 to 500	1
501 to 600	1
Above 600	2 (b)
Total	11 (c)

(a) 205, 219, 251.

(b) 953, 1161. The factory employing 953 persons had one plant in Macclesfield, another in Congleton. The factory employing 1,161 persons had five plants in Macclesfield.

(c) These 11 factories employed a total of 4,698 persons, or an average of 427 persons. The median, however, is 251.

Outside the limits of the two counties, Chester and Lancaster, there were factories whose magnitude either equalled or exceeded that of Chester and Lancaster factories. For instance, Joseph Grout, the famous Middlesex crape manufacturer, had several plants of his own located in different parts of the Eastern counties, the capital investment of which in machinery, buildings and land was estimated at the huge sum of £143,546 in 1832. Despite the decline of business from 1822 to 1831, in the number of persons as well as of spindles employed, this remained the largest of all silk factories that we have yet come across in the silk manufacture.[8]

Table 96. Statistics of Grout's Factory, 1822-1831

Year	No. persons employed	No. spindles Employed	No. spindles Unemployed
1822	2,747	59,754	None
1823	2,974	59,754	None
1824	3,594	63,756	None
1825	3,908	86,664	None
1826	3,516	77,384	9,280
1827	3,164	78,522	8,142
1828	2,818	83,252	3,312
1829	1,540	60,024	26,040
1830	1,624	59,616	27,048
1831	1,871	59,616	27,048

After 1831 the factory began to recover, and in 1840 its employees had increased from 1,871 in 1831 to 2,630.[9]

Employment of Power Machinery

Motive power. In 1838 the percentage of steam power employed in silk factories was 71.33, which stood between that of 61.3 in woolen factories and that of 81.1 in cotton factories. In the following table, the increase of steam motive power over water since 1838 is shown.

Table 97. Motive power in silk factories, England, 1838-56

(In horse power)

Year	Steam motive power Amount	Steam motive power Percent.	Water motive power Amount	Water motive power Percent.
1838	2,309	71.3	928	28.7
1850	2,718	76.0	856	24.0
1856	4,238	83.9	816	16.1

Another table is given below to show the increase in the amount of motive power per silk factory from 1838 to 1856.

Table 98. Amount of motive power per silk factory, England, 1838-56 (in horse power)

Year	Total number of factories	Total amount of motive power	Motive power per factory
1838	263	3,237	12.3
1850	272	3,574	13.1
1856	460	5,054	11.0

Spindles. In 1838 the majority of spindles—about 54 percent.—were employed in the silk factories of the Northern Counties.

Table 99. County distribution of spindles in silk factories, England, 1850-56

County		1850		1856
Northern		635,697		625,046
Chester	287,575		292,849	
Lancaster	162,988		117,614	
York	128,808		117,164	
Derby	56,326		97,419	
Eastern		197.803		237,917
Essex	80,250		107,898	
Norfolk	63,692		78,690	
Hertford	39,459		36,639	
Suffolk	14,402		14,690	
Western		280,472		145,049
Somerset	178,798		41,548	
Dorset	55,360		60,232	
Gloucester	26,014		22,653	
Wilts	20,300		20,616	
Other		74,936		55,543
England		1,188,908		1,063,555

The silk factories in 1838 and since then were employed chiefly in throwing,[10] and partly in throwing and weaving. In 1850 as well as in 1856 more than two thirds of the total number of spindles were employed in throwing factories.

Table 100. Percentage of spindles distributed between silk throwing and throwing-and-weaving factories, England, 1850-56

Type of factory	1850	1856
Throwing	74.5	76.1
Throwing-and-weaving	35.5	33.9

However, the throwing-and-weaving factory had a larger number of spindles per factory than the throwing factory.

Table 101. Number of spindles per silk throwing (or, throwing-and-weaving) factory, England, 1850-56

Type of factory	No. factories	No. spindles	Spindles per factory
Throwing			
1850	192	886,417	4,617
1856	249	809,755	3,252
Throwing-and-weaving			
1850	30	302,491	10,083
1856	36	253,800	7,050

The change in size as shown in the above table can be brought out more clearly in the following arrangement.

Year	Spindles per throwing factory	Spindles per throwing-and-weaving factory
1850	4,617	10,083
1856	3,252	7,050

Power looms. In the period under consideration, the power loom in silk manufacture was yet in the experimental stage, and ranked after the cotton, woolen and worsted manufactures in the extent of its employment. In 1831, Porter wrote that "it becomes very doubtful whether the use of power looms, however they may be modified, is susceptible of much extension in any save the commonest branches of the silk manufacture."11 To this statement we may add that besides "the commonest branches of the silk manufacture," narrow ribbons and crapes were also to be included. The reasons for the slow introduction of power loom weaving into the silk manufacture were plain enough. One manufacturer, Joseph Grout, remarked that there was no "advantage, at the present [1832] price of hand weaving, in using the power loom, for inasmuch as we save in the price of weaving done by power, we lose by interest of capital on the first cost of the power looms, the additional waste of silk, and heavy expense of wear and tear, and wages of mechanics employed to keep them in order".12 Other reasons given were more technical than economic. In the first place, broad silk goods must have dyed silk for weaving, but in the process of dyeing silk became tender or brittle through chemical reaction, and thereby unfit to resist the rough and quick treatment of power loom weaving. Secondly, it would be difficult to introduce variety of colors or stripes for broad silk, as simplicity rather than intricacy in mechanical construction constituted the chief advantage of power looms at that time. Thirdly, a power loom, on account of its quick movement in weaving, might make it impossible for the weavers to prevent any imperfection in the goods woven before its discovery. Silk being a valuable material, the waste thus resulted might cause a material loss to the manufacturer. Fourthly, in silk if not in cotton weaving, tedious trimming of the warp consumed a large portion of labor, while weaving itself actually took less time. The economy of labor possible in cotton power loom weaving was absent in silk. Finally, the fickleness of fashion, especially in the figured branch of silk manufacture, resulted in frequent cessation of the looms at work at different seasons of the year, and thereby increased the overhead cost through disuse of fixed capital.13 All these reasons, technical and economic, accounted for the slow introduction of power loom weaving into the silk manufacture, which could be illustrated by the following table on the number of silk power looms in England from 1835 to 1856.

Table 102. Power looms in silk factories, England and three important counties, 1835-56

Year	England	Lancaster	Chester	Norfolk
1835	1,714	366	414	300
1850	6,092	1,977	955	951
1856	9.260	3,770	1,125	1,029

In Lancashire the 366 power looms in 1835 were employed by five factories in Manchester, Salford and Eccles as follows.

Table 103. Power loom factories in Lancashire, 1835

Name of factory	No. power looms
Royle and Crompton	40
William Harter	184
Smith and Thorp	60
B. Williams and Co.	22
J. and J. Clegg	60
Total	366

The average size of silk weaving, or of silk throwing-and-weaving factory in 1850 and 1856, is shown in the following table. As is true with the average factory size according to the number of spindles employed, the throwing-and-weaving factory had a larger number of power looms per factory.

Table 104. Number of spindles per silk weaving (or, weaving-and-throwing) factory, England, 1850-56

Type of factory	No. factories	No. power looms	Power looms per factory
Weaving			
1850	40	2,122	53.1
1856	129	5,119	39.7
Weaving-and-throwing			
1850	30	3.970	132.3
1856	36	3,536	98.2

The change in size as shown in the above table can be brought out more clearly in the following arrangement.

Year	Power looms per weaving factory	Power looms per weaving-and-throwing factory
1850	53.1	132.3
1856	39.7	98.2

Among the effects arising from the employment of power machinery we may first mention the displacement of labor. The silk factories, like the cotton and worsted factories, made little or no demand on muscular effort, and therefore employed a very small proportion of males or adults. In view of our previous discussions on child and woman labor, we may dismiss the subject by quoting the following statistics.

Table 105. Child labor in silk factories in England and Chester, 1838-47

(in percentage)

Year	England	Chester
1838	59.54	61.84
1847	47.31	48.78

Table 106. Woman labor in silk factories in England and Chester, 1838-56

(in percentage)

Year	England	Chester
1838	68.19	58.84
1847	68.48	59.76
1850	69.99	60.55
1856	69.73	63.46

The second effect is the presence of overhead cost. According to one silk manufacturer, the difference between the fixed property of the throwster, and the fixed property of the manufacturer, "is so great that I can hardly state it; the manufacturer begins trade with his outlay, with his silk and bobbins, and his reeds and harness, etc. etc.; and I apprehend if a man has one thousand weavers in Spitalfields, his fixed capital in those materials will be perhaps two or three hundred pounds. I apprehend that the worst silk mill [i.e. silk throwing mill] is worth more as old materials than all the reeds and harness in Spitalfields."[14]

The third effect is the increased productivity. In 1838 Mr. Mitchell of the Assistant Handloom Weavers' Commission reported that "the same work [narrow ribbons], which, in Spitalfields in 1820, was paid 9d., is now done for 3½d or 4d."[15] In Newbury, Berks, the efficiency of power loom in crape manufacture had so outrivalled that of the handloom that its introduction was thought to be the only "prospect of relief for the weavers".[16]

Statistical Summary

Table 107. Factory statistics on silk manufacture, England, 1838-56

Year	No. factories	No. employees	Amt. motive power (in horse power)	No. spindles	No. power looms
1838	263	33,565	3,237	....	....
1847	...	43,690	....	....	....
1850	272	41,703	3,574	1,188,908	6,092
1856	460	55,300	5,054	1,063,555	9,260

If we take 1838 statistics as a base of 100 percent., we may express the above table in relative terms below.

Year	No. factories	No. employees	Amt. motive power (in horse power)
1838	100.0	100.0	100.0
1847	....	130.2	....
1850	103.4	124.2	110.4
1856	174.9	164.8	156.1

From the above table it is clear that for the two years 1838 and 1856 the different percentage of increase in the number of factories on the one hand, and employees and motive power on the other, has given us a factory smaller in number of persons and amount of motive power employed, in 1856 than in 1838. Foreign competition, as has already been observed, is the chief factor for such decrease in factory size.

Handloom Factories

In silk manufacture scattered references to the existence of handloom factories are not uncommon.[17] In 1838 at Derby, Messrs. Moore had two factories with twenty handlooms each, while Messrs. Robinson had one factory containing thirty-five handlooms. Both manufacturers also gave out work to be done by weavers in the neighboring towns and villages.[18] At Chippenham, Wilts, there was another factory employing fifty-six handlooms.[19] At Sudbury, Suffolk, "Hill had a factory in which many of his weavers worked; he liked the factory, as the work was more under his own eye than when the men worked at home; there was less pilfering, and men, in case of any difficulty, could easily get information how to go from more experienced hands."[20]

Merchant Employer System

The merchant employer system which was typical of the woolen manufacture in West Country, had a fuller development in the silk manufacture. In the latter manufacture, about fifty percent. of the total employees worked under the merchant employer system, while the other fifty percent. were subject to the factory system.

Table 103. County distribution of silk employees under merchant employer system, England, 1840

County	Number	Percent.
North		
Lancaster	12,999	69.9
Chester	6,891	36.8
East		
Middlesex	7,109	97.9
Other	6,066	27.6
Total	33,065	49.6

In the above table it is to be observed that a large majority of the silk employees under the merchant employer system resided in the Northern counties, and that Middlesex was the county most completely organized under the system, with Lancaster occupying a middle position and Chester at the lowest end. The other counties were more or less organized under the merchant employer system, but on the whole the factory system prevailed.

Social classes. The merchant employer system in silk manufacture had a more elaborate development than that in woolen manufacture. In woolen manufacture there were fewer social classes through whose hands the finished product must pass before reaching the consumer from the producer. In silk manufacture such classes were more numerous, and we may first refer to the following diagram for a comparative analysis.

Consumer	←	Distributor of product 1. Wholesale 2. Retail	←	Employer ("manufacturer")	←	Distributor of work (undertaker)	←	Producer (weaver, warper and winder)

In the above diagram the presence of the two classes of distributors constitutes the main difference between the merchant employer system in woolen manufacture and that in silk manufacture. In the woolen manufacture these two classes of distributors were generally absent; while in the silk manufacture in 1840 the distributor of work was being slowly eliminated and the distributor of product had become more capitalistic than his predecessor. The distributors of work, generally known as the undertakers, performed the function of winding and warping the raw silk and giving it out to the weavers to be woven into silk goods at their own homes.[21] The distributors of product were divided into two classes—the retail "drapers" and the wholesale "warehousemen." The drapers were the original middlemen in the distribution of silk products, while the warehousemen sprang up only in the second quarter of the nineteenth century.[22] The employer, generally called "manufacturer," was the pivotal figure of the merchant employer system, although he had now lost his "merchant" function largely to the warehousemen. To quote a contemporary description of his role,

> "the manufacturer who procures his thrown 'organzine' [silk warp] and 'tram' [silk weft], either from the throwster or from the silk importers, selects the silk necessary to execute any particular order. The weaver goes to the house or shop of his employer and receives a certain quantity of the material, the 'tram' being generally wound on bobbins, and the 'organzine' in the form of what is called *cane* (derived from the French word 'chain', and so called from the silk being taken off the warping-mill in loops or links): This cane or warp varies from one or two hundred

yards in length. The weaver takes the material home to his own dwelling and weaves it at his own looms, or sometimes at looms supplied by the manufacturer. He is paid a certain rate per ell for his labor; but, as the weavers are not remarkable for provident habits, even in the best of times, they are accustomed to 'draw' money on account while the work is in progress, and to receive the remainder when the woven material and overplus material are returned to the manufacturer."[23]

The above quotation is typical of the conditions in Spitalfields around 1840, although the direct contact between the employer and his workmen did not exist in all cases. When the weavers were beyond the reach of the town limits and resided in outlying country districts, either undertakers or employer's own agents distributed the work to the weavers. In Manchester, the Committee on Manufactures, Commerce and Shipping were told by one witness that

> "the manufacturer at Manchester sends his clerk with warps and with wefts to a village, perhaps three or four, or five or ten miles off; he sends him to a public house, probably, and he gives out his warps and his wefts to the weavers that come, and he goes again and collects (in a fortnight, or three weeks or one month) his goods when they are ready; so that the weavers do not alter their locality, but the manufacturer takes the labor to them."[24]

These employers, besides, may also be manufacturers as we truly understand the term. In Spitalfields, for instance, they generally resided in the town, but had their factories built at Coggeshall, Sudbury, Andover, Taunton, Reading and some parts of Derbyshire, these factories being chiefly throwing or spinning factories.[25] Other employers combined both throwing and weaving in their factories.

The weavers, with a few winders and warpers, constituted the last but largest class of the silk manufacturing population. In 1840 there were two groups of weavers, namely, the town and the country groups. As Professor Fay has pointed out, in analyzing Hickson's Report on Handloom Weavers, [26] their employment conditions varied in accordance with the group to which they belonged, and could be compared in respective cases with the intensive and extensive margin of cultivation of land.

> "The town workers represent the intensive margin of cultivation, the country workers the extensive margin. First of all the Spitalfields weavers, who have been short of work, have more work given to them. The weavers' wives also get work, and their boys and girls who never were on a loom before are now put to the trade. Fresh hands are introduced. From the metropolis the demand for labor pushes outwards over the country. Recourse is had to 'inferior soils.' Old weavers in the villages get work, together with their wives and families. Even farm laborers are impressed. Blemishes for which at other times deductions would be claimed are overlooked. Carts are sent round to the villages and hamlets with work for the weavers so that time may not be lost in going to the warehouses to carry home or take back work. Then comes the ebb. 'The immediate effect is that all the less skillful workmen, the dissolute and disorderly, are denied work, the third and fourth looms, those worked by the sons and daughters of the weavers, are all thrown out of use.' The intensiveness of cultivation has been reduced in the towns, the least remunerative 'doses' no longer pay. The ebb of the tide, which reduces the quantity of employment in the towns, leaves the country districts high and dry. 'At such times the country towns and villages, to which work is liberally sent,

> when there is a demand for goods, suffer still more. A staff or skeleton only is kept in pay, and that chiefly with a view to operations when a demand returns.' A skeleton—well said!"27

Home work. The chief feature of home work is family labor, which, in 1850 and thereabouts, was well described in the personal recollections of a clever, shrewd and aged weaveress, born in 1844.

> "I never went to school, and cannot remember beginning to wind and weave. I always had to work and sleep among the looms in my father's workshop. There were six of us children, and we were all taught to wind quills for the shuttles as soon as we could talk, and to weave as soon as we could sit in the loom. My mother used to weave as well, and only left off to bring up our food to us, so that we should not lose more time than could be helped in eating...... My father hardly ever did any work himself after he had taught me to weave fancy silks with a Jacquard machine. When I was, I think, about twelve, I could do the work as well as he could. He used to come in and put the machine right when a needle got bent or anything else went wrong, but mostly he was talking with other men. He used to pick and look over all our work when it was finished, and take it to the warehouse. My mother used to make plain satins, and the younger children used to weave low quality plain silks. Sometimes I used to get fidgety and want to get up and move about. To prevent this, father used to tie me to the loom in the morning, before he went out, and dare me to leave till he came back."28

When the weaver's family did not have any or sufficient number of its own children, it hired the children of others to meet the need, out of which a peculiar situation arose. In Bethnal Green Road, in the neighborhood of Spitalfields, a "children's labor market," for instance, was held every Monday and Tuesday morning between the hours of six and eight.

> "There is an open space, to which children of both sexes, from the age of 7 and upwards, resort to be hired by the week or month by any person who may require their services. When trade is slack as many as 300 children may be found in the market, seeking employment, but when trade is brisk there will not be more than fifty to one hundred. These children are chiefly hired by weavers; the boys as substitutes at the loom, or for winding, and the girls as little domestic servants or maids of all work. Many of the girls who accept these situations are between seven and nine years of age, and are paid from 1s. 3d. to 2s. per week. In a weaver's family where the wife is employed at the loom, she requires a servant to nurse the child, make the beds, and help cook the victuals; and as a strong girl of fifteen would be too expensive for the purpose, a mere child of 7, 9, 10 and 11 is hired instead. A child thus hired comes to work before 8 in the morning, having had her breakfast, returns home to dinner, and at night again returns to sleep. On Saturday night she receives the 20d. she has earned, and gives it to her parents."29

Abuses

In the silk manufacture, official reports indicated a smaller extent of embezzlement than the economic conditions of the industry warranted. The high value of silk materials in small bulk was a temptation to embezzlement. The distressed conditions of the weavers added a further motive. In Spitalfields there was a weavers' society—a benefit society—that undertook to indemnify the employers for any embezzlement by their own members, but in the course of eight years there were only two cases reported! Public flogging, it is said, was a punishment for embezzement in Spitalfields, although the little excitement it caused among the passers-by indicated the frequency of its occurence.[30] In the South and West of England, Commissioner Austin reported that there was a temptation to embezzlement, but rather little of it. "It is certainly quite possible to abstract (without immediate detection) half a yard from the end of a piece, but he [a manufacturer] does not think that it is done here."[31]

Persistence and decline

With the coming of the foreign competition the merchant employer system in silk manufacture gave way rapidly to the factory system. The former system lingered on only in some branches of silk weaving where the handloom performed a service that the power loom by its technical nature was unable to replace. It is also possible that in such cases, the domestic weavers may have been employed by the factory owners as outworkers. In 1924 one writer remarks that

> "the handloom is still used as a commercial machine for the finest silks and linens, which are unable to withstand the severe strain, friction and high speed of the power loom; for the most expensive and elaborately ornamented fabrics such as are used for ceremonial purposes, in which the cost of weaving forms a comparatively small proportion of the total cost and in which the increased output of the power loom is counterbalanced by the greater risk of faultily woven cloth; and for pattern weaving, owing to the facility with which changes can be made."[32]

B. Flax and hemp

The flax and hemp manufactures of England in 1841 may be defined as including (1) all persons who were fully employed in preparing and spinning flax or tow fibre into yarn, and weaving the flax or tow yarn into linen cloth, and (2) all persons who were fully employed in preparing, spinning, or twisting the hemp fibre into rope or cord. The combined treatment of these two manufactures is necessary for a comparison between the Census and the factory statistics, as the factory statistics, unlike the census statistics, do not distinguish the flax from the hemp manufacture. These two manufactures, as here defined, employed in 1841 a total of 30,764 persons,—19,474 persons[33] in the flax manufacture and 11,290 persons in the hemp manufacture.[34]

In the local distribution of these two manufactures the contrast was notable. The flax manufacture was concentrated in the northern counties of West Riding and Lancaster, while the hemp manufacture was highly scattered. The latter industry was so situated as to meet the local demand for its product, and on that account it was found in nearly all parts of England, especially in London, Liverpool, Hull, Plymouth, and other principal seaport towns.[35]

Table 109. County distribution of persons employed in flax and hemp manufactures, England, 1841

County	Flax		Hemp	Flax & Hemp	
	Flax	Linen		Number	Percent.
York, W. R.	7,424	4,747	587	12,758	41.4
Lancaster	1,316	1,177	2,111	4,604	14.9
Other	2,214	2,596	8,592	13,402	43.7
England	10,954	8,520	11,290	30,764	100.0

Industrial Organization

The industrial organization of the flax and hemp manufactures of England in 1840 was unique, as it represented the three systems at the same time. These three systems, namely, master craftsman, merchant employer and factory systems, varied with the technical processes as well as the raw materials employed in the flax and hemp manufactures. With regard to the technical processes employed, flax spinning was chiefly organized under the factory system in 1840.[36] The system was first introduced with the commencement of flax spinning by power machinery in 1790, by John Marshall.[37] Its rapid extension since that date was accompanied by a similar decline of hand spinning under the old merchant employer system. Such extension, furthermore, concentrated the industry within the two northern counties of West Riding and Lancaster.

Flax weaving, on the other hand, was organized under the same system in England as in Scotland, namely, merchant employer system.[38] At Barnsley, Yorkshire, for instance, the employers (generally called "manufacturers") purchased flax yarn from the factory spinners, and gave it out to the handloom weavers. The latter wove it into linen cloth at their own homes, brought it back to the warehouse of the employer, and received payment for their labor.[39] This system of having to go to the employer's warehouse to receive flax yarn for weaving worked well with the town weavers, but with the country weavers, of whom there must have been a great number in 1840, it involved a heavy load for a long trip. With some of the warehouses it often happened that the weaver had to go several times before he got the whole of such load home.[40]

With regard to the raw materials employed, hemp spinning, unlike flax spinning, was organized under the master craftsman, as well as under the factory system. Under the master craftsman system no machinery was employed. Spinning or twisting of the hemp fibre into rope or cord was generally carried on inside the master's workshop or under the open air outside. These "roperies" and "spinning ways," as they were called, were found in every considerable town in England. Their owners—chiefly master craftsmen—sold the rope or cord at the very spot where it was spun or twisted, and on that account had a limited extent of operation.[41] In Lancashire in 1841, Mr. Austin of the Children's Employment Commission visited 25 roperies located within the three towns of Preston, Ormskirk and Liverpool,[42] of which three-fifths did not employ more than 14 persons per ropery, nor did they employ any machinery other than hand wheels turned by the human power of little children. The following table shows the sizes of these 25 roperies in Lancashire.

Table 110. 25 roperies in Lancashire, 1841

Persons per ropery	No. roperies
1 to 10	9
11 to 20	5
21 to 30	6
Above 30	5
Total	25

Under the factory system hemp spinning was generally carried on with the aid of power machinery as invented during the period 1793 to 1833.[43] Such power machinery, however, was not so extensively employed in hemp as in flax spinning. The factories where it was found at work were scattered throughout the important seaport towns of the country, but compared poorly with the flax spinning factories as to the number of persons employed per factory. Messrs. Jackson & Co., of Liverpool, had probably one of the largest factories in hemp spinning, with a total labor force of 233 persons in 1841. [44]

In the following table the extent of each of the three systems of industrial organization in the flax and hemp manufactures in England in 1840 is shown. [45]

Table 111. Persons employed under different systems of industrial organization in flax and hemp manufactures, England, 1840

System	Process		Persons employed: Number	Persons employed: Percent.
Master craftsman	Hemp spinning		5 671	18.4
Merchant employer	Flax weaving		8,520	27.7
Factory	Flax spinning	10,954		
	Hemp spinning	5,619	16,573	53.9
Total			30,764	100.0

According to the above table, the factory system was the predominating system of industrial organization in the flax and hemp manufactures in England in 1840, employing 53.9 percent. of the total employees, while the merchant employer system, with 27.7 percent. of the total employees, ranked next.[46] The rest, 18.4 percent., were organized under the master craftsmen system. In the following paragraph we may direct our attention to a discussion of the factory system in flax and hemp spinning.

The Factory System

York and Lancaster, together with four other Northern counties including Westmoreland, Durham, Cumberland and Northumberland, had 84.6 percent. of the total factory employees in the flax and hemp manufactures in England in 1838. This high degree of concentration in the North was an unusual feature; it suggested as one explanation the close proximity of the English flax and hemp manufactures to the Scotch and the Irish. In the following table the geographical distribution of the factory employees in the two manufactures among different counties is shown.

Table 112. County distribution of factory employees in flax and hemp manufactures, England, 1838

County		Persons employed Number	Percent.
Northern		14,024	84.6
York	9,654		
Lancaster	2,881		
Westmoreland	449		
Durham	422		
Cumberland	366		
Northumberland	252		
Other		2,549	15.4
Total		16,573	100.0

The two elements of a factory—labor concentration and employment of power machinery—and their effects are briefly summarized in the following paragraphs.

Labor Concentration

In following our previous practice in treating the cotton, woolen, worsted and silk manufactures, the index of labor concentration as found in the number of persons employed per flax or hemp factory in the important flax and hemp counties of England in 1838 is given below.

Table 113. Number of persons employed per flax or hemp factory in important counties, England, 1838

County	No. factories	No. employees	Employees per factory
York	91	9,654	106.1
Lancaster	16	2,881	180.1
Other	72	4,038	56.1
England	179	16,573	92.6

The average number of persons employed per flax or hemp factory in England in 1838, according to the above table, was 92.6. This average is hereafter referred to as average no. 1. Disregarding the unimportant counties, in which most of the smaller sized factories were found, Lancaster had unusually large factories with an average of 180 persons per factory. York, with that of 106, more closely possessed the typical factory. The varation in factory sizes among the important parishes where the flax and hemp manufactures were carried on is shown in the following table.

Table 114. Number of persons employed per flax or hemp factory in important parishes, England, 1838

County	Parish	No. factories	No. employees	Employees per factory
York	Leeds	44	6,430	146.1
Lancaster	Preston	6	1,392	232.0

The average of 92.6 persons per factory for England, an arithmetic mean, is subject to the influence of extremes, and therefore not representative. In following our previous practice in treating the cotton, woolen, worsted and silk manufactures, another average, a median, is compiled after an enumeration of all factories located in parishes having one or two factories per parish. Omitting all details, the enumeration is summed up in the following table.

Table 115. 89 flax and hemp factories located in parishes having one or two factories per parish, England, 1838

Persons per factory	No. factories
1 to 10	7
11 to 20	8
21 to 30	13 (a)
31 to 40	8
41 to 50	10 (b)
51 to 60	7
61 to 70	6
71 to 80	5
81 to 90	2 (c)
91 to 100	3
101 to 110	3
111 to 120	2
121 to 130	2
131 to 140	1
141 to 150	0
151 to 160	2
161 to 170	1
171 to 180	0
181 to 190	0
191 to 200	2
Above 200	7 (d)
Total	89 (e)

(a) 21, 23, 23, 23, 24, 25, 26, 27, 27, 28, 28, 29, 29.
(b) 41, 41, 42, 44, 45, 45, 50, 50, 50, 50.
(c) 87, 88.
(d) 209, 261, 271, 339, 397, 801.
(e) These 89 factories employed a total of 7,187 persons, or an average of 80.8 persons per factory.

According to the above table, the median is 50, the lower quartile, 27, and the upper quartile, 88. The quartile coefficient of deviation is, therefore, 0.27. This average is hereafter referred to as average no. 2.

Still another average, also a median, is compiled after an enumeration of the sizes of 56 flax or tow spinning factories in Inspector Horner's Districts in 1841, viz. 20 factories in York, 15 in Lancaster, 14 in Cumberland, and 7 in Durham. The full enumeration is presented in the following table.

Table 116. 56 flax and tow spinning factories in Yorkshire and neighborhood, 1841

Persons per factory	No. factories
10 or below	6
11 to 20	6
21 to 30	9 (a)
31 to 40	5
41 to 50	3 (b)
51 to 60	2
61 to 70	2
71 to 80	1
81 to 90	0
91 to 100	2
101 to 200	10 (c)
201 to 300	5
301 to 400	2
401 to 500	2
501 to 600	1
Total	56 (d)

(a) 21, 23, 23, 26, 28, 28, 30,30.

(b) 42, 46, 50.

(c) 110, 145, 147, 150, 155, 158, 174, 177, 180, 199.

(d) These 56 factories employed a total of 6,184 persons, or an average of 110.4 persons per factory.

According to the above table, the median is 46, the lower quartile, 23, and the upper quartile, 158. The quartile coefficient of deviation is, therefore, 0.75. This average is hereafter referred to as average no. 3. In the following table, the two averages, average no. 2 and average no. 3, are compared with respect to their medians, quartiles, and quartile coefficients of deviation.

Average	Lower quartile	Median	Upper quartile	Quartile coefficient of deviation
no. 2	27	50	88	0.27
no. 3	23	46	158	0.75

The close correspondence between the two medians is striking. In flax and hemp manufactures, as well as in cotton, woolen, and worsted, such correspondence suggests that the median average of factory size is more representative than the arithmetic. In flax and hemp manufactures, the median average is 48 (which is derived by averaging the median averages no. 2 and no. 3), while the arithmetic average is 92.6.

In the following table, the growth of average factory size in England from 1838 to 1850 is given, using the arithmetic averages as the only data yet available for a comparative analysis of this nature.

Table 117. Persons employed per flax or hemp factory, England and two important counties, 1838-56

Year	England	York	Lancaster
1838	92.6	106.1	180.6
1850	140.8	191.9	302.7
1856	143.1	197.6	266.0

The increase of average factory size in flax and hemp manufactures from 1838 to 1856 is largely a result of the increasing application of power machinery not only to flax and hemp spinning, but also to flax weaving.

The two returns below, compiled respectively from the Factory Returns of 1838 and 1841, [47] give some notion of the sizes of the "first-rate" flax and hemp factories.

Table 118. "First-rate" flax and hemp factories in 1838 and 1841

Persons per factory	No. factories 1838	No. factories 1841
96 to 100	2	0
101 to 200	13 (a)	10 (d)
201 to 300	4 (b)	5 (e)
301 to 400	2	2
401 to 500	0	2
501 to 600	0	1
Above 600	1	0
Total	22 (c)	20 (f)

(a) 102, 104, 105, 112, 112, 122, 127, 139, 154, 154.
(b) 170, 200, 200.
(c) 209, 261, 271, 271.
These 22 factories employed a total of 4,544 persons, or an average of 206.5 persons per factory.
(d) 110, 145, 147, 150, 155, 158, 174, 177, 180, 199.
(e) 218, 234, 240, 264, 274.
(f) These 20 factories employed a total of 5,020 persons, or an average of 251 persons per factory.

In the following table, the medians, quartiles, and quartile coefficients of deviation for the two averages are compared.

	Lower quartile	Median	Upper quartile	Quartile coefficient of deviation
1838	112	154	200	0.28
1841	155	199	274	0.28

The largest factory, however, belonged to the Marshalls of Yorkshire. John Marshall, the first to introduce flax spinning by power machinery around 1790, took a leading role in the development of the industry. In his evidence before the Parliamentary Committee on the Linen Trade of Ireland, in 1825, he gave detailed statistics concerning the extent of his factory in flax spinning. He had four plants at Leeds and Shrewsbury in Yorkshire, with a fixed capital of £100,000 in buildings and machinery. These plants employed 300 steam horse power, and 900 to 1,000 flax spinners.[48] In 1841 the younger Marshal, James, gave even more startling statistics about the famous factory that was generally pronounced to be the largest in England and in the world.[49] According to his evidence before the Parliamentary

Committee on the Export of Machinery, the Marshall plants at Leeds and Shrewsbury employed altogether 2.000 persons in flax spinning, 330 steam horse power, and 25,000 spindles. These 25.000 spindles, calculated at a capital cost of 4 pounds per spindle, were valued at a fixed capital of £100,000, and produced annually a total output of flax yarn valued at £200,000.

Employment of Power Machinery

Motive power. In 1838 the percentage of steam power employed in flax and hemp factories was 73.5, which was next only to the percentage of steam power employed in cotton and worsted factories, viz. 81.1 and 81.7 respectively. In the following table, the increase of steam motive power over water from 1838 to 1850 is shown, while in 1856 there was a small decrease.

Table 119. Motive power in flax and hemp factories, England, 838-56 (in horse power)

Year	Steam motive power. Amount	Steam motive power. Percent.	Water motive power. Amount	Water motive power. Percent.
1838	3,134	73.5	1,131	26.5
1850	3,616	80.6	871	19.4
1856	3,639	78.4	1,005	21.6

Another table is given below to show the increase in the amount of motive power per flax or hemp factory from 1838 to 1856.

Table 20. Amount of motive power per flax or hemp factory, England, 1838-56 (in horse power)

Year	Total no. factories	Total amount of motive power	Motive power per factory
1838	179	4,265	23.8
1850	135	4,487	33.2
1856	139	4,644	33.4

Spindles. The North of England, including York, Lancaster and four minor counties, had the largest percentage of the total number of spindles, 90.5 percent. in 1850 and 95.1 percent. in 1856.

Table 121. County distribution of spindles in flax and hemp factories, 1850-56

County		1850		1856
Northern		240,446		419,907
York	82,768		214,547	
Lancaster	117,356		137,220	
Cumberland	19,986		30,897	
Westmoreland	11,808		692	
Northumberland	6,508		36,551	
Durham	2,020			
Other		25,122		21,852
England		265,568		441,759

In view of Yorkshire's leadership over Lancashire in flax spinning in 1838, it is doubtful whether the smaller number of spindles in Yorkshire than in Lancashire in 1850, not in 1856, can be accepted as being accurate. It is still more doubtful whether in the course of six years (from 1850 to 1856) Yorkshire had increased its number of spindles from 82,768 to 214,547,—an increase of more than two and a half times. It is more likely that the factory inspectors, in summing up the returns, made an error in reporting 82,768 spindles in 1850 instead of 182,768.[50]

In the following table, the increase in the number of spindles per flax or hemp spinning factory in England from 1850 to 1856 is shown.[51]

Table 122. Number of spindles per flax or hemp spinning factory in England, 1850-56

Year	Total no. spinning factories	Total no. spindles	Spindles per factory
1850	100	197,564	1,975.6
1856	96	366,019	3,812.7

Power loom. In flax manufacture, the employment of power looms was yet in its infancy in 1840,[52] and for many years since that date. In the following table the number of linen power looms in England from 1835 to 1856 is shown.[53]

Table 123. Number of linen power looms, England, 1835-56

Year	No. power looms
1835	41
1850	1,083
1856	1,987

In 1850 there were only ten linen weaving factories employing altogether 782 power looms in England; in 1856 there were nineteen such factories employing 1,452 power looms. The rest of the power looms in 1850 and 1856 were employed in the spinning-and-weaving factories.

Among the effect arising from the employment of power machinery we may first mention the displacement of labor. The displacement of male adult by child and woman labor is shown in the following tables, in which the proportion of child and woman labor in flax and hemp factories from 1835 to 1856 is given.

Table 124. Child labor in flax and hemp factories in England and Yorkshire, 1838-47 (in percentage)

Year	Engalnd	Yorkshire
1838	52.33	54.95
1847	42.79	45.27

Table 125. Woman labor in flax and hemp factories in England and Yorkshire, 1838-56 (in percentage)

Year	England	Yorkshire
1838	67.55	66.44
1847	64.42	63.28
1850	67.71	68.55
1856	68.84	68.46

The second effect was the increased productivity in flax spinning, which was accompanied by the better quality of yarn produced. Machine spun yarn was stronger and more even than hand spun yarn, and therefore commanded a better market.[54] In 1825 it was estimated that machine spun yarn could be sold at a price ten or fifteen percent. higher than the hand spun yarn.[55]

Table 126. Factory statistics on flax and hemp manufactures, England, 1838-56

Year	No. factories	No. employees	Amt. motive power (in horse power)	No. spindles	No. power looms
1838	179	16,573	4,265		
1847	...	19,840			
1850	135	19,001	4,487	265,568	1,083
1856	139	19,787	4,644	441,759	1,987

If we take 1838 statistics as a base of 100 percent., we may express the above table in relative terms below.

Year	No. factories	No. employees	Amt. motive power
1838	100.0	100.0	100.0
1847	....	119.7	.....
1850	75.4	114.7	105.2
1856	77.7	119.4	108.9

Comparing the factory statistics in the various years, it is clear that the disproportionate percentage of increase between the number of factories on the one hand, and the number of employees and amount of motive power on the other, resulted in a larger factory in 1856 than in 1838, with regard to the number of persons and amount of motive power employed per factory.

Merchant Employer System :

In flax weaving the merchant employer system prevailed in general. Exceptions, however may be made to the existence of a distinct class known as "custom weavers," or wage workers in Bücher's terminology. These custom weavers were mentioned only in flax and cotton handloom weaving, although they might have been found also in the handloom weaving of woolen, worsted or silk fabrics.[56] In flax weaving, according to Commissioner Mitchell,

> "These domestic artizans were at one time numerous in the counties of Lincoln and Nottingham, and were employed to weave the yarn spun in private houses. Their number is now much reduced, there being far less domestic manufacture. The cheapness of cotton good is considered the principal cause. Some custom-weavers employed themselves partly in agricultural labor and partly in weaving, that is, when any is to be obtained.
>
> "Dr. William Ellis of Caistor, in Lincolnshire, states, that there is a remnant of custom-weavers in that part of the country, but some of them have ceased on account of not obtaining employment, and others of them can get weaving for only a part of their time.
>
> "In a commercial point of view these men are totally insignificant, and can excite no interest but as a remnant of a body once very numerous in the generations now gone past."[57]

Besides this system of wage work, a few power loom factories might also be found.[58] But on the whole, the merchant employer system was predominant in flax weaving. In the following paragraphs, the elements and abuses of the merchant employer system in flax weaving are sketched in brief.

Social classes. There were two soc employer purchased the yarn and gave it the outworking weavers. The merchantial classes, viz. the merchant employer and out to be woven by the weavers on looms hired from the merchant employer. In many cases the work was done at the weavers' houses, while in some, the merchant employer also provided the shop that was inadequately called "factory".[59] When the weavers worked at their own houses, such houses, or cottages, were in a few cases built by the merchant employer and rented out to the weavers. In Newark, Nottingham, for instance, Mr. George Scales, the only merchant employer told the Handloom Weavers Commission that,

> "the system is for the manufacturer to build cottages adapted for weavers, and filled up with looms, and to let those cottages at a moderate weekly rental; and every weaver taking a house and not having a family to employ the looms, then sublets the use of looms and lodging to two or three other weavers, which renders his own rent moderate. There are 32 such cottages belonging to Messrs. Scales".[60]

These merchant employers had varying number of looms, Mr. Scales in Newark, Nottingham, for instance, Mr. George Scales, the only merchant employer, told the Others employed only a few, and usually had to travel about the neighboring country in order to obtain orders for their goods.[61]

Home work. In flax weaving the work was carried on in cellars attached to the weavers' homes. In Barnsley, "the cellars in which they work are not more damp than is desirable for carrying on the trade."[62] Child and woman labor, however, was not so commonly employed in flax as in other branches of handloom weaving.

Abuses

Truck was neither frequent, nor resorted to by the large merchant employers, but it was said to have existed in Barnsley. William Ashton, a weaver, told the Handloom Weavers' Commission,

> "I can name an instance where a manufacturer named Tyne refused to pay the weaver for the first piece before he brought home the second, as he found fault with 4 yards of the first piece, and when the man brought home the second he would not pay him otherwise than in cloth, and gave him 13 yards of 4-4 duck at 1s. 2½d. per yard; he took the cloth and tried to sell it at different shops in the town, but found he could have bought it at 1s. a yard."[63]

At Haverhill, Suffolk, one of the merchant employers stated that the act against the truck system was much evaded.[64]

Chapter V. Paper, Pottery and Glass

A. Paper

The paper manufacture of England in 1841 employed a total of 5,526 persons,[1] who were scattered throughout every county. Some counties, however, may be called the centres of paper manufacture, relatively speaking.[2] The eastern counties of Kent, Buckingham, and Hertford were early known for the paper manufacture. In Hertford the first English paper mill was said to have been erected by John Tate the Younger, son of the Lord Mayor of London, around the year 1500.[3] In Kent, William Gill started the first paper mill for the county in the year 1719.[4] The early localization of the paper manufacture in this eastern region is explained by several factors. The immigration of the Huguenots into Kent and the neighborhood introduced the art of paper making;[5] while the presence of water supply furnished the necessary motive power as well as the desired quality of water indispensable to converting the rags into pulp. Finally, the proximity of the region to the Metropolis was an important advantage from the viewpoint of marketing. In the north, paper mills were not always located near the streams, as steam power replaced water power in many instances. The element of market, however, was a determining one; and consequently, Lancashire and West Riding of Yorkshire were the leading centres of paper manufacture.

Factory System

A paper mill in 1840 approached a factory where labor was concentrated under the employer's supervision, and where machinery, driven by water or steam power, was employed. In some paper mills, it is true, machines were still not in use, and papers were made by hand.[6] Hand mills, however, supplied the smaller part of the product. In 1840, out of a total output of 47,000 tons of paper made in the United Kingdom, only 9,122 tons, or 19 percent., were made by the hand process.[7]

A paper mill, unlike other factory, had a comparatively small average of employees. A typical paper mill might employ from 40 to 45 hands, while a paper mill with 120 employees was considered as an "extensive works" by the Children's Employment Commission in 1841.[8] The largest paper mill, however, was that of Messrs. Hilton at Over Darwen, Lancashire, where 301 persons were employed in 1841, including 249 adults and 52 children.[9] This paper mill was different from many others insofar as it was integrated. Besides making paper by machinery, it also operated several coal pits to supply coal for the steam engines, and erected immense reservoirs for the reception of water. The total number of employees, including the colliers, was given by one visitor at upwards of 450. According to the same visitor,

> "the horse power of their engines is computed at 300—they consume 80 tons of coal per day, and manufacture upwards of 30 tons of paper weekly! ...... The position of the mills is eminently favorable for an extensive manufactory, and when we add, that the proprietors have availed themselves of all the modern improvements in machinery, and execute extensive government orders, we have adduced sufficient reason for our belief that here the art of paper making may be seen under the most favorable aspects."[10]

In the following table, the sizes of paper mills are cited, for which scattered statistics on the number of persons employed were given by the Children's Employment Commission in 1841.[11]

Table 127. Size of selected paper mills in England, 1841

Name and location of mill	No. employees
Messrs. Hilton, Over Darwen, Lanc.	301
Mr. Joynson, Foot's Cray, Kent	120
Mr. Woodfull, Foot's Cray, Kent	60 to 70
Mr. Nash, Foot's Cray, Kent	60 to 70
Messrs. Magnay, Surrey	65
Scotswood, Northumberland & Durham	43
Messrs. Annandales, North. & Durham	42

The history of paper manufacture by machinery may be traced back to the year 1770, at which date Hollander's beating machine was first invented. It was worked by hand at the beginning, but later on by steam and other mechanical power. By the end of the eighteenth century Hollander's beating machine was fully established. At the same time, Robert, a French artisan, invented the automatic paper making machine. Didot, by placing his mill, his workers, and his financial resources at Robert's disposal, made Robert's invention a success.[12] The French invention was soon taken up by the English through Didot's brother-in-law, Gamble, who in turn interested Henry and Scaley Fourdrinier, two prosperous London stationers, to take out an English patent and to engage a capable and ingenious engineer, Donkin, for the production of the first English paper making machine.[13] Two patents were taken out in the years 1801 and 1803; in the latter year the first English machine successfully making a continuous reel of paper was started at Frogmore, Hertford.[14] Other inventions, including that of Dickinson in 1809, quickly followed until 1830, the year which marked the end of an important epoch in the history of machine paper manufacture. Later improvements of the century were chiefly based upon inventions already discovered before 1830.[15]

In 1837 Fourdrinier appeared before the Parliamentary Committee on his patent, a Committee appointed to find out if the famous patent deserved public compensation on account of its usefulness, he made a statement on the number of machines at work in the 168 paper mills in the United Kingdom as follows.[16]

Table 128. Machine paper mills in the United Kingdom, 1837

Kind of mill		No. mills	No. machines
1	machine	120	120
2	machines	36	72
3	,,	5	15
4	,,	4	16
6	,,	1	6
7	,,	2	14
	Total	168	243

According to one writer for the Penny Magazine in 1833, "a paper mill with only one machine, and no vats, is held to carry on a respectable business, employing about 40 hands"[17] In 1841, Messrs. Magnay's paper mill in Surrey had two machines and employed 65 hands.[18] From these two estimates, it is reasonable to conclude that each machine in 1837 would employ about 35 hands. The 120 mills employing one machine each in 1837 might have an average of about 35 to 50 hands per mill, as it was not uncommon for a paper mill to install a paper machine in addition to the vats already in use.

The rapid triumph of paper manufacture by machinery over that by hand has been explained in a previous paragraph, for which several reasons may be given. First and foremost must be mentioned the reduced cost of paper making by machinery. "The expense of making paper by hand," according to the Parliamentary Committee on Fourdrinier's Patent, "is 16s. per cwt., whereas by machine it is only 3s. 9d."[19] Secondly, with machinery skill was not an important element, and any workman not regularly educated to the trade could learn it in a very short time, "the most uninstructed person being competent for the purpose."[20] Finally. "with regard to the superiority of paper manufactured by machinery, it is necessary to say but little, as that can best be appreciated by an examination of the different specimens."[21]

B. Pottery

The pottery manufacture of England in 1841, in a broad sense, included the manuufacture of such articles as pottery, earthenware, china and porcelain; it employed a total of 23,468 persons. Out of this total, 17,822 persons,[22] or 76 percent, were concentrated in the famous county Stafford, or, more strictly speaking, in that region of the county known for centuries as the "Staffordshire Potteries." This high degree of concentration was a joint outcome of several factors. The conscious determination to plant the industry, which was early carried out through the governmental policy of encouraging the alien immigration into the district, gave the pottery manufacture an initial start.[23] The immigration first took place in Burslem, "the most early and chief seat of the manufacture in this district."[24] The bounty of nature constituted another important factor. According to Shaw,

> "the situation, on the ridge of the low part of the Moorlands, is so much diversified with hills and dales, that however little might be its value for agricultural purposes, or black and barren its aspect, no inconvenience is ever experienced by the inhabitants from the volumes of smoke arising from the coal consumed in making bricks, tiles and earthenware. ...... Here also the potters enjoy four of the chief material benefits which may be connected with human existence—air extremely salubrious, water of tolerable purity, the sun seldom obscured by fogs, and an entire freedom from dampness. Other advantages for the manufacure, which have contributed to make this district a permanent seat, and enabled it to triumph over every attempt at its removal, are the coals and clay for the several purposes, being readily obtained on moderate terms."[25]

Finally, access to the best market for the pottery produced, like nearness to the raw materials, was further facilitated by the ready means of communication at disposal. Canals as well as railroads connected the district known as the "Staffordshire Potteries" with Manchester, Birmingham, Liverpool, London and other important towns of England.

Factory System

In 1840 the pottery manufacture of England was chiefly organized under the factory system. A pottery factory, like an "ironworks," comprised under single ownership and management many constituent units that were scattered throughout an extensive area of land, but were connected with one another in accordance with the order of technical processes performed. Indeed, in approaching the pottery districts where factories were congregated side by side,

> "a stranger might be tempted to believe [that] he saw a vast line of fortifications rising before him. The surrounding hills are all crowned with the lofty columns and the huge pyramids of the chimneys, and with the great rounded furnaces, of which dozens are often seen close together, looking like colossal bomb-mortars. The high roofs of the drying houses, the magnificent warehouses, and the massy walls that enclose the whole great establishment, or 'workhouse bank,' with the piles of clay, flints, bones, cinders, and other matters, serve rather to strengthen the illusion. . . . . Between the great workhouse banks lie scattered the small houses of the shopkeepers, the workmen, the painters, the engravers, the color-men, and others."26

In short, the buildings of a pottery factory were very different from those of a cotton factory, or other factory in which machinery supplied the motive power. They were very irregular, very much scattered, and, in general, only of two stories in height.

In accounting for the scattered character of a pottery factory, two reasons in particular may be given. In the first place, despite the fact that in a pottery factory labor was concentrated under the employer's supervision, neither machinery nor power was commonly employed. The different processes in the pottery manufacture, in general, were still performed by manual labor, although in some large factories, steam engine had been applied to the processes of grinding and turning. Secondly, the minute division of labor could be more efficiently carried out in separate buildings for a bulky product like pottery than under one roof. The pottery products varied in size as well as in kind; consequently, a separate building was generally set aside to the making of those articles that were of one kind if not of one size. In Messrs. Copeland & Garraths' factory, for instance, there were "19 ovens, 272 working rooms, 19 slip kilns, 42 warehouses, and 33 other offices."27 These 272 working rooms were used for carrying on the different processes of dipping, printing, painting, scouring, turning, and throwing, etc. If we were to go back half a century from the period of our investigation, we find that, in 1790, in the factory of the famous pottery manufacturer, Josiah Wedgwood, such division of labor was already carried out to an extreme nicety. Of the 160 workers in the factory, there were more than eighteen distinct classes of workers. namely, 8 slip makers, 2 clay beaters, 9 throwers, 18 platemakers, 4 dishmakers, 11 hollow ware pressers, 13 flat ware turners, 14 hollow ware turners, 6 handlers, 6 biscuit oven keepers, 10 glost oven keepers, 3 glost warehousemen, 11 enamellers and painters, 1 coal getter, 4 odd men, 12 warehousemen, 5 counting house clerks and 23 others, making in all a grand total of 160 workers.28

In view of the scattered character of a pottery factory, several workrooms were generally under the control of one foreman; these formen, in turn, carried out the orders of the factory superintendent. In some workrooms, such as painting rooms, a great number of persons—chiefly women,boys and girls—were congregated; while in others there were only one or two. In the first case the duty of a foreman was primarily supervisory; while in the second, a foreman was himself a workman. In both cases, the size of a workroom varied in accordance with the nature of process employed and the number of workmen therein. The workrooms for painting, burnishing, gilding, flower-making, moulding, figure-making and engraving "constitute literally schools of art, under the superintendence of masters and mistresses. . . . . . They [appretice boys and girls] are seen sitting at their clean tables, at a comfortable distance from each other, in an airy, commodious, and warm room, well ventilated, and heated by a stove or hot plate, on which they dress their meals. The women who superintend their work are generally selected from among the rest on the premises on account of their good moral conduct and long servitude."[29] These rooms, painting rooms for instance, oftentimes employed a considerable number of workmen. In John Ridgway's factory at Shelton, there were three groups of painting rooms employing a total of 160 persons.[30]

Kind of room	No. rooms	No. employees
Women's painting room	4	84
Apprentice girls' room	2	52
Apprentice boys' room	2	24
Total	8	160

These 160 painters constituted one third of the total employees of the factory. This proportion between the painters and other workmen in a pottery factory was undoubtedly too large, for in Messrs. Copeland and Garrath's factory there were only 76 painters out of 780 workmen.[31] The different proportion in different factories must be explained in accordance with the kind of pottery produced. The coarser articles, in general, required little or no painting.

Labor Concentration

In order to measure the size of a pottery factory, we may resort to the number of persons employed per pottery factory as the index for which comparatively complete statistics can be obtained. According to this index, a pottery factory tripled in size in the course of seven or eight decades. In 1762 it was stated in a petition for a Turnpike Act that "in Burslem and its neighborhood are near 150 separate potteries for making various kinds of stone and earthenware, which together find constant employment and support for near 7,000 people."[32] The average size per pottery factory, according to the petition, was 47 persons. In 1836, seven or eight decades later, there were 130 pottery factories employing a total of 20,100 persons in the "Staffordshire Potteries."[33] The average number of persons employed per pottery factory, therefore, was 155, an average about three and half times higher than that for 1762.

An average leaves a good deal untold. The Children's Employment Commission of 1841, however, cited the individual sizes of certain pottery factories in a casual but useful fashion. Out of 94 pottery factories for which evidences were taken, the individual sizes for 26 factories were given. In the following table, the individual sizes of 14 pottery factories that may be termed "first-rate" are quoted.[34]

Table 129. Size of 14 "first-rate" pottery factories in "Staffordshire Potteries", 1841

Name and location of factory	No. employees
Wm. Davenport & Co., Longport	1,200 to 1,400
Copeland & Garraths, Stoke	780
Adams & Sons, Stoke	648 (a)
F. & R. Pratt & Co., Fenton	upwards of 500
J. Ridgway, Shelton	500
T. & J. Mayor, Longport	450 to 500
G. Phillips, Longport	450 to 500
C. J. Mason & Co., Fenton	400 or more
Ridgway, Morley, Wear & Co., Shelton	348
Meller, Venables, Pindar & Co., Burslem	300 (ca.)
Charles Meigh, Hanley	300 (ca.)
Z. Boyle, Stoke	245
Daniel & Sons, Stoke	240
T. Dimmock, Hanley	236

(a) There was another factory belonging to the same owner, for which the number of persons employed was not given.

In the above table, the factory belonging to William Davenport & Co. at Longport was the largest of all, employing from 1,200 to 1,400 persons. The factory was first established in 1800, and was one of the largest potteries known in 1829.[35] It had five plants engaged in the manufacture of china, earthenware, and glass, and was one of the few factories that employed machinery for grinding and other purposes. It was said to have produced earthenware and china alone to the value of nearly £100,000 in 1836;[36] while its glass output was also considerable. In Messrs. Copeland & Garraths' factory at Stoke, the second largest in our table above, "there are 454 men, 249 women, and 77 children, the latter being under 13 years of age. The premises extend nearly 14 acres of land, all enclosed, and consist of 19 ovens, 272 working rooms, 19 slip kilns, 42 warehouses, and 33 other offices; all well drained by sewers, which empty themselves in running water and cess-pools away from the premises. ...... One steam engine of 45 horse power, which is applied to grinding clays, flint stone, bone, glaze, paint, and other materials used on the works; it also drives two blungers, ten sieves, and two stampers."[37] Indeed, the factory was very extensive, and in the opinion of the Children's Employment Commission, appeared

> "to have been erected many years, as there is neither conformity, order, nor system of arrangement, in any or either apartment (sic), except in the dipping house, green house, and one or two others, which are comparatively modern structures. The printing rooms, throwing rooms, pressing rooms, and painting rooms, in which numbers are working together, are close, low, small, and inconvenient places for the purposes to which they are applied—even at this season when the Ther. is 38, they are hot and unhealthy. There are no means of ventilation except at the doors or small opening at the windows, although there are hot-stoves and plates in every room. The premises are well drained, and great order and regularity is observed by the workpeople. The boys and girls are in separate premises, and are under good discipline."[38]

Messrs. Adams & Sons, the third largest factory, had four plants, three at Lower and one at Upper Stoke. The three plants at Lower Stoke employed a total of 648 persons. The number of persons employed for the plant at Upper Stoke was not known, but it is probable that were we to include these persons also, the factory as a whole might have exceeded Messrs. Copeland and Garraths.' These plants "are of the most ancient in the district, extending over 12 or 14 acres of ground, and situate in the lowest part of the town; the rooms throughout are in a very dilapidated condition, as well as close, damp, hot, dirty, and uncomfortable."[39] A "Sick Society" was organized among the workmen of the factory. Each workman paid 1s. per month, for which payment he received 7s. per week in case of sickness.[40]

Details on other factories listed in the above table may be given, but they are unnecessary in view of the great similarity that prevailed among the pottery factories except in size. Commissioner Scriven observed in 1841, "when first I began my inspection of the several works, it was my intention to report on each separately, but I found there was so great a similarity in their character, that it would have occupied my time uselessly, and been attended with no advantage."[41] The Commissioner, however, divided the whole 127 pottery factories in 1841 into three classes, unfortunately "not according to their magnitude and extent, but to their merits as to drainage, ventilation, and convenience."[42]

The table above given, it must be observed, did not include all the "first-rate" factories in the "Staffordshire Potteries," as statistics on the number of persons employed per pottery factory were not always given by the Commissioner. Messrs. Minton and Boyle's china and earthenware factory at Stoke, for instance, was minutely described as to its various departments, but nothing was said by the Commissioner on the number of persons employed. We must look elsewhere for such statistics on labor concentration. Kohl, writing in 1846, was told by the proprietors of the factory that Messrs. Minton and Boyle employed from 800 to 900 workmen in the factory;[43] while thirteen years earlier, they employed only 700.[44] Messrs. Allcock's china and earthenware factory at Burslem, to take another instance, was considered by the Commissioner as "one of the largest and best conducted in the Potteries,"[45] but no notion of its extent was given at that date. In 1833, however, it employed 400 persons. In the same year, other factories, those of R. Stevenson and of Hicks, Meigh & Co. at Shelton, employed respectively a greater number of persons than Allcock's, namely, 600.[46]

C. Glass

The glass manufacture of England in 1841 was divided into four branches, viz. bottle, crown, flint and plate glass. The relative importance of each branch may be seen in the following table showing the amount of output.[47].

Table 130. Output of glass manufacture of England in 1841 classified according to branches.

Branch	Amount of output (in cwts.)
Bottle glass	414,907
Crown glass	146,613
Flint glass	89,905
Plate glass	32,941
Total	684,366

These four branches of the glass manufacture of England in 1841 employed a total of 6,799 persons, who were distributed in the leading counties in the following table.

Table 131. County distribution of persons employed in glass manufacture, England, 1841

County		Number	Percent.
Northern:		2,391	35.2
Lancaster	1,073		
Durham	858		
Northumberland	460		
Midland		2,291	33.7
Warwick	975		
Stafford	930		
Worcester	386		
Eastern, etc.:		2,117	31.1
Middlesex, Surrey	1,098		
Other	1,019		
England		6,799	100.0

Master Craftsman System

In the glass manufacture of England in 1841, the factory system was the prevalent form of industrial organization. The only exception to this statement was found in the flint glass branch, in which the master craftsman system was adopted by the small glass makers as a means to evade the glass duties levied under the excise law. Under the excise law, every "glass house" must be licenced and supervised by the excisemen from the government with a view to register the amount of output as a basis for the assessment of excise duties. Not every "glass house," however, was licensed and paid taxes to the government, as the motive for evasion was strong under a system of heavy taxation—a system that almost crippled the industry but for its repeal in 1845.[48] According to Mr. William Powell, "an extensive and experienced manufacturer from Bristol,"[49] there were in 1835 "thirty manufacturers in this city [London] ...... unknown to the government, employed in melting up what we call callet, or broken glass, such as the stems of goblets, bottoms of tumblers, the thick parts of decanters, and so on; they can be melted in a garret or a cellar, and made up into salt-cellars, cruets and castors, bird boxes, smelting-bottles, and a variety of articles used by perfumers, and it is done to a very large extent, and my belief is that there is no legislative precaution [that] can prevent it."[50] This estimate of thirty unlicensed glass manufacturers in the flint glass branch was undoubtedly too low, for one of the surveyors-general in the London Excise Office, Mr. Hetherington, gave a minimum number of 98 unlicensed glass manufacturers. Mr. Hetherington told Mr. Hawkes, a glass manufacturer,

> "I think you are considerably under the mark; I take them at 98. They formerly melted the glass with coal that made a very great smoke, and you might detect the illicit work from the smoke, and, therefore, they work with coke, which makes not so much smoke as the common chimney, and the heat is much more intense; they are enabled to melt their glass much more quickly; and the consequence is, that they produce a much finer article than they formerly did." [51]

In other glass making centres, similar conditions were found.

> "At Sheffield there used to be an immene trade carried on in cruets, and goods of that kind. For that market there is scarcely an order [that] comes into the country for such manufacturers from that quarter, because they are made in the unlicensed houses. At Birmingham fraud is carried on to an amazing extent; and in that town they have such a variety of furnaces, that if you take them, the consequence is that there is a complaint made of the hardship of taking up those poor men, and they have nothing to arrest, and if you seize the implements, they are worth very little."[52]

On the whole, however, such unlicensed glass manufacturers were not important, for in the first place, they were limited only to the flint glass branch, for technical if not for other reasons. According to Mr. Harrison, "bottle glass, from the nature of the material, can only be manufactured in glass houses expressly built for the purpose, and upon a scale that cannot evade observation, and consequently easily kept under the control of the Excise; whereas flint glass can be, and is frequently, manufactured in private houses, and escapes the duty altogether."[53] "Flint houses," as a result, "are of different sizes, but not so with crown house or bottle houses."[54] Secondly, flint glass branch produced only 13 percent of the total glass output of England in 1841, a large portion of which must have been produced by the licensed glass manufacturers in the factories. A licensed glass manufacturers employed a large number of persons; while an unlicensed one working under the master craftsman system required but a handful of workers. Thirdly, the disadvantages under which an unlicensed glass manufacturer had to labor were sufficient to check the growth of his class on an extensive scale. "Those [i.e. unlicensed] manufacturers," said Mr. Shortridge, "a very extensive crown glass maker",[55]

> "work at a greater disadvantage than the regular manufacturers. . For instance, if a man who makes a small quantity of glass, perhaps half his time is expended in going about to sell it, and that he must do in an underhand way; he is watched by the Excise, and it is well known that a man has been watched for miles and miles in London, and has not entered into a shop, because he knew that he was pursued by the Excise; and fire, and wages, and everything is proportionably higher. And they work at great waste of materials. . . They cannot produce so good an article in the first instance, and in every possible way they work at greater disadvantage, and they have to pay persons in their employ greater wages. They must pay a sort of hush-money."[56]

Factory System

In the glass manufacture a factory was not infrequently composed of several plants or houses. The Cookson Brothers, for instance, had eleven glass houses in South Shields and Newcastle, which formed the largest glass factory in 1833. John Bower, Joseph Lamb, and the Ridley Brothers had respectively four glass houses; while William Chance, W. Featherstonhaugh, John Biddle, and Joseph Price had respectively three glass houses. Nine glass manufacturers had two glass houses each; while the other fifty-three had one glass house each. In the following table, the 70 glass manufacturers of England in 1833 were classified according to the number of glass houses that each glass manufacturer owned and managed.[57]

Table 132. 70 glass manufacturers of England in 1833 classified according to number of glass houses per glass manufacturer

No. glass houses per manufacturer	No. glass manufacturers	No. glass houses
11	1	11
4	3	12
3	4	12
2	9	18
1	53	53
Total	70	106

In the above table, the 70 glass manufacturers in 1833 had a total of 106 glass houses. In the following table 17 of the 70 manufacturers, being those having two or more glass houses each, were listed in order of the amount of glass duties paid to the government by each manufacturer for that year.[58]

Table 133. 17 leading glass manufacturers with amount of excise duties paid, England, 1833

Name & location of manufacturer		No. glass houses	Amt. glass duties paid (in £)
Cookson Bros., S. Shields and Newcastle		11	69,304
William Chance, Stourbridge		3	53,982
Ridley Bros., Newcastle		4	43,400
Couthupe & Co., Bristol		2	39,190
Charles Atwood, Durham		2	37,922
William Richardson, Newcastle		2	33.939
Joseph Lamb, Newcastle		4	29,943
Richard Shortridge, Durham		2	22,379
John Bower, Leeds		4	16,762
A. Fenwick & Co., Durham		2	12,986
Joseph Price, Newcastle		3	12,150
W. Featherstonehaugh, Durham		3	11,243
John Biddle, Salop		3	10,783
J. B. Falkner, Warrington		2	9.923
Henry Ricketts & Co., Bristol		2	8,392
John Hubbard, Durham		2	7,761
John Carr, Newcastle		2	6,481
Total	17 manufacturers	53	426,540
Add other	53 manufacturers	53	253,544
	70	106	680,084

According to the above table, seventeen manufacturers, or less than one fourth of the total number, paid £426,540 in glass duties, or 63 percent. The glass manufacture, in other words, was highly concentrated in the hands of a few capitalist manufacturers on a large scale and was organized under the factory system.

Statistics for later years tended to confirm the conclusion that, to a large extent, the amount of glass duties paid by each manufacturer in 1833 was an approximate index of the extent of the manufacturer's business. In order to substantiate this statement, another table like the one given above is compiled, in which only those manufacturers, for whom information other than glass duties paid can be secured, are included.[59]

Table 134. Selected list of leading glass manufacturers, England, 1833

Name & location of manufacturer	No. glass houses	Amt. glass duties paid (in £)
Cookson Bros., S. Shields and Newcastle	11	69,304
William Chance, Stourbridge	3	53,982
Ridley Bros., Newcastle	4	43,400
Thomas Cockburn & Co.	1	15.395
George Bacchus, Birmingham	1	11,016
Apsley Pellatt, London	1	7,852
Rice Harris, Birmingham	1	7,259
John Carr, Hartley Pans, Newcastle	1	6,481
Thomas Molineux, Manchester	1	5,200
John Davenport, Longport	1	4,211

In 1841 the glass bottle houses at S. Shields employed 118 adults and 38 children. At Newcastle the glass bottle houses employed 19 children. Were the proportion between the adults and the children at Newcastle glass houses the same as that at S. Shields, the Newcastle glass bottle houses would have employed 57 adults. Besides, the plate glass houses at Newcastle employed 200 adults and 30 children. As a whole, Cookson Brothers' glass factory employed a total of 462 persons in 1841.[60] In 1835 the firm of Chance and Hartley, at Birmingham, "is stated to be the greatest crown glass work now in England. The business of this house is confined to window glass, the principal denominations of which, namely, crown glass and German sheet glass, are manufactured to a great extent, both for home consumption and for exportation."[61] This was the opinion of the Commissioners of Excise Inquiry, to which McCulloch, writing in 1839, concurred in his remark that "the largest manufactory of crown glass in the Empire is that of Messrs. Chance and Hartley, near Birmingham."[62] In 1843 Mr. William Chance, the proprietor, told the Midland Mining Commission that in his glass factory "we employ about 550 men, and 150 women and boys, and we may safely say that from 1,500 to 2,000 are dependent on the works."[63] In 1841 the broad and crown glass factory of the Ridley Brothers employed a total of 379 persons, including "77 men, 27 male apprentices, 64 wives of men, 107 sons, and 104 daughters."[64] The employment of the workman and his family members presented an interesting feature that was not uncommon in the glass factory. "Free house and firing"[65] was provided for by the factory, and constituted a part of the wages. Thomas Cockburn & Co., or, British Plate Glass Co., Lancashire, was incorporated in 1773, when it erected its extensive works at Ravenhead, near St. Helen's, Lancashire.[66] In 1835 Head said that "this extensive establishment consists of iron foundries, as well as the manufactory [plate glass manufactory] aforesaid, and contains within its walls an area of 22 acres of land."[67] In that year, this factory was the only one of its kind besides the plate glass houses of Messrs. Cockson Brothers.[68] In the glass factory of George Bacchus, Birmingham, Mr. Stone, the superintendent, told the Children's Employment Commission that in 1841 "about 220 mechanics are employed, of which number 21 are women, and the remainder men and boys; no young girls are employed. . No machinery is employed where the boys work, but there is a steam engine of 14 horse power for the polishing and cutting."[69] Mr. Pellatt's flint glass factory was "decidedly the largest in London"[70] in 1835, at which year he had already given up the manufacturing part but not the mercantile part, on account of the crippling effects of the heavy excise duties.[71] In 1841 Mr. Pellatt must have resumed the operation of his factory, as his factory was at that year considered "a large establishment"[72] by the Children's Employment Commission. The large scale upon which the factory was carried on had been effectively told by Dodd in 1843. The premises of the factory, said Dodd, "comprise the various buildings necessary for the production of flint-glass ware; such as a horse mill, for grinding old melting pits, as one of the ingredients in the manufacture of new ones; a room wherein ground or powdered clay is mixed or kneaded into a working state; another in which the pots are made; others for drying the manufactured pots; rooms for storing, washing, and preparing the alkaline salts; others for washing and drying the silicious sand; a mixing room, wherein the sand, alkali, and oxides are combined; two coking ovens, or furnaces for converting coal into coke; the glass house, with its working furnaces, pit-furnace, and annealing oven; glass-cutting and glass-engraving shops; and others for subsidiary purposes; the whole occupying an area of three fourths of an acre."[73] The factory of Rice Harris, Birmingham, employed 540 persons in 1841. "The premises are large, and the shops tolerably spacious. The dangerous parts of the machiney are guarded."[74] The factory of John Carr, Hartley Pans, Newcastle, made glass bottles, and employed in 1841 a total of 129 persons, including 97 adults and

32 children.[75] In 1841 Thomas Molineux's factory in Manchester employed 32 children under 18 at the glass house, and 15 or 16 at the grinding shop.[76] A grinding shop, however, was not invariably attached to a glass house, and was sometimes conducted as a separate trade. Whereas the proportion between adults and children in a glass house was three to one, and whereas that in a grinding shop it was one to one, the factory of Thomas Molineux would have employed 111 adults and 50 children. Lastly, John Davenport's factory at Longport, owned by the famous pottery manufacturer, John Davenport, employed 64 hands in 1841.[77]

Chapter VI. Metal: I

A. Iron

The iron manufacture of England in 1841 defied accurate definition, a task that even the Census authorities had failed to accomplish. The latter did not differentiate the iron from the steel manufacture; and instead, merged together the two manufactures under the single heading "iron manufacture" in the summary return for England.

The difficulty of defining the iron manufacture was due to its highly integrated character. A typical "ironworks" in 1841, which in our terminology may be called an iron factory, included under single management and ownership all or a part of such constituent units as coal and ironstone mines,blast furances, forges, rolling mills and foundries. These different units carried on different processes, some of which did not properly belong to the iron manufacture. For instance, mining must be eliminated as a process in the iron manufacture, while founding could be included only in a restricted sense. In view of this difficulty, our definition is justified insofar as it enables us to make use of the Census statistics of 1841. In our definition, the iron manufacture of England in 1841 included all persons fully employed in the following three processes: (1) smelting iron from the ore; (2) manufacturing the pig or cast iron into wrought iron, such as bars, rods, sheets etc.; and (3) casting the pig iron into products for domestic and building purposes.[3]

Table 135. Number of persons employed in iron manufacture, England, 1841

Branch of manufacture		No. persons employed
Iron (a)		21,872
Moulder (b)		3,316
Founder		946
Iron products (c)		4,039
Pipe (d)	2,691	
Crate	546	
Fender	462	
Grate, range & stoves	211	
Fire iron	129	
Total		30,173

These 30,173 persons employed in the iron manufacture were distributed among the leading counties as follows.

Table 136. County distribution of persons employed in iron manufacture, England, 1841

County		Number	Percent
Midland		11,464	38
Stafford	7,601		
Warwick	1,444		
Worcester	1,334		
Salop	1,085		
Northern		9,923	33
Lancaster	3,701		
York, W. R.	3,535		
Durham	1,181		
Northumberland	796		
Derby	710		
Western etc.		8,786	29
Monmouth	3,042		
Other	5,744		
Total		30,173	100

The concentration of the iron manufacture in the midland and northern counties of England in 1841 was a phenomenon that had its origin in the latter part of the eighteenth century. Before the general use of coke smelting of iron, the industry was dispersed wherever wood was available. The dependence upon water power to drive the wheels for generating the blast was another factor in scattering the iron furnaces along the rivers or streams. The "tyranny of wood and water" was overthrown only after the inventions of Darby and Watt. The concentration of the iron manufacture was rapidly brought about now that iron could be smelted in the coal-producing regions, and the application of steam engines freed the industry from being dependent upon water power. In 1841 the localization of the industry was based on different principles, of which proximity to ironstone, coal and limestone on the one hand, and nearness of market for iron and iron products on the other, were the most important. Besides, facility of transport must also be considered.[4]

Integration in iron factory

That the iron manufacture in 1840 and much earlier had been conducted under the factory system needs no further proof after Mr. Ashton's study into the original sources on the subject. "From the earliest period of which we have exact information," says Mr. Ashton, "iron making in this country [Great Britain] has been conducted on capitalistic lines—capitalistic not only in that the workers are dependent upon an employer for their raw materials and market, but also in that they are brought together in a 'works', are paid wages, and perform their duties under conditions not dissimilar to those of almost any large industry of modern times. The scale of operations has increased enormously; the sapling has become an oak, deep-rooted and widespread; technique has been revolutionized. But in structure and organization there is no fundamental change."[5] In this quotation the key elements of a factory were described, namely, labor concentration in the employer's establishment and the use of considerable capital. In almost similar terms, another authority on the subject, Mr. Lloyd, remarked that in the British iron industry "the adoption of large-scale production must be dated at least as far back as the seventeenth century, and indeed long before that time the relatively large capital and extensive labor force required for the operation of furnaces and forges of even the moderate size then employed, gave the industry a character far removed from the category of handicraft production."[6] Take, for instance, the enterprize carried on by Sir Ambrose Crowley and his son John, at Swalwell and Winlaton in the county of Durham, which "was surprizingly modern. In its intergrated structure, its internal administration, its methods of ensuring industrial peace, and even in such minutiae as the provision of standardized gauges, the concern appears properly to belong to the twentieth, rather than the seventeenth or eighteenth century. . . . In the foundry, in the converting of steel, and in the heavy forge processes, the men worked in buildings and with machinery provided by the capitalist; they were sometimes paid time wages, and were in other respects similar to the modern factory works."[7]

Indices of Factory Size

Output. A better index of the factory size in the iron manufacture is the output of pig iron, for which there exists more definite information than for other features of an iron factory such as capital and labor. Ever since 1844 the Commissioner of Mining Population has made annual report on the "ironworks" in Great Britain, although with no systematized plan in gathering facts. For our purpose of analyzing the factory size in the iron manufacture, the most satisfactory statistics available in these and other reports were for the years 1843 and 1848, from which the following two tables are compiled, showing the output of pig iron for each of the 20 leading "ironworks" in the years 1842 and 1848. In this connection, it may be repeated that an iron factory, like a textile factory, is a unit of ownership and management. Several "ironworks", for instance, were scattered in the different counties of England, or in the different towns or parishes of a county, and yet each formed a constituent part of the iron factory if owned and managed by a single proprietor.[8]

Table 137. Size of 20 first-rate "ironworks" of England as measured by output of pig iron per "ironworks" in 1842

Name of owner	Location	No. blast furnaces	Output (in tons)
Lilleshall Co.	Salop	not given	19,030
Coalbrookedale Co.	Salop	not given	18,360
W. & B. Botfield	Salop	not given	15,639
Blackwell & Jones	Stafford	7	14,940
P. Williams & Sons	Stafford	6	14,240
Sir H. St. Paul	Stafford	6	14,240
G. Jones	Stafford	5	13,280
B. Gibbons	Stafford	7	12,400
Low Moor	York	7	12,348
Butterley Co.	Derby	6	11,208
W. & J. Sparrow	Stafford	5	10,680
J. Bradley & Co.	Stafford	4	10,680
Parkfield Co.	Stafford	4	10,680
Pelsall Iron Co.	Stafford	2	10,680
Chillington Co.	Stafford	4	10,340
Matthews & Dudley	Stafford	4	9,840
W. & J. Firmstone	Stafford	3	8,960
Madeley Wood Co.	Salop	not given	8,768
J. F. Wombridge	Salop	not given	8,498
Bowling	York	4	7,642
Total			242,453

Table 138. Size of first-rate "ironworks" of England as measured by output of pig iron per "ironworks" in 1848

Name of owner	Name of "ironworks"	Location	No. blast furnaces In blast	No. blast furnaces Total	Output (in tons)
British Iron Co.	Dudley Wood & Northerton	Staf.	2	6	35,360
	Congreaves	Staf.	2	2	
	Abersychan	S. Wales	4	6	
Derwent Iron Co.	Consett & Crook Hall	Durham	7	14	29,120
Colebrooke-dale Co.	Colebrooke-dale Co.	Salop	8	8	24,960
J. Foster	Shutend	Staf.	3	4	21,900
	J. Foster	Salop	3	3	
P. Williams & Son	Wednesbury Oak	Staf.	3	3	20,900
	Union	Staf.	2	3	
Butterley Co.	Butterley Hall	Derby	3	3	20,800
	Codnor Park	Derby	3	3	
Heathcote	Apedale	Staf.	4	4	18,720

W. & P. Botfield	Old Park	Salop	6	8	18,720
Lilleshall Co.	Lilleshall Co.	Salop	6	7	18,720
G. Jones	Cosely	Staf.	2	2	16,720
	New Duffield	Staf.	2	3	
	New Birchills	Staf.	—	2	
Bagnall & Sons	Gold's Green	Staf.	2	3	16,720
	Cappon Field	Staf.	2	3	
W. & J. Sparrow	Slowheath	Staf.	—	5	16,680
	Osier Bed	Staf.	2	3	
	Lane End	Staf.	2	3	
Hird, Dauson & Co.	Low Moor	York	5	6	14,560
Kinnersley	Kidd's Grove	Staf.	3	3	13,520
Daws & Sons	Crookhay	Staf.	3	3	12,540
	Oldbury	Staf.	—	2	
Weardale Iron Co.	Tow Law	N. & D.	2	3	12,480
	Stanhope	N. & D.	1	1	
Witton Park Co.	Witton Park Co.	N. & D.	3	4	12,480
Sturgess & Co.	Bowling	York	4	5	10,400
Yates & Co.	Wengerworth	Derby	2	3	10,400
Smith & Co.	Stanton	Derby	3	3	10,400
	Total		94	131	356,100

In the following table, the total number of "ironworks" and total amount of pig iron output in the two years 1842 and 1848 are summarized.

Table 139 "Ironworks" and their pig iron output in England, 1842 and 1848

Number of "ironworks"	Total pig iron output (in tons) 1842		1848	
20 first-rate "iron works"	242,453	(53%)	356,100	(57%)
51 other " "	215,626	(47%)	....	
69 other " "	....		263,220	(43%)
71 " "	458,079	(100%)	....	
89 " "	....		619,320	(100%)[9]

According to the above table the 20 first-rate "ironworks" produced 53% and 57% of the total pig iron output in the years 1842 and 1848 respectively. Or, to speak in another term, in 1842, 28% of the total number of "ironworks" produced 53% of the total pig iron output; in 1848, 22% of the total number of "ironworks" produced 57% of the total pig iron output. The total pig iron output per first-rate "ironworks", therefore, has increased in the course of six years. The measurement of factory size according to the output of pig iron factory or "ironworks" was the only index applicable to all the "ironworks" in England in 1840 and thereabout. This application, however, was limited, and, for various reasons, must be qualified by the employment of other indexes. In the first place, the output varied a great deal with the business conditions of the iron industry in 1841. That the actual output of an "ironworks" did not always correspond with its potential output could be easily seen in an analysis of the number of blast furnaces "in blast" and "out of blast". In 1842 such statistics were not complete, and therefore omitted from Table 137. But in 1848 such statistics were given for each of the twenty first-rate "ironworks". Thus according to Table 138, of the 131 blast furnaces existing in 1848, only 94 (or, 72%) were "in blast". In the second place, the integrated character of an "ironworks" renders our analysis of the above two tables somewhat misleading. We measure the size of an "ironworks" in these tables by the output of pig iron alone, while in many cases such a measure indicated very little about the actual size of an "ironworks". Not every "ironworks" carried on the process of smelting iron as the only employment; indeed, the majority were also engaged in mining and founding processes.[10] When an "ironworks" carried on the mining and smelting processes together the mining process generally employed a greater number of persons than the smelting process. One witness, Mr. Best, estimated in 1843 that each blast furance employed a total of 157 persons, of whom only 41 were employed in the smelting process, while the other 116 persons carried on the mining process.[11] In other words, the proportion between smelting and mining as regards the number of persons employed was one to three.

With these qualifications in mind, we may quote a few details on labor concentration in the leading "ironworks" of England, some of which are listed in the two tables for the years 1842 and 1848 respectively.

Labor. The Consett Iron Works, which was owned by Derwent Iron Co., was first established in 1841. In 1846 the "ironworks" was considered the second largest in the United Kingdom, and had a population of 12,000 dependent upon it for means of livelihood. Perhaps it furnished the most perfect example of an integrated industry. It owned and managed 35 coal and ironstone pits, 14 blast furnances, 12 refineries, 2 rolling mills turning out a weekly output of 900 tons of bar iron, and 22 steam engines. The pits that the "ironworks" owned and managed extended over a radius of 28 miles.[12] Another example of a highly integrated "ironworks" was that belonging to Mr. James Foster and carried on under the firm name of J. Bradley & Co. Commissioner Horne told us in 1841 that Mr. Foster's "ironworks", situated in the town of Stourbridge, Worcester, was the largest in England. It comprised "forges, refineries, rolling mills, foundries, etc. Mr. James Foster has also blast furnaces, and extensive colliers and works near Kingswinford [in Stafford], and likewise in Shropshire [i. e. Salop]. There are as many as 5,000 persons employed by this one gentleman.....To attempt any detailed account of these stupendous works would occupy a space far beyond the limits of the report [i. e. Report for the Children's Employment Commission] to admit; I shall merely say, therefore, that the good order, rate of wages, and general good treatment of the young persons employed, appears to me to leave little to be desired in those respects."[13]

In the following table further statistics on the employees per "ironworks" are given for the years 1844 to 1847. They are fragmentary in character, but may help to illustrate the varying size of different "ironworks" during that period.[14]

Table 140. Size of "ironworks" as measured by number of persons employed, 1844-1847

Name of owner of "ironworks"	Location	No. employees
J. Barker	S. Staf.	1,500
Tow Law	N. & D. (a)	1,200
Kinnersley	N. Staf.	1,000
Apedale	N. Staf.	700 to 800
Silverdale	N. Staf.	500

(a) Northumberland & Durham.

Without further statistics conclusions on the average size of an "ironworks" cannot be drawn. It may be said, however, that "ironworks" of 1,500 employees each were frequently found if we count the miners of raw materials together with the other works. Commenting on the size of an "ironworks" in the first quarter of the nineteenth century, Mr. Ashton says that "it was by no means uncommon to find 300 or 500 men employed in a single works, apart from the colliers and others connected with the supply of raw materials."[15] The "colliers and others connected with the supply of raw materials", as we have shown in a previous paragraph, outnumbered the smelters, moulders and refiners etc. in a blast furnance by threefold.[16]

Capital. The integrated character of an "ironworks" necessitated the employment of large capital. Even before the end of the eighteenth century, two out of the "four great typical concerns", as Mr. Ashton told us, had already employed a capital amounting respectively to 100,000 pounds and 213,393 pounds.[17] In 1812, according to Thomas Attwood, a complete "ironworks" could not be constructed for less than 50,000 pounds;[18] in 1833 the estimate was tripled by William Matthews, a witness before the Committee on Manufactures, Commerce and Shipping. The Committee asked Mr. Matthews,

> "do you think, even in the present time, that an ironwork of the capability of manufacturing 300 tons of iron per week can be erected for £150,000, consisting of a sufficient quantity of ironstone and of coal mines opened, and a sufficient number of furnaces, with materials for the working of those engines; with refineries, with forges, and with mills, with the necessary quantity of rooms, and all other appendages?"

The reply from Mr. Matthews was,

> "yes, I do; in Staffordshire I think it might be done, but of course under favorable situations and circumstances; and there are many such situations and circumstances that would enable a man to erect an ironwork in Staffordshire of the capacity stated for less than £150,000."[19]

One effect resulting from the employment of large capital in an"ironworks" was the heavy overhead cost, which consisted chiefly in interest, depreciation charges, and expenses for upkeep and maintenance. The overhead cost varied, in each case, with the extent of integration in an "ironworks". The more highly integrated an "ironworks" became, the greater overhead cost it incurred. As a consequence, the policy of an "ironworks" had always been to avoid any stoppage of operation, whether partial or complete. In a period of business depression, it was not uncommon for the proprietor of an "ironworks" to go a step further by increasing the output "so as to reduce the common charge there is upon his produce."[20] The loss consequent upon such a period of business depression was thus aggravated by ruinous competition among the ironmasters themselves; and it is not strange that combinations were early formed among these ironmasters to reduce output and to maintain price.[21] Such combinations, however, tended to keep the inefficient producers in business.

Industrial Organization and Technique

The close connection between industrial organization and technique can be best witnessed in the iron manufacture. In the words of Mr. Ashton, "the new coke-smelting process led to a transformation of what had previuosly been relatively small-scale production into those forms of great industry, which for want of a better phrase, are generally termed the 'factory system'. This had existed in the metal trades even in the seventeenth century, as the examples of the Crowley and Sitwell undertakings clearly show. But the new processes of ironmaking were to turn what had been exceptional and sporadic manifestations into a dominating tendency of industry."[22] Besides, each invention in technique generally resulted in an increase in the size of "ironworks". The process of coke-smelting was a leading example, but other inventions such as those of the steam engine and hot blast were equally important in this respect. In the following table, it may be observed that the increased output per blast furnace from 1740 to 1839 indicated not only the improvement in technique, but also the growth of industrial unit in the iron manufacture.[23]

Table 141. Output of pig iron per blast furnace, Great Britain, 1740-1839

Year	No. furnaces		Total output (in tons)		Output per furnace (in tons)	
	Charcoal	Coke	Charcoal furnace	Coke furnace	Charcoal	Coke
1740	59		17,350		294	
1788	26	59	14,500	53,800	545	909
1796	12	122	16,900	124,179	450	1,018
1806	11	162	7,800	250,406	709	1,546
1823		266		455,166		1,730
1830		372		677,417		1,824
1839		396		1,248,781		3,153

B. Engine and Machine

The engine and machine manufacture of England in 1841 may be tentatively defied as including all persons fully employed in making and repairing the following products: (1) general (primarily textile) machinery and millwork, (2) steam engines (stationary, locomotive and marine), (3) hydraulic machinery, (4) machine tools, and (5) subsidiary parts of machines and machine tools.[24] Under this definition the manufacture employed a total of 32,667 persons.[25]

Table 142. Number of persons employed in engine and machine manufacture in England and two important counties, 1841

Branch	England	Lancaster	York, W. R.
Trades			
Engineer (a)	10,175	2,253	947
Millwright (b)	6,622	1,313	784
Products (staple)			
Engine and machine (c)	6,083	2,058	1,188
Boiler	2,673	873	194
Frame	1,149	43	4
Pump	526	21	110
Products (subsidiary)			
Card	2,050	378	1,459
Bobbin	1,004	403	101
Roller	595	455	22
Reed	515	339	89
Shuttle	364	179	118
Spindle	239	35	126
Heald	199	....	199
Die	266	6	8
Sinker	207	....	....
Total	32,667	8,356	5,349

In the engine and machine manufacture, the two leading counties were Lancaster and West Riding of York, which employed a total of 13,705 persons, or 41 percent of the whole. The proximity of demand from the textile and metal industries was an important factor for the localization, while the nearness to raw materials such as iron and coal constituted another.

Industrial Organization

Before analyzing the industrial organization of the engine and machine manufacture, it is necessary to note that the two classes of workers enumerated by the Census as "engineers" and "millwrights" in counties other than Lancaster, the West Riding of York, and probably Middlesex (with Surrey and Kent), Stafford, Durham and Northumberland, were, on the whole, not employed as factory workers, but as master craftsmen or wage workers. These two classes, according to Professor Day, had a provincial instead of national distribution with regard to the market of their products or services. An industry was national in its distribution "if more than half the persons engaged in it were concentrated in districts comprising less than one quarter of the total population."[26] The failure of the "engineers" and "millwrights" to meet this test of national distribution indicated their dispersion, a phenomenon explained by the nature of their occupation. In every county of England,[27] whether large or small in population, "engineers" and "millwrights" were needed either for the erection of engines and machines, or for their repair. Bedford, for instance, had 3 "engineers"; Westmoreland, 3; Hereford, 5; Cambridge, 6; Northampton, 10; etc.

Again, with regard to the 6,083 persons enumerated by the Census under "engine and machine makers" Professor Day remarks that "there is evidence that this group still included many petty artisans, scattered through the agricultural districts and constructing their simple mechanisms largely of wood. It is significant that this group, with a function so important in the development of modern manufacture, was not merely small, but was so widely dispersed. More than half the workers, it is true, were gathered in Lancashire and the West Riding, but the coefficient of deviation, calculated for the county as a whole, was barely larger than unity. The boiler makers show a somewhat higher degree of concentration probably because the trade was one in which a petty craftman could hardly be enumerated."[28]

Factory System

In the eight counties where 22,129, or 68 percent of the total persons employed in the engine and machine manufacture, were concentrated, the factory system was the prevalent form of industrial organization except in the manufacture of subsidiary products such as healds and cards.[29] These eight counties are listed in the following table with the number of persons employed in each county.

Table 143. County distribution of persons employed in engine and machine manufacture, England, 1841

County	Number	Percent.
Lancaster	8,356	25.6
York, W. R.	5,349	16.3
Middlesex. Surrey & Kent	4,331	13.3
Stafford	1,657	5.1
Durham	1,352	4.1
Northumberland	1 014	3.1
Other	10,608	32.5
Total	32,667	100.0

For Lansaster, the most important of all counties in engine and machine manufacture, reliable statistics that illustrated the average size of an engine and machine factory in 1841 were given by Mr. W. Jenkinson before the Committee on the Export of Machinery. Mr. Jenkinson's statistics, however, were estimates, which, concerning the number of persons employed, would have to be discounted by at least two fifths, if not one half. The reason for such discounting is obvious. After having given his estimates on "the principal mechanical establishments in eleven of the manufacturing towns of Lancashire, with the number of horses' (sic) power attached to each, the number of hands which each is capable of employing, and the amount of capital invested in each",[30] Mr. Jenkinson remarked that "the convenience now arising and the facilities afforded for producing machinery are so great, that I have no hesitation in saying that, within the next 12 months, we could produce double the quantity that we have produced in the last 12 months if it was required; the establishments now in existence, I am sure, could produce it."[31] Mr. Jenkinson's estimates for Lancashire are as follows.

Table 144. Statistics on principal mechanical establishments in Lancashire, 1841

Town	(A) No. factories	(B) Amt. motive power (in H. P.)	(C) No. hands each factory is *capable* of employing	(D) Estimate of capital invested (in £)
Manchester	26	404	5,430	400,000
Salford & Patricroft	9	244	2,520	240,000
Stockport	8	50	500	40,000
Ashton & Neighborhood	8	122	1,155	85,000
Rochdale	13	113	720	80,000
Burnley	5	40	210	30,000
Bury & Heywood	7	76	740	70,000
Bolton	7	225	2,250	220,000
Preston	7	127	930	100,000
Blackburn	6	73	720	50,000
Oldham	19	337	2,207	200,000
Lancashire	115	1,811	17,382	1,515,000

If we discount column (c) two fifths for the reason already stated, and calculate the average size per engine and machine factory as regards the motive power, employees and capital, we get the following table for the eleven manufacturing towns of Lancashire.[32]

Town	Motive power per factory (in H. P.)	Employees per factory	Capital invested per factory (in £)
Manchester	15.5	125.3	15,385
Salford & Patricroft	27.1	168.0	26,667
Stockport	6.3	37.5	5.000
Ashton & Neighborhood	15.3	86.6	10,625
Rochdale	8.7	33.2	6,154
Burnley	8.0	25.2	6,000
Bury & Heywood	10.1	63.4	10,000
Bolton	32.1	193.0	31,429
Preston	18.1	79.7	14,286
Blackburn	12.2	72.0	8,333
Oldham	17.7	69.6	10,526
Lancashire	15.7	90.7	13,174

According to the above table, the average engine and machine factory in Lancashire employed 15.7 horse power, 90.7 employees and £13,174 capital. Bolton, where an average engine and machine factory employed 32.1 horse power, 193 employees and £31,429 capital, had the largest average size factory among all the manufacturing towns in the county.

Mr. Jenkinson's estimate for Lancashire was to a certain extent confirmed by Mr. Fairbairn's with regard to the average size of an engine and machine factory except in the number of persons employed. Mr. Fairbairn calculated that in 1841 there were in Leeds 18 factories, employing a total of 255 horse power, 2,950 persons and £305,000 capital. In other words, each engine and machine factory in Leeds employed, on the average, about 14 horse power, 164 persons and £17,000 capital.[33]

With the average factory for Lancashire and Leeds as our basis, we may now proceed to examine a few of the first-rate factories in the engine and machine manufacture in the leading counties of England. In the following table, 9 factories employing varying numbers of persons are enumerated from the different sources.[34]

Table 145. Details on 9 "first-rate" factories in engine and machine manufacture, England, 1841-46

Name and location of factory	Number of employees	Branch of manufacture
Acramans, Morgan & Co., S. Glou. (a)	1,200	Engineers, millwrights, &c.
Hibbert & Platt, Oldham, Lanc. (b)	900	Machines
Sharp & Roberts, Manchester, Lanc. (c)	800	Locomotive engines
Wm. Fairbairn, (e) Leeds, York (W. R.)	550	Textile machinery
J. & S. Seawards, London (f)	500	Marine engines
Nasmyths Gaskell & Co., Patricroft, Lanc. (g)	500	Machine tools
R. Stephenson & Co., North. & Durham (h)	466	Locomotive engines
R. & W. Hawthorn, North. & Durham (d)	440	Locomotive engines
Wolverton works, Bucks, (i)	400	Locomotive engines

Some of the engine and machine factories, like many of the "ironworks",[35] were foundries also, making machines and engines with other products. In the factory of Acramans, Morgan & Co., for instance, "besides the usual operations of engineers and millwrights, those of iron shipbuilding, chain cable and anchor making, are carried on by this spirited company to a vast extent."[36] In the above list, Acramans, Morgan & Co. furnished the only example, but in Lancaster and the West Riding of York, many factories known under the general name "iron foundries" made engines and machines together with other metal products.[37]

Among the "first-rate" factories, those of Nasmyths, Gaskell & Co., Wm. Fairbairn, and Sharp & Roberts were cited by the Children's Employment Commission as examples of establishments working "on a very large scale and fitted up with great elegance", with great subdivision of labor.[38] Fairbairn's factory, which employed only 60 to 70 persons in 1824,[39] employed 550 in 1841. It had a capital of £50,000 to £60,000 in the manufacture of machinery for flax and waste silk spinning, and produced £60,000 to £70,000 worth of machinery per annum.[40] The Atlas Works of Sharp & Roberts in Manchester, which employed 800 persons in 1846, had 1,200 to 1,500 persons four years later. According to one visitor to the Works in 1850, "a spirit of universal satisfaction appears to reign, and a finer or more muscular army of men than those swarthy mechanics, with their strong limbs and firm gait, it would be hard to select."[41] In 1847, it was said, the Atlas works made and sent out 87 locomotive engines, and orders were already on hand which it would take until 1850 to execute.[42] In the factory of J. & S. Sewards in London, the large extent of operation carried on may be seen from the description of a foreign visitor. "The total or aggregate power they make," said this visitor, "is about 2,000 horses annually besides repairing old engines. The consumption of coals in 12 months is 3,000 tons, of iron 2,500, of copper 200, and of brass 100 tons...The wages they pay weekly is about 800 pounds, and the expenses of the establishment, rent and taxes, about 9,000 pounds per annum, so that wages, rent and taxes all amount nearly to 50,000 pounds a year."[43]

C. Button, Screw, Steel pen, Pin & Needle

The five small manufactures of metal in England in 1841, viz. button, screw, steel pen, needle and pin, employed a total of 9,605 persons. Despite their small extent, they furnished excellent examples of the factory system. Three of them, viz. button, screw, and steel pen, were chiefly Birmingham manufactures, as Birmingham alone employed 3,632 out of a total of 5,484 persons, or 66 percent. In the following table the number of persons employed in each of the five manufactures in England and Birmingham in 1841 is summarized.

Table 146. Number of persons employed in five small metal manufactures, in England and Birmingham, 1841

Manufacture	England	Birmingham
Button	3,915	2,888
Screw	1,242	438
Steel pen	327	306
Needle	2,815	13
Pin	1,306	120
Total	9,605	3,765

Button

The button manufacture of England in 1841 employed a total of 3,915 persons, and was chiefly carried on in the two counties of Warwick and Dorset. In Warwick, 2,913 persons, or 74 percent. of the total employees in the manufacture, made various sorts of buttons, namely, metal, gilt, and plated buttons; pearl, horn, bone, and glass buttons; Florentine and silk buttons; etc. In Dorset 205 persons, or 5 percent. of the total employees in the manufacture, made pearl and other buttons. The Census return for Dorset was probably an under-estimate, as many button workers in Dorset were part-time family craftsmen including women and children, and were therefore not included in the Census.

Two systems of industrial organization were found in the button manufacture in 1841, namely, factory system and family craftsman system. In Warwick where 74 percent. of the total employees in the button manufacture were located the factory system prevailed; while in Dorset the family craftsman system was predominant. According to the Children's Employment Commission in 1841, there were in Birmingham, the button metropolis of Warwick,[44] six button factories employing varying numbers of persons per factory.[45] Messrs. Turner and Sons, "one of the principal establishments in the button trade",[46] employed about 500 persons and had branches at London, Manchester, Leek and Paris.[47] Mr. Elliot's button factory, "one of the largest manufactories in the town", also employed about 500 persons "when trade is tolerably good, in the manufactory and out of it".[48] Mr. Chatwin's, a third button factory, employed upwards of 200 persons; Mr. Hasluck's employed "a large number of workmen—as many as any other manufacturer in this business in the town."[49] The button factories of both Mr. T. W. Ingram and Mr. J. Morton employed "a considerable number of children and young persons".[50] These factories were mentioned because they gave evidences before the Commission, but there were undoubtedly others of the same or smaller size that were omitted. For instance, one witness, A. B., "from the nature of his employment has free and unconstrained access to one of the largest manufactories of the town, where upwards of 200 mechanics are employed. Also knows several of the smaller manufactories where from ten to eighteen or twenty persons are employed."[51] These "smaller manufactories" were shops rather than factories—shops of the "small masters" who were either master craftsmen working on their own account, or outwork masters employing journeymen and apprentices to execute orders given out by the factory manufacturers. They were, however, insignificant in the proportion of persons employed, and would not affect our conclusion that the Birmingham button manufacture in 1841 was organized chiefly under the factory system.[52]

In Dorset the making of shirt wire-buttons was carried on from Blandford to Shaftesbury and from Shaftesbury to Sturminster; it was carried on more in the villages than in the towns.[53] The family craftsmen, including the farm laborers' wives and children were occupied in this employment all the year round. Girls and boys, before they went into field work after eight years of age, were employed in buttoning at home. "The importance of this employment to the family," it was said, "is perceptible in the unwillingness of the parents to let their children attend the schools where the work might be interrupted."[54] The buttons were brought to the drapers by the makers after they were made. "Upon the makers offering the buttons for sale, a price is fixed upon, half of which is paid down by the draper in money and the other half is applied in the purchase of fresh materials, and in purchasing articles of linen or clothing from the draper. This mode of dealing between the draper and the button maker has been in existence as long as the trade of button making in these parts,- more than a hundred years."[55] In Blandford in 1841, "the introduction of the pearl-button," however, "has made a serious difference to the button makers; it has very considerably diminished the demand for the wire-buttons, which were the most profitable to make, whilst it has increased, perhaps, the demand for the coarser articles, upon which the earnings are small."[56]

Screw

The screw manufacture of England in 1841 employed a total of 1,242 persons, who were concentrated chiefly in the Midland and Northern counties as follows: Stafford, 496; Warwick, 438; Worcester, 57; Derby, 97; West Riding, 92; other counties, 65.

Before 1760 screws were generally made by hand, but at that year Brothers Wyatt of Birmingham, by the use of a screw cutting machine, rendered the process of screw-making to a certain extent automatic.[57] In 1819 steam power was introduced in driving the screw making machines at Birmingham.[58] In 1841 we are told that screw making required five processes, namely, foorging, turning up, nicking, worming and topping. In all these processes except forging, power machinery was employed.[59]

Before the general use of machinery, screw-making was in the hands of "small master". In the early years of the nineteenth century, when screws could not be profitably made without mechanical arrangements somewhat beyond the means of a humble workman, the factory system became the prevalent form of organization. In the county of Stafford there were in 1833 three screw factories, one at Wolverhampton, and two at Bilston.[60] In 1841 screw factories were found in Wolverhampton, Bilston, Sedgley and Darlaston.[61] In Bilston Messrs. Hawkesford Brothers' factory employed about 200 persons when at full work;[62] in Sedgley, Mr. Baker's factory was considered "one of the two large factories in town".[63] In Birmingham, where according to the Census all the 438 persons of the county of Warwick were employed, a larger total, about 800 persons, were reported for the four factories by the Children's Employment Commission. Mr. James' factory was the largest employing 60 men and 300 women;[64] Messrs. J. Hawkins and Co.'s ranked second with 194 workmen, employing 39 men, 138 women, and 17 children[65] The other two factories, those of Messrs. F. H. Ryland and Co. and Messrs. Daniel and Joseph Ledsam, employed respectively 100 and 140 hands.[66] In these Birmingham screw factories the preponderance of woman labor was made possible by the employment of power machinery. Woman labor, it was estimated, constituted "80 or 90 percent. of the whole number employed"[67] in a screw factory. One visitor to the screw factory in Birmingham "found a considerable number of women employed, in one room as many as 60, all busily engaged in turning the screws at small iron engines placed before them on a bench. The cutting of the lengths, heading screw and notch cutting, and other parts of the process, although performed by the agency of steam power, appeared throughout to be anything but work for women. Some of these females can turn out 24 gross of screws in a day, and at this rate they will realize from 6s. to 8s. weekly."[68] In Derby, the third largest county for screw making, there was in 1829 "a curious and interesting mill for the manufacture of common screws, at Hartshorn, belonging to Messrs. Smith, Port, Wood & Co. It was originally a branch of a similar manufactory at Burton-upon-Trent. Numerous hands are employed, and many hundred gross are made per week, by means of engines and lathes, turned by a water-wheel."[69]

Steel Pen

The steel pen manufacture of England in 1841 employed a total of 327 persons, of which number Warwick claimed 307 persons. The manufacture was insignificant in view of the small number of persons employed in it, but for the analysis of industrial organization, it offered an excellent example of the factory system.

In the steel pen manufacture the employment of machinery was the primary factor for the introduction of the factory system. The making of metallic pen was first put into practice by Mr. Harrison around 1780,[70] and fifty years later machines were invented to make them by Mitchell, Gillot and Mason.[71] Mitchell's claim was much earlier than Gillot's, as Gillot improved upon Mitchell's invention only after he had learnt the art from Mitchell's daughter, whom he married after a brief courtship.[72] Mason, Mitchell, and Gillot were all manufacturers in Birmingham, while in London James Perry also took out several patents on the making of steel pens by machinery.[73] From 1829 on, Perry was so successful in his business that Mason supplied his own pens to Perry under the latter's name. Perry's pens, "pushed into the market with remarkable energy, were known and used throughout the country before the great Birmingham penmakers, Mitchell, Gillot, and Mason, created a revolution in the trade, and ensured its enormous development, by the use of machinery in the production of pens."[74]

In 1830 Gillot advertised at Birmingham in the "Penny Cyclopaedia" that "the number of pens produced in this factory in the year ending December 31, 1836, was 250,000 grosses (sic) or 3,000,000 dozens, or 36,000,000 pen."[75] At about the same time an article on Birmingham in the Penny Cyclopaedia said that "the principal manufacturer of steel pens [at Birmingham] employs 250 individuals, and consumes annually upwards of 40 tons of fine sheet-steel, each ton of which will make nearly 10,000 gross of pens!"[76] By the process of deduction, if a factory producing 400,000 gross of pens employed 250 workpeople, then Gillot's factory producing 250,000 gross of pens would have employed 155. Mitchell's factory in 1841 employed "a large number of children under 18, boys and girls; also many young women and some few men as tool makers".[77] The number of persons employed in the factory was not given, but we may infer that the factory must have been considerable in size, as there were 14 or 16 tool makers employed.[78] The Commissioner, after inspecting Mitchell's factory, remarked that "the premises are very much crowded. The shops are of the common character in this town. In some the number of workpeople is very great, so that they are greatly crowded. The privies are used separately; but in a small yard like this, and with the great number of mechanics, it is impossible that any effective restriction as to communication between the males and females can be carried into operation."[79]

Needle

The needle manufacture of England in 1841 employed a total of 2,815 persons,[80] of whom 2,150 persons, or 76 percent., were located in the two Midland counties of Worcester and Warwick, Worcester having 1,322 persons and Warwick having 828 persons. The hosiery making counties of Leicester, Nottingham and Derby, where knitting needles were consumed to a large extent, employed a total of 477 persons, or 17 percent. In the counties of Worcester and Warwick, Redditch, Alcester and Studley formed the trio district for the needle manufacture, Redditch having long been the leading source of supply for needles.

In 1762 the art of needle-making was so sub-divided as to include thirteen separate processes for the completion of a single needle.[81] This minute division of labor gave the first impetus to the concentration of labor inside the factory, under the constant supervision of the employer. Later on, the introduction of machinery for some of the needle-making processes such as piercing, clearing (i.e. polishing) and pointing brought the industry under the complete operation of the factory system. In 1840 all the processes were carried on inside the factory with the exception of pointing. "Some of the needle pointers work at their own homes while some work at the factories; but the process is the same in either case."[82]

In 1843, there were in Redditch about a dozen manufacturers, each of whom conducted the needle manufacture on a large scale under the factory system, and employed a considerable number of persons. The subdivisions of the famous factory, the British Needle Mills at Redditch, corresponded with the technical processes employed. Accordingly, we find that while some of the shops were occupied by men, others contained only women or children.[83] At Hathersage, Derby, there was "a celebrated manufactory of needles"[84] in 1829, conducted by Messrs. Cocker and Sons. In 1841 it was said that the proprietor, Mr. Henry Cocker, "has taken great and meritorious pains to Christianize and instruct the working classes, and the absence of town vices, and the temptations incidental to the congregation of large bodies, have materially aided Mr. Cocker's efforts; and the state of morals at Hatersage is very satisfactory, and the prospect of further improvements most gratifying."[85]

Pin

The pin manufacture of England in 1841 employed a total of 1,306 persons, who were chiefly distributed in the three counties of Gloucester (524), Lancaster (469) and Warwick (120).

Ever since the time of Adam Smith the pin manufacture has always been cited as a classic example of the principle of division of labor.[86] In this manufacture, "one man draws out the wire, another straights it, a third cuts it, a fourth points it, a fifth grinds it as the top for receiving the head; to make the head requires two or three distinct operations; to put it on is a peculiar business, to whiten the pins is another; it is even a trade by itself to put them into the paper; and the important business of making a pin, is, in this manner, divided into about eighteen distinct operations, which, in some manufactories, are all performed by distinct hands, though in others the same man will sometimes perform two or three of them."[87] On account of this minute division of labor, the pin manufacture was more economically carried on under the factory than under any other system of industrial organization.[88] John Tilsby, whose improvement on the art of pin-making in 1626 laid the true foundation for the English pin manufacture, was said to have employed about 1,500 people in his factory, in the county of Gloucester.[89] Taylor, writing in 1826, said that there were 9 factories employing a total of 1,500 hands in the same county.[90] The manufacture declined, however; and in 1833 there were only 3 factories left, the first employing 150 to 160 persons, the second employing 130 persons, and the third employing 150 persons.[91] In 1841 there were but two pin factories in South Gloucester, those of Robert Charleton at Two Mile Hill, and James Dobson at Soundwell. Mr. Charleton's factory had 160 persons on the premise, while another 500 persons, chiefly women and children, were "partially employed at their homes, heading and sticking pins".[92] Mr. Dobson's factory had 34 children and young persons, and a few adults.[93]

In the county of Lancaster there were in 1841 4 factories employing a total of about 1,300 persons.[94] This total, which was given to the Children's Employment Commission by the proprietors of these factories, was almost three times higher than the total given for the county of Lancaster by the Census, viz. 469.[95] Three of the four factories belonged to the Edelsten brothers, under the firm names of J. M. Edelsten, Samuel Edelsten, and James Edelsten (or Edelsten and Price), employing a total of 1,034 persons. These factories were all located at Warrington and neighborhood. Each factory, however, was not a unit in itself; it had several shops scattered in the different parts of the town or country. Mr. J. M. Edelsten, for instance, reported himself as a "pin manufacturer in Warrington and Latchford, at the Mersey Pin Works in the latter place for the heading-rooms; the warehouse, patent heading-rooms in King-Street; and a heading place at Newton, five miles from here; and sundry other heading shops in Turner's Fold and Oliver street (sic)".[96] Mr. J. M. Edelsten employed "about 400 hands altogether, principally children and young people" [97] who were congregated in rooms of different sizes, varying from 20 to 60 persons in each room.[98] The other two factories of the Edelsten brothers, namely, those of Samuel Edelsten and James Edelsten, employed about 600 persons. Messrs. Johnson and Co.'s pin factory at Manchester, the fourth factory in the county of Lancaster, employed 150 to 200 children and young persons on the company's premises, and 80 to 90 persons on those of the New Bailey Prison. The premises of the New Bailey Prison were superintended by one Peter Webster, who was in the direct employ of the company.[99]

At Birmingham, Warwick, there were in 1841 two pin factories employing a total of about 300 persons. This total, again, was two and half times higher than that given by the Census, viz., 120. The first factory, that of Messrs. Phipson, was "one of the largest pin manufactories in the Kingdom",[100] and employed about 200 persons.[101] This factory was also composed of shops, and in two of them, used for the purpose of heading, about 90 to 100 persons were congregated. The second factory was much smaller in size, employing about 50 persons in the heading shops and probably not more than 100 in all. This factory belonged to Messrs. Palmer and Holt of Birmingham.[102]

In 1841 the processes of pin manufacture included wire drawing, straightening, cutting and pointing; making and cutting heads for pins; heading or attaching heads to pins; cleaning and whitening pins and sheeting them into packs of convenient sizes.[103] With the exception of wire drawing and pointing, no power machinery was employed. In heading, machines were employed, but were driven either by steam power or by the feet.[104] Hand machines for heading were employed in general, and constituted the major part of the machinery in a pin factory.[105]

Chapter VII. Metal: II

A. Clock and Watch

The clock and watch manufacture of England in 1841 employed a total of 13,118 persons. With the exception of the leading counties such as Middlesex, Lancaster, and Warwick,[1] etc., the manufacture was primarily a local one, and was scattered throughout every county in England. Rutland, for instance, had 7 clock and watch makers, being the smallest number; next in order were Westmoreland, 22; Hereford, 31; etc.

Table 147. County distribution of persons employed in clock and watch manufacture, England, 1841

County	Number	Percent
Middlesex, Surrey, Kent(a)	4,655	35.6
Lancaster	2,446	18.6
Warwick	1,184	9.0
Other	4,833	36.8
Total	13,118	100.0

(a) The merging together of the three counties is necessary because of the presence of London, London being located in a part of each of the three counties. London alone employed 4,290 persons out of a total of 4,655 persons for the three counties.

In the clock and watch manufacture, the diversity of industrial systems was a notable feature. The three systems under review, namely, master craftsman, merchant employer, and factory systems, were found in parallel existence. First of all, the master craftsman system prevailed in those counties where the manufacture existed primarily to meet the local demand. "Every town, indeed, has its clock-makers, whose business is generally to finish the works sold by the first fabricators in a certain state, put them into appropriate cases, and afterwards clean and keep them in repair."[2] Such a clock and watch maker was in all cases a master craftsman—a petty entrepreneur working on his own account, perhaps with the assistance of a few journeymen and apprentices. Furthermore, he was a shopkeeper, selling clocks and watches to his customers—the ultimate consumers. In our table on the county distribution of persons employed in the clock and watch manufacture, the total of 4,833 persons employed in the counties other than those where the manufacture was more or less concentrated, or 37 percent., would properly fall under this class of master craftsmen. It was also possible that even in the counties where the manufacture was more or less concentrated, such master craftsmen could also be found to a small extent.[3]

Secondly, in the leading counties of the manufacture,—Middlesex, Lancaster and Warwick,—the merchant employer system was the general form of organization. This system was preferred to the factory system for obvious reasons. The watch manufacture, long before Adam Smith wrote on his pin manufactory, was cited by Sir William Petty as an excellent illustration of the principle of division of labor.[4] In 1817, the making of a watch involved 102 separate processes, and each class of workmen, in general, carried on two or three processes at a time.[5] This high degree of specialization, as Marx had remarked, "permits little use of the instrument of labor in common, and the capitalist, by scattering the work, saves the outlay on workshops etc."[6] It also replaced the master craftsman system, for the manufacture, being highly specialized, required the presence of a class of capitalists to bring together the products of these sub-divided branches either for finishing (i.e. assembling) or for sale to the master craftsmen scattered throughout the towns of England.[7] The want of connection between different processes gave rise to the merchant employers, whose chief function consisted in organizing the industry along capitalistic lines. The majority of the clock and watch makers carried on two or three processes at a time at their homes or shops, receiving raw materials from the merchant employer and working them up into finished products.[8] It was not unusual, however, to find that some of these artisans were "men possessing sufficient capital to employ in their workshops a considerable number of workmen, among whom they can carry out the principle of the division of labor to a still greater extent."[9] The workshops of these artisan employers formed what we may better term "immature factories."

Finally, the factory system was also found in the clock and watch manufacture, but to a very limited extent. In finishing, the general practice was either for the master craftsman to purchase from the big mercantile establishment the movements, that is, the parts of a clock or watch, and then to finish them himself or with the assistance of journeymen and apprentices,[10] or for the merchant employer to make the movements by one class of outworkers, and have them assembled by another. However, back in 1780 and thereabouts, a Birmingham manufacturer introduced the new system of finishing. For the first time in the history of Coventry clock and watch manufacture, this manufacturer established a watch factory, where he employed outdoor apprentices to finish the movements under the direction and supervision of some experienced journeymen. From that time on, his factory employed about 30 to 40 apprentices at different years, and in 1817 was employing 30.[11] This system was introduced primarily to economize the cost of labor by substituting apprentices for journeymen, and was made possible by the facility with which finishing was commonly performed. In 1817 when one Piercy was asked by the Parliamentary Committee as to the progress of the factory system in Coventry, his reply was, "it has been for 35 years augmenting, to the amount of 12 or 14 manufactories, and each of them has retained on their separate account from 12 to 30 or 40."[12] According to another witness before the same Committee, "12 manufacturers have amongst them about 150 to 160 at a time; two houses have from 25 to 30 each."[13] Despite this comparatively small number of persons employed under the factory system, the alarm among the outworking artisans was already great, and equally so was their opposition. "Such apprentices," claimed the opponents to the factory system, "are never properly taught, nor even afterwards employed as journeymen by those to whom they were apprenticed; and the trade is over done with the too great number of working people, who are half starved, because they cannot obtain any steady employ; the whole number of clocks and watches wanted for the use of all parts of the world, being very limited, as they are not articles of consumption, but very durable, even for centuries."[14] "It is," said another witness, "a great aggravation to the distresses of the journeymen who have served seven years apprenticeship to the trade, that they should be driven out of it by the conduct of persons not brought up to the business, and know nothing of it practically; and that, too, by means that deprive regular bred artisans of all hopes of ever getting into it again."[15] Furthermore "the practice of Coventry must tend to depreciate the soundness of the article manufactured,"[16] as the apprentices were generally less skilled in finishing than the journeymen. This lack of skill, although skill was not an important factor in finishing, might cause expensive accidents in the opinion of the opponents.[17] On the other hand, the manufacturers told a different story. To them it was "absolutely necessary that the apprentice system be carried on in the watch trade."[18] One manufacturer said, "I think none so likely to continue that skill and general knowledge, that are essential to our maintaining the preference which I think the British manufacture still has in Europe and in the world to all others, [than the apprentice system]."[19]

In 1841 the Coventry system of employing outdoor apprentices in the factory, in view of its past persistence, must have in all probability been in existence still, although its extent might not have been very significant. At Clerkenwell, Dodd said that

> "there are not 2 or 3 hundred men employed in a large factory, to make a watch throughout, but there are 30 or 40 distinct classes of tradesmen, comprising, perhaps, three times that number of minor subdivisions, all living and working at their own homes, and contributing the various parts of a watch, which is finally completed by the 'watch-manufacturer.' "[20]

Meantime, factories making the accessory products had sprung up. In Christchurch, Southampton, there were two factories in 1841 making watch chains. Messrs. H. T. Jenkins & Co. "have one factory, in which are conducted all the processes of watch chain making; but two of these processes are also carried on by hands working at their own homes. These are the wire pointing and the riveting of the links together."[21] Messrs. Cox, the other manufacturer, had "more hands employed out of the factory than in it; all at riveting and wire making."[22] These factories, however, were probably sporadic growths, and therefore unimportant. Bevan, writing in 1876, said,

> "the factory returns [of 1875?] throw no light upon these numbers, giving only six factories for England and Wales, employing 385 persons; but it must be remembered that clock and watch-making is a solitary occupation, employing only the owner of the shop and perhaps another, and that consequently it does not come under the Factory or Workshop Act, which latter requires a minimum of five workers."[23]

B. Gun

The gun manufacture of England in 1841 employed a total of 5,385 persons, and was located chiefly in the three counties of Warwick, Stafford and Middlesex.[24]

Table 148. County distribution of persons employed in gun manufacture, England, 1841

County	Number	Percent.
Warwick	2,332	43.3
Stafford	739	13.7
Middlesex	1,109	20.6
Other	1,205	22.4
England	5,385	100.0

In certain respects the gun manufacture compared very closely with the clock and watch manufacture. In both manufactures, the majority of employees were concentrated in a few counties, Warwick and Middlesex having long been celebrated for the gun manufacture; although the gun smith and maker, like the clock and watch maker, could be found in every county of England except Rutland. Because of this difference in local distribution of employees, different forms of industrial organization were commonly adopted in both manufactures. In the counties where the gun manufacture was not concentrated, the master craftsman system prevailed. Thus, in the words of a contemporary writer in 1833,

> "besides the numerous reported manufacturers in London and Birmingham, there is hardly a town of any size in the Kingdom that is without its gunmaker. These provincial artists, although not infrequently individuals of first rate ingenuity as workmen, never entirely fabricate the articles sold by them. Obtaining the barrels and the locks from the regular houses, the business of the country gunmaker is mostly to stock and finish the piece in a superior manner; and more especially to clean, repair, and alter the guns of gentlemen in the immediate neighborhood. He is likewise a dealer in all the necessary accoutrements of the sportsman,-powder, shot, percussion caps, and so forth."[25]

In the counties where the gun manufacture was highly concentrated, including Warwick, Stafford and Middlesex, the merchant employer system, with a few exceptions, prevailed for the same reason as in the clock and watch manufacture. In the gun manufacture the division of labor was quite elaborate,[26] for which reason "there is no such trade as a gun maker, properly so-called, in Birmingham."[27] A want of connection between the different classes of gun workers consequently resulted, to which the merchant employer supplied a necessary connecting link. He was, in the gun as well as in the clock and watch manufacture, the capitalist entrepreneur furnishing both capital and organization to a highly specialized industry. Meantime, he was not subject to heavy losses arising from the fluctuating nature of his business under this system,

> "time was when Birmingham gun-workers were employed only fitfully. The master or his traveller went out seeking orders, and when he returned, or an order came into the town, the workpeople were summoned, the work was begun and continued until the order was executed; the workers were then discharged until wanted again to execute another order, just as seamen are engaged for a particular voyage and are paid off when it is finished."[28]

This system of industrial organization—the merchant employer system—persisted several decades after the period of our investigation, 1840. In 1866, the merchant employer was still a factor to be reckoned with, although in some cases he had also become a factory manufacturer. Goodman,a leading authority on the gun manufacture, said in 1866 that

> "gunmaking, or setting up, is again much subdivided. It is only in the more important establishments that all the branches are carried on on the premises of the gun-maker. More or less, outworkers are engaged in every branch. This system makes it difficult to obtain a correct estimate of the number of workmen employed in the trade. Probably no master can tell how many hands he is employing at any given time, and the number varies, from month to month, with the demand."[29]

This outwork system, the same authority continued,

> "leads to the employment of a considerable number of young boys, who are employed mainly in carrying the work from one to another, as it passes through its several stages; we regret to say that many are employed in this work too young for the burdens they are called upon to bear."[30]

In the counties where the gun manufacture was more or less concentrated, the factory system, although relatively insignificant as compared with the merchant employer system, was also in operation. With regard to the factory of the Westley Richards started in 1812, Taylor said that

> "from the earliest times it was customary to keep those separate processes apart, but the Westley Richards' policy has ever been one of centralization. Colonel Hawker, for instance, it may be remembered, records that he would have needed to take a journey of twenty miles through London in order to see so many departments in the art of gun-making as were gathered together under the Westley Richards' roof, 'the splendid factory' of 1849." [31]

In Birmingham, other factories were found in 1841. According to the Children's Employment Commission, Messrs. Tipping and Lauder, gun makers, had "one of the principal manufactories in this trade. The premises are open, there being a spacious yard in front of the workshops. The latter, which are light and airy, are built on a plan common in this town; each is only wide enough for a single row of vices, etc., so that each workman faces the windows. A few men only work in each shop."[32]

C. Cutlery

The definition of "cutlery" is highly elastic, and varies with time, place and author. Cutlery in 1841 included many branches that were not conceivable in the seventeenth century;[33] cutlery in Sheffield had a much narrower scope than was generally given it elsewhere.[34] One author, again, defined cutlery as designating "all kinds of sharp cutting instruments";[35] while still another, after having defined cutlery in a rather broad sense, added to it such manufactures as were preparatory, subsidiary, and contributory to the cutlery trades.[36] The latest authority on the subject defines "cutlery" as including all implements involving "the operations of forging, stamping or pressing, hardening or putting together,"[37] but includes under "cutlery and allied manufactures" such products as files, tools and saws.[38] For our purpose, the cutlery manufacture may be defined in an extended sense, from the viewpoint of similarity in products made as well as in technical processes employed. Cutlery manufacture of England in 1841, in our definition, included all persons who were fully employed in the forging, grinding (including polishing), cutting or hafting processes for the making of cutting implements such as cutlery (in a restricted sense), files, and tools (including saws, edge tools, etc.). According to this definition the cutlery manufacture of England in 1841 employed a total of 25,099 persons, who were distributed in the different branches in England, West Riding of Yorkshire, and Sheffield parish as follows.[39]

Table 149. Number of persons employed in different branches of cutlery manufacture in England, West Riding of Yorkshire, and Sheffield Parish, 1841

Branch	England		York, W. R.		Sheffield	
Cutlery (restricted sense)		16,167 (a)		11,486		10,006
1. Products						
Knife (b)	1,944		1,944		1,795	
Scissor	1,042		945		924	
Razor	581		581		555	
Fork	571		568		268	
Blade	331		316		287	
Sword	116		—		—	
Haft & scale	759		729		596	
Spring	331		82		78	
Agricultural implement (c)	1,111		195		121	
2. Processes (d)						
Cutler	5,982		4,240		3,824	
Grinder	2,825		1,879		1,553	
Polisher	574		7		5	
File		4,259		2,855		2,325
Tool		4,673		1,441		1,210
Saw (e)	1,019		709		522	
Edge tool	985		559		553	
Etc. (f)	2,669		173		135	
Total		25,099		15,782		13,541

The prominence of the West Riding of Yorkshire as the great cutlery district in 1841 is seen in the above table, which shows that the county employed 15,782 persons out of a total of 25,099 for all England, or 63 percent. Of the 63 percent., 54 percent. were located in the Sheffield Parish,[40] the cutlery metropolis. This high degree of concentration had its origin in the sixteenth century, when the alien refuges established their new homes under the patronage of English royalty.[41] The new trades of cutlery which these artisans brought over to Sheffield grew rapidly in the course of the next three centuries; while London, the original centre of English cutlery, declined at an equally rapid rate during the same period. This superiority of Sheffield as a cutlery centre was not unaccounted for. "Sheffield," the City Council tells us,

> "must have possessed peculiar advantages over other districts, such as the availability of raw materials, including rock suitable for grindstones, whilst the numerous hill streams, by supplying water power to drive grinding wheels, both quickened and cheapened output. Besides those things the local water is said to contain properties valuable for hardening and tempering steel.......... It is also probable that the Sheffield workmen excelled in acquiring the higher mechanical skill called for by the demand for knives of more complex patterns, while later they developed artistic gifts to no small degree in producing the modern types of penknife."[42]

Furthermore, Sheffield, being so near to the great textile centres, had a ready market and excellent means of transport for the output. Coal mines abounded in the neighborhood, providing plenty of fuel for power purposes.

In the cutlery manufacture, the predominant form of industrial organization was the so-called "little master" system. The factory system, however, was also found to a small extent. In 1841 six factories of varying size were manufacturing cutlery articles of different descriptions in Sheffield, which included the following.[43]

Table 150. Six cutlery factories in Sheffield, 1841

Name of occupier	Description of manufacture	No. employees	Amt. motive power (in H. P.)
Rodgers & Sons	Cutlery	494	16
Kenyon & Co.	Files, saws, steel	166	70
S. Cocker & Son	Steel, files, wire, needles, etc.	134	66
Whitham & Co.	Files, steel, screws & spindles	75	30
Gilbee & Co.	Table-cutlery	61	9
Moorhouse	Penknives	45	—

The minute division of labor in the largest of the above six factories, namely, that of Rodgers & Sons, was already described by the historian of Yorkshire in 1831. The latter, after having given an account of the various processes, proceeded to observe that "in this progression every table-knife, pocket-knife, or penknife, passes, step by step, through no less than 16 hands, involving, in the language of Mr. Rodgers, at least 144 separate stages of workmanship in the production of a single penknife."[44] Such factories required a good-sized capital investment; it was said that some of them cost £30,000 to £40,000.[45]

"Little Master" System

The predominant form of industrial organization in the Sheffield cutlery manufacture in 1840 was the so-called "little master" system. The term "little master" is misleading, as it is composite in character. In one sense, a "little master" is a middleman—an outwork master; in another, he represents the small merchant employer.[46] In whichever sense, however, a "little master" may become a master craftsman if the economic conditions in the trade so demand.[47] Indeed, the instability and flexibility in the economic status of a "little master" defies any clear-cut distinction. In our study, the term "little master" is retained in deference to the common usage, but it must be understood that the systems which this term "little master" represents are essentially identical with the merchant employer and middleman systems.

Social classes. Under the "little master" system there were three social classes, namely, the merchant employer, the middleman and the outworker. Of these three classes, two were commonly known in the Sheffield cutlery manufacture as the "little masters," namely, the small merchant employer and the middleman. As a small merchant employer, the "little master" was an entrepreneur whose enterprize, on account of its nature or scope, involved a substantial share of commercial risks and liabilities. To quote a recent description not inapplicable to the conditions in 1840, such "little masters" "occupy a small office and warehouse, and are really factors on a small scale, having little or no work done on the premises, but utilizing the labor of outworkers—generally those who are neediest and the poorest workmen. These men buy the parts of the article stamped out by machinery, and having had them worked up, travel round to country towns and watering places selling at prices at which they can defy the larger establishments who have heavy fixed charges and higher wage rates to meet."[48]

As a middleman, the "little master," whether grinder or cutler, employed a team of several workmen in his own forge[49] or in a "tenement factory" where power and space could be hired at a fixed rental. He secured the work either from the small merchant employer, the "little master" *par excellence*, or from the large merchant employer carrying on an extensive business;[50] and distributed it among his own workmen. The latter received a piece wage for the work done from which were deducted the various shop charges incurred by the middleman. This practice here was similar to that in the hosiery manufacture, and constituted a source of sweating by the middleman.

Shop work. In the Sheffield cutlery manufacture a shop was sometimes located near the residence of a "little master" (middleman), or an outworker. Such an arrangement was generally true of the forgers and hafters, the number of whose hearths in 1846 reached 2,535 in Sheffield.[51] But the grinders adopted a different plan in order to facilitate the effective utilization of steam power. According to the Children's Employment Commission in 1841,

> "grinding is generally, indeed almost entirely, done at what are called public wheels—large buildings of two or three stories, to which steam engines are attached, and where power is let out, together with the room for a stone and apparatus, to each individual grinder, who pays 7 pounds to 8 pounds per annum for its use, supplying his own stone, together with the bands which connect the stone with the shaft, etc.; from 10 to 15 stones and glazers will occupy one room, or hull as it is called, each grinder being unconnected with the others, and all of them independent of the manufacturer, whose work they have got to grind. These public wheels belong generally to persons or companies who have no connexion with the grinding or other trades. The individual grinder is, therefore, the only person who is concerned in the arrangement or improvement of the grinding apparatus."

Such "public wheel" constituted what was generally known as the "tenement factory." According to our definition, it could not be a factory. The concentration of labor was apparently present in a "tenement factory," but such concentration carried no significance. In a "tenement factory," there might be a dozen or more tenants, each working independently of the other and with the assistance of his own workmen. The space rented to each tenant was practically equivalent to a small shop, and a "tenement factory" constituted, in fact, a group of small shops located together in one building for the common purpose of effectively utilizing the steam power.

For the use of 1,553 grinders in Sheffield in 1841 (or 3,106 grinders if we included both grinders and assistants, each grinder requiring one boy assistant), there were 90 "grinding wheels" in operation, 50 of which employed steam and 40 employed water power. Each "grinding wheel" contained a varying number of troughs; and its size could be measured accordingly. In 1841 a census of troughs in the different "grinding wheels" (i. e. "tenement factories") employing steam power was taken, with the following results. Of the 2,222 troughs, 423 were unoccupied at the time when the Census was taken by Commissioner Symons.[53]

Table 151. 50 "steam wheels" (or "tenement factories") in Sheffield as measured by number of troughs per "steam wheel", 1841

No. troughs	No. wheels
1 to 10	9
11 to 20	6
21 to 30	10
31 to 40	5
41 to 50	3
51 to 60	3
61 to 70	3
71 to 80	3
81 to 90	3
Above 90	4
Unknown	1
Total	50

What we have described so far applied to the "steam wheels," or "tenement factories." In the rural districts, the "water wheels," or grinding establishments employing water power, presented a different picture. According to one account in 1831,

> "these grinding wheels being, in most cases, situated beside a fall in the stream, for the convenience of working an overshot water wheel, their roofs are frequently but little elevated above the superior level of the adjacent ground, or of a large dam, which is constructed to collect the water during the night in seasons of scarcity, as well as for the ordinary convenience of allowing it to flow into the pentrough with regularity when the machinery is in motion."[54]

The life of a grinder, too, was much harder working on "water wheels" than on "steam wheels." It was vividly described in a doggerel ditty on "The Grinder's Hardships," dating from the early nineteenth century.

"It happened in the year eighteen hundred and five,
From May Day to Christmas the season was quite dry,
That all our oldest grinders such a time never knew,
For there's few who brave the hardships that we poor grinders do.

"In summer-time we cannot work till water does appear,
And if this does not happen the season is severe.
Then our fingers are numbed by keen winter frosts or snow;
And few who brave the hardships that we poor grinders do.

"When war is proclaimed our masters quickly cry,
'Orders countermanded; our goods we all lay by;.
Your prices we must sattle, and you'll be stinted too'.
There's few suffer such hardships as we poor grinders do.

"There seldom comes a day but our dairy-maid[55] goes wrong,
And if that does not happen perhaps we break a stone,
Which may wound us for life, or give us our final blow.
For there's few that brave such hardships as we poor grinders do.

"There's many a poor grinder who's thus been snatched away;
Without a moment's warning, to meet the Judgment Day:
Before his Judge he must appear his final doom to know;
There are few who brave such hardships as we poor grinders do.

"Thus many a poor grinder whose family is large,
That with his best endeavor cannot his debts discharge,
When children cry for bread how pitiful the view:
Though few can brave such hardships as we poor grinders do.

"So, now I must conclude these few humble rhymes
With 'success to all grinders who suffer in hard times':
I wish them better fortune—their families the same,
And may we ne'er experience such hardships again,
By being further stinted, and paying discount too.
There are few who brave such hardships as we poor grinders do."[56]

Abuses

Truck. Truck was practiced chiefly by the small merchant employer upon his outworkers. The "stuffing system," as truck was generally called in the cutlery manufacture, "had been an abuse of old standing and of formidable extent. It had risen into prominence in the seventeenth century, and though periodically efforts were made to repress the custom it constantly broke out afresh."[57] By the end of the eighteenth century, "the factor's ware-house was the store whence food and raiment were exchanged with the poor cutler for his labor on such terms as the factor chose to impose."[58] The rise of the trade societies or unions among the journeymen might even have intensified the evil, for, as one master (i.e. merchant employer) wrote in 1814: "If the men charge double what is proper for their labor, a master may perhaps justify himself by saying that he in self-defence is compelled to pay them, not in money, but in coffee, cloth, etc., for which he charges double or treble its value."[59] About 1840 a master at Sheffield was fined by a magistrate for having obliged his workmen to receive as wage, a piece of cloth, at the rate of 35s. per yard, which was not worth more than 11s.[60]

Embezzlement. There was little evidence of embezzlement in the cutlery manufacture from the official reports around 1840. Judging by the prevalence of the practice among the workmen in other manufactures organized under the merchant employer system, and the extension of embezzlement legislation to the cutlery manufacture in 1817,[61] it is not unlikely that embezzlement was also common in the cutlery manufacture.

Persistence

The persistence of the "little master" system could be attributed to three factors,—personal, technical and industrial. The personal factor lay in the character of the Sheffield artisans. These artisans were independent from the early days of the manufacture. Working as they did with solitary grinding wheels on the banks of the streams, it was impossible for the rules and regulations governing ordinary factory life to become traditional among them. Freedom from restraint was to them a time honored habit. They were conservative, and were on that account an obstacle to progress. A "little master," for instance,

> "would not be convinced that the markets of the world were changing, and that what suited the grandfather and father would not please the son and grandson. The head of one large firm remarked, 'we spend considerable sums of money in sending our travellers to the four quarters of the globe for orders, and when we have received their reports and samples of what was wanted, the 'little master' has stood in the way of the suggestions being carried out.' "[62]

The requirement for few and inexpensive tools in production, and the demand for a great variety of designs in the cutlery products, constituted the second factor—the technical factor. Variety of design, it is said, "determines the sale of articles which are of the nature of luxuries, and in which custom and personal idiosyncrasy are determining factors, so that a pocket knife, for instance, which sells with a horn handle in one district must be supplied with a bone or wooden handle in another. This makes standardization difficult, if not impossible, and compels the production of goods in such small quantities as to render the adaptation of machinery a matter of great difficulty."[63] Finally, the industrial conditions constituted the third factor for the persistence of the "little master" system. The facility with which the small merchant employers could sweat their workmen by means such as truck made it impossible for the large factory owners to compete with them. The former could oftentimes undersell the latter and drive them out of the trade. Business depression, again, prolonged the existence of the "little masters" as it had caused their rise centuries ago.

D. Nail

The nail manufacture of England in 1841 may be defined as including all persons who were fully employed in making nails by hand or by machinery. Under the Census definition which excluded the part-time outworkers, chiefly women and children, it employed a total of 18,040 persons.[64] In the following table, the county distribution of these 18,040 persons is shown.

Table 152. County distribution of persons employed in nail manufacture, England, 1841

County		Number	Percent.
Midland		11,997	66.5
Worcester	6,226		
Stafford	3.992		
Salop	1,274		
Warwick	505		
Northern		4,118	22.8
Lancaster	1,825		
York, W. R.	1,262		
Derby	1,031		
Other		1,925	10.7
Total		18,040	100.0

The nail manufacture of England was highly concentrated in the Midland counties, especially Worcester and Stafford; it was generally called the "Black Country" manufacture on that account.

Spread of Factory System

In the nail manufacture the industrial organization varied in accordance with the technical process employed. In the hand nail branch, in which the majority of workers were employed, the merchant employer system prevailed. However, the invention of nail making machinery in 1790[65] introduced a new system of industrial organization into the manufacture—the factory system. Despite the opposition of the hand nailers,[66] the system had grown rapidly since then because of its superior economy in production. The quality of nails was improved with a simultaneous increase in quantity; consequently, the price of nails was reduced.[67] The machine nails began to replace the hand nails. "As showing the effect upon the trade of the introduction of machine-made nails, it is stated that while before machine-made nails were introduced, one of the great London Dock Companies used 67 tons of hand-made nails, chiefly for use in the construction of tea chests, whereas after machine-made nails were made available for the purpose their orders were for 50 tons of machine made nails and only for 17 tons of the hand-made. The orders of all the large docks, of the Admiralty, and of all the large contractors showed a similar falling off in demand for the hand-made nails."[68]

Machine nails were made in two ways. "The production by the *patent* machine wrought system [since 1790] is limited, the nails being formed while the iron is heated. In *cut* nails [since 1813] the metal is operated upon cold; pointing forms a special operation in both hand and machine wrought nails; but in cut nails, the production of shank and point is simultaneous."[69] Cut-nail factories were chiefly located in Birmingham and leading seaport towns; patent nail factories in Worcester and other places. At one cut-nail factory in Birmingham "a steam engine of 50 horse power moved long ranges of machines, before each of which stood a lad with a rod of iron in his hand, and the duty of this operative consisted in little else than holding the rod to the machine, which chopped off and stamped a head upon the nails at the rate of probably 50 in the minute. The nails dropped in a finished state from a hopper into a basket. About 100 men and boys were employed in this work ....; and the produce of nails, large and small, is from fifteen to twenty millions weekly. It is not unusual to receive an order for a million nails, all of one size."[70] In Worcester, Messrs. Rowland and Hughes, a patent nail factory, employed altogether 84 persons, including men, women and children.[71]

Merchant Employer System

Merchant employer and middlemen. There were two forms of the merchant employer system in hand nail manufacture in 1840. Under the first, the merchant employer gave out work directly from his warehouse to the nailers; under the second, the middlemen, known as "factors" or "foggers," were employed. Under the first the merchant employers, generally called "nailmasters" or "nail ironmongers,"

> "have in some part of the nail district a large central warehouse, but in addition they have lesser branch warehouses in other of the nail making villages. These warehouses form the centres of the local trade. Some of the nailmasters have recognized sets of outworkers who work for no one else, but in many cases several warehouses may issue work to the same people. Iron of all the kinds required for the production of the nails the firm deals in is kept at all these warehouses, and is generally stored in 'bundles' of 56 pounds. In some cases iron is sent out to the workers by the firm, but the vast majority of outworkers come to the warehouse for their own iron, and carry back their finished proportion of nails. Thus there is no commoner sight in the nail district than that of men, women, or young people carrying through the streets, on Mondays especially, their 'bundles' of iron, or on Saturdays chiefly, returning to the warehouses their loads of finished nails."[72]

Under the second form of the merchant employer system in the hand nail manufacture, a new class, the middlemen, were introduced in different capacities. The merchant employer, in the first place, might employ the middlemen to give out work to the nailers, and thereby dispense with the expenses of having to maintain central or branch warehouses.[73] Secondly, he might purchase from the middlemen nails that were either bought up by the middlemen from the nailers, or wrought up by the nailers employed by the middlemen. It is easy to classify the middlemen under the merchant employer system in the first case, while in the second, difficulty arose from the fact that a relationship of buyer and seller, not one of employer and employee, existed between the merchant employer and the middleman, or between the middleman and his workman. Here as elsewhere in other manufactures, the boundary line between one industrial system and another is vague and indefinite, and any attempt to a clear-cut distinction is futile and confusing. Theory, it may be repeated, does not always agree with fact. For our purpose, we are perhaps justified in classifying the middlemen in the second case under the merchant employer system, as their primary function consisted in giving out work to nailers on the merchant employer's account rather than in buying and selling nails as commercial middlemen.

The reasons for the rise of the middleman, or "fogger," may next be explained. According to one authority,

> "his existence is in the first place due to an ill-regulated or over-crowded labor market. Under the special conditions of the nail trade there must always be floating about a certain proportion of unemployed labor in the form of workers not regularly employed by any nailmaster, or of nailers whose regular employers may not for a few days be able to provide them with full work. In such cases the nailer, in order to earn something, works up into nails iron he has saved from his allowances [i. e. employers' allowance for waste], or which he has bought himself, or which he has obtained from the 'fogger's' at the 'fogger's' price. The nail thus made the nailer must dispose of as best he may. If his usual employer can't take them the nailer is driven back upon the 'fogger', from whom he need not expect any approach to list price. The 'fogger' in his turn disposes of the nails so obtained either to nail masters who like to buy below list price, or to general customers whom he may find in the towns beyond. The 'fogger' also executes orders for nailmasters, or for outside merchants who send him the iron, which he distributes to the lower class of workers, who convert the material into nails, and return the finished article at less than list price."[74]

Outwork. Outwork included shop work or home work. In the nail manufacture, shops were generally attached to the homes of nailers.[75] In a few cases, however, nailers worked at the shops owned by others or by a group including himself, so as to "economize coals, shoproom, etc." by "using the same fire and the same bellows in turn."[76] The equipment of the shops was exceedingly simple. A small hearth and bellows for heating the iron rod, a small anvil, a few dies and hammers, and an "oliver," were all the tools required. This simplicity and inexpensiveness of shop and tools accounted for their ownership by the nailers. It was also responsible for the domestic character of the manufacture. In these shops, which were also homes in the majority of cases, one account tells us that "men, women, boys and girls, all work at it [nail]; they have sometimes assistants in the work, but no regular apprentices, only some agreement, generally a verbal one. Carries on the business in the same way as the Spitalfields weavers; believes that similar laws apply to both."[77] These nailers, according to another account, "are a very rough set. Nothing tends to demoralize the people more than women and men working together in the nail shops. The consequence. . . .is that there is no comfort at home, and consequently, when they are earning good wages, both men and women go to the public houses and drink and sing together."[78]

Abuses

Sweating. Under the merchant employer system the "fogger" had become the middleman between the merchant employer or nailmaster and the nailers, making his profits chiefly by underpaying the workers for the work they did—a clear application of the sweating system. These "foggers," in the words of a nailer in 1840,

> "are generally bad characters; they are no better than rogues; they have their weights and scales wrong, so that if they buy 50 pounds of nails, they will have 51 for the 50, and sometimes more than that, and they are generally bad, which hurts the trade again, and they pay in truck, and they sometimes have other manoeuvres besides; so that the men are oppressed in three, if not in four ways, in every transaction, and it is said that they have not more than 15s. or 16s. in the pound in reality."[79]

The truck system referred to in the above quotation prevailed to the greatest extent in the iron and coal trades of the Midland district, to which nail-making was by no means an exception.[80] In a letter written by one nailer, the truck system was considered as "a nefarious robbery to the workman and a disgrace to the trade." "This worm," continues the nailer,

> "has been growing at the root a great number of years till he has assumed the form of a giant. When the workman goes to the warehouse of this monster [i. e. 'fogger' or truckster], he has to submit to an extra balance on the weight side, and sometimes he robs him of his tale, a practice known only too well by the workman. He comes to the books and then he has to suffer very often another injustice. Having done this business there, he has to find his way to the tommy shop, and there he meets the giant, who compells him to buy his tommy at 10 sometimes 15 percent. above market price of inferior quality. Some places this giant keeps a public house, and then the workman is highly blessed when seduced into the top room and is riddled again. He tells him he must come on Monday for his iron (another trick). He goes accordingly to order, but no iron—you must go into my castle and have some beer today. He is like the fly and the spider which he cannot extricate himself from. He is bound hand and foot by this modern Goliah."[81]

These abuses of a truckster were sufficient to depress the economic conditions of a nailer to the lowest level, but the worst still had to be told. Price, in his address to the nailers in 1833, remarked that

> "the truckster thus paying, brings down the unwilling money-paying master in order that he may meet him, though with distressing odds, in the market; then it is that wages are step by step brought down so low, as just to enable the workmen to keep body and soul together, lower than this the most steel hearted and brazen faced do not venture to go."[82]

Embezzlement. In view of the wretched economic conditions and of the low level of moral habits, embezzlement was not uncommon among the nailers receiving work from the merchant employer directly. One Moore from Wigan, a merchant employer, told the Children's Employment Commission in 1841 that

> "nailers are quite proverbial for dishonesty; they will cheat you in every possible way they can; they are very quick at calculating; and if by mistake you make an account out more in their favor than it should be, there is not one in ten will tell you of it. Colliers are said to be bad, but nothing can beat nailers for dishonesty. Sometimes they will exchange good iron which they have received for bad, and keep the difference.

Sometimes they will make away with iron, and never give any account of it. Sometimes they will make it into nails, and dispose of it to carpenters and ironmongers on their own account. We are frequently obliged to issue out warrants for their apprehension on that score. This is a printed form of notice which we keep by us. and I believe the same thing is done by every nail master in the trade. It being printed, will give some idea of the frequency it is requisite to use them:

> If the iron you took from us, to work into nails, is not immediately brought back, either worked or unworked, you will be apprehended and brought before a magistrate without any further notice.
>
> Yours etc.
>
> Wigan Thomas Moore & Sons"[83]

E. Anchor and Chain

The anchor and chain manufacture of England in 1841 employed a total of 1,672 persons,[84] of whom 902 persons were engaged in making chains, and 770 persons in making anchors and chains.

Broadly speaking, the anchor and chain manufacture of England in 1841 may be divided into two branches, namely, that making the anchors and heavy chains used mostly for ships, and that making the small chains. Because of this division, the local distribution of employees as well as the industrial organization also differed. The small chains, like the nails, were manufactured in the Midland counties of Stafford and Worcester—the "Black Country"; and were organized under the merchant employer system. Indeed, the two manufactures—those of the small chain and the nail—"have much in common, and are, in some respects, carried on under conditions so identical, that to many people the mention of the chain and nail trades conveys the notion of one industry only."[85] On the other hand, the anchors and heavy chains used mostly for ships were manufactured in the counties of Durham and Northumberland, Lancaster, Middlesex and Gloucester, where the leading ports such as London, Liverpool, Newcastle, Bristol, etc. were located; and were organised under the factory system. The factory system prevailed primarily on account of the technical requirements. "There is perhaps no kind of smith's work more gigantic than that of anchor-making: the large masses of iron employed and the mechanical forces necessary for welding the several parts together render the employment of powerful apparatus indispensable." [86] In the manufacture of heavy chains about the same kind of smith's work was required. In general, heavy chains and anchors were manufactured in the same factory at the leading ports.[87] At North Shields, Northumberland, for instance, Messrs. Pow and Fawcus was listed as a "chain and anchor works" in 1841.[88] In Liverpool, we are told that there were in 1838 "numerous extensive manufactories of iron chain cables, anchors etc., in and near to the town".[89] The "manufactories" might have been "extensive", but certainly could not have been "numerous" in view of the small number of persons employed in the manufacture for the county, viz., 96 persons according to the Census.

F. Lock and Key

The lock and key manufacture of England in 1841 employed a total of 5,408 persons, of whom 4,208 persons, or 79 per cent., were located in the county of Stafford. In this county, the leading centres of lock and key manufacture were, in order of importance, Willenhall, Wolverhampton, Walsall and Wednesfield.[90]

In 1840 the lock and key manufacture was organized under the master craftsman system.[91] In the county of Stafford where four fifths of the total employees in the manufacture were concentrated, the master craftsman sold his products to the merchants known as "factors," who in turn distributed them among the retailers in the different parts of England. In other counties the master craftsman sold his products directly to the consumers, without the intermediation of "factors." Furthermore, he sold not only his product, but also his service in repairing locks and keys.[92] The master craftsman system that prevailed in the county of Stafford is described in the following paragraphs.

Master Craftsman System

Social classes. Under this system of industrial organization there were three social classes, namely, the merchant (generally called "factor"), the master craftsman (generally called "small master"), the journeyman and apprentice. The rise of the first class to a prominent position was of a rather recent date. In the seventeenth century the merchants were but chapmen who travelled from place to place with packhorses, so as to purchase the locks from the master craftsmen of Willenhall and Wolverhampton. "At the commencement of the eighteenth century, however, the merchants began to establish store-rooms in Wolverhampton and Birmingham, whither the locks and other hardware products were conveyed by the smiths, in wallets,—or as they have been facetiously termed, 'human saddlebags' a system not even yet quite obsolete" in 1866.[93] These merchants or "factors" increased in number until they reached 16 in 1841 in Wolverhampton, a town which was then considered by one leading factor as "an extremely rich town." "Scarcely any of the larger warehouses," remarked the same factor, "were worth less than £20,000 capital; at least a dozen were worth that; some £40,000 and £60,000, etc."[94] At these warehouses the factor gave the orders to the small masters for execution at a price, to be returned within a limited time. This, nevertheless, was not always the case. The small master could sell his locks and keys to the factor without having received the orders beforehand, the price varying with the market conditions, and more frequently, with the factor's shrewdness to strike a bargain. The second class, the master craftsman, or "small master," varied in the extent of his manufacture. In a few cases, the small master employed both journeymen and apprentices to work up his raw materials into finished products, at his own shop. In the majority of cases, however, he had very little assistance from that class, but worked them all by himself.[95] The third class, the journeymen and apprentices, were often hired on contract terms by the small master; they constituted the wage-earning class. Their conditions of employment were greatly abused by the small master, to which we shall return in a later section on "loans."

Shop work. The shops of small masters in the lock and key manufacture were generally located in the back of the street, and "as much out of sight as birds' nests." "You might pass along a street fifty times, up the passages and court yards of which there were shops containing nests full of young children, and never know it."[96] Furthermore, "whatever variation there may be in the size of some of these small shops, there is none as to the relative space of shops and operatives. The operative has what is aptly called 'a standing', for he seldom has more than just sufficient room for his work. Each shop, therefore, usually contains as many as it will hold, with barely room for one person to pass to and fro while the rest are at work, though the space for doing even this is far from a common circumstance."[97] These shops, in general, were filthy beyond description. "Many of the shops are extremely dirty and dilapidated, sometimes presenting the appearance of little broken sheds. Their floor is generally the earth, a few are boarded, many bricked. Their windows are often a mere hole in the wall, into which a shutter is fitted at night. They must be wretched places in winter, especially for the feet of all those who stand from day to day at the vice."[98]

One distinct feature of the shop work was its "systematic irregularities," which were "carried to the utmost excess by the host of small masters."[99] According to the Children's Employment Commission,

> "the majority of the working classes do no work at all on Monday. Half of them do not work much on Tuesday. Wednesday is the market-day, and this is an excuse for many to do only half a day's work; and in consequence of attending the market they are often very unfit for work on Thursday morning. Lights are seen in the shops of many of the small masters as late as ten and eleven o'clock at night, on Thursday. During the whole of Friday the town is silent in all the main-streets and thoroughfares, and seems to have been depopulated of all its manufacturers. Lights appear in the workshops to a late hour in the night—sometimes till morning. All Saturday morning the streets present the same comparatively barren and silent appearance. Everybody is working for his life. Among the small masters, their wives, children, and apprentices are being almost worked to death. Kicks, cuffs, curses, and blows, are abundantly administered to the children at this crisis of the work. ...... About two o'clock ...... some of those who did some work on Tuesday begin to appear in the streets; and large masses issue forth between 4 and 5 o'clock. The wives and elder girls go to market; the husbands and other adults to the beer-shops. By 7 or 8 o'clock the market is full; the streets are all alive; the beer-shops and ginshops are full; and all the other shops are full The manufacturers are stretching their limbs expanding their souls to the utmost, and spending their money as fast as they possibly can. No one ever thinks of saving a shilling."

This description applied to the Wolverhampton men, but more so to the Willenhall men. "The Willenhall men," reported Commissioner Horne, "habitually devote themselves to almost unexampled labor, and abandon themselves at intervals to an equally habitual debauchery."[101]

Abuses

Discount. Out of the general custom of the small masters selling their finished products to the factors at the warehouses, arose the practice of discounting, a means to dodge the frequent fluctuations of prices in the lock and key trade. Under this practice the price for locks and keys had remained unchanged for almost one century, while the percent. of discount allowed might have varied considerably. Such variation often depended upon the kinds of articles made, the demand for them, and above all, "the factor's power to apply 'the screw' "[102] A discount of 50 or 60 percent. was not uncommon,[103] and a factor frequently demanded a higher one in case the small master was in financial difficulty. In the case of one padlock maker, William Kempsen, "the factors know he's so poor that he must have the money somehow, and must sell for whatever they choose to give; he must, for he canna stand agen them—no, nor can those big men here."[104]

Underselling. The evil of discount was further aggravated by the practice of the small masters to undersell one another. In Willenhall the small masters "are in the habit of accepting orders for any quantity of work that may be desired, and at almost any price." "The cheapness of their work," said one Mr. Wright, "beats everything of the kind anywhere else. A padlock, which is sold at ironmonger's in London for 1s., is made and sold by the Willenhall men for 1 3/4 d., including all the material, labor, everything, or 1s. 9d. per dozen. Other things are in the same proportions.

The price of iron is now [1841] double what it was formerly, in 1830 and 1831, yet the Willenhall men sell for even less money than formerly."[105] Indeed, "they will work the greatest number of hours for the smallest amount of profit, and by these means compete with and undersell the lock and key manufacturers of every other town. Besides this, they all compete with each other."[106]

Truck. The system of truck payment by the merchant prevailed less generally in the towns than in the rural parishes surrounding Wolverhampton. For the majority of the small masters who were used to drunkenness and debauchery at the week-ends, the system had its good side inasmuch as it prevented them from getting cash and thereby visiting the beer-shops. But the evils of the system outweighed the good side by far. According to Commissioner Horne, such evils were five in number:

> "First, it obliges the man to go to a market where there is no competition, and where he consequently depends upon one individual for the quality, right weight and measure, and price of every article. .... Secondly, it encourages improvidence by preventing the chance of a habit of saving, for nobody can save food. Thirdly, it prevents the family from obtaining a sufficient supply of clothes, and more comfortable furniture,—in proportion of the possession of which it is always found that the working man becomes more steady, careful, industrious, and respectable. ...... Fourthly, it enables the employer to lower wages imperceptibly at any time, by lowering the quality of the tommy goods without abating their price, or by raising the price upon an inferior article, or by his weights and measures. Fifthly, it enables the employer to save the interest of the money he would have expended in the payment of wages in cash, instead of the more steady operative having the chance of doing the same for himself by depositing a portion of his wages in a savings' bank."[107]

Loan. The practice of the small master lending money to the journeyman or apprentice at the time of employment offered a fertile field to the former to retain the latter's service at a lower wage rate. At the time of employment the small master agreed to lend a certain sum to the journeyman, to be paid back by weekly installments spreading over a period of one or two years. These weekly installments generally amounted to one or two shillings, and were not always paid up regularly. As a consequence, when the two year term ended, the loan remained partly unpaid. In fact, the small master might have advanced him another loan before the end of the term, and the unpaid loan was therefore very great in amount. The journeyman must pay the loan or face a jail term, for the contract was strictly enforced. If he sought to avoid the jail, then the only way out was to continue his employment at the small master's for perhaps another one or two years. The evils of this arrangement of making loans and then deducting them from the wages of the journeyman are obvious. In the first place, the practice incited the journeyman to a system of borrowing, and of extravagance. Secondly, it prevented a journeyman from availing himself of any opportunity that might occur for bettering his condition. He had bound himself to the small master under a virtual slavery.[108] These arrangements and evils applied equally to the apprentice as well as to the journeyman.[109]

Chapter VIII. Clothing: I

A. Ribbon

The ribbon manufacture of England in 1840 may be defined as including all persons who were fully employed in the winding, filling, warping and weaving of silk yarn into ribbon. Under the Census definition which excluded part-time outworkers such as winders, fillers, warpers,[1] and some weavers, it employed a total of 6,826 persons in 1841. Warwick, with the important city of Coventry, was the leading county, and employed 6,264 persons.[2]

In 1838, out of a total of 13,219 ribbon looms in Warwick, only 598 (or 4.5 percent) were employed in the 27 factories of varying sizes. These 27 factories were operated mostly by hand power, only two factories with 53 looms being moved by steam.[3] In many cases, especially at Bedworth, they were owned by small merchant employers, who had twice as many looms working under the merchant employer as under the factory system.[4]

This small extent of the factory system was surprising in view of its early application. In Coventry the handloom factory system was introduced before 1818, when a few enterprising merchant employers collected the handlooms in the factories, and employed halfpay apprentices to operate them. At that date, it was already complained that "on account of these factories, His Majesty's revenue is much injured, but to put a good aspect on the factory proceedings, they keep the most profitable part of our business inside their own manufactories, while the father of a family, living on a tolerably large house, having poor rate and taxes to pay, commonly receives the refuse or the most unprofitable part of the work, which is one reason why he that lives in a large house, under heavy rent, etc. is not able to pay king's taxes."[5] In 1831 when one Mr. Beck of the same city, a competent manufacturer and inventor of 'peg batten', erected a steam power factory filled with ribbon looms, the workmen, after a meeting on November 7,

> "rushed down to the new factory, forced their way in, brutally treated Beck, cut out the warps, threw the silk into the river, commenced at once to demolish the machinery, and ended by setting fire to the building. . . . The destruction of Beck's factory had serious consequences for the trade. Mr. Timmins says that it put back the employment of power in Coventry for five years. In 1832 and again in 1838, it was confidently asserted that steam power could never be economically applied to the manufacture of good ribbons, and it was not until Coventry felt the competition of Congleton, Leek and Derby, where steam power had been employed, that the manufacturers began to use it generally."[6]

The reasons advanced against the introduction of the factory system were diverse in nature, but many of them were the same as given in almost every other manufacture. It was argued, for instance, that "the men have no centre for their habits in a home, and in the power of forming an individual character."[7] In a factory, again, a skilled weaver was reduced to an unskilled worker; and the dignity of his profession was thereby lost. Other reasons, more fundamental than these, set a real check to the introduction of the factory system. The factories that were in existence in 1838 had been established chiefly by the small merchant employers. The large ones, however, were not yet in a position to follow similar courses. Those whose extent of business operation in ribbon manufacture was rather extensive required a considerable amount of fixed capital to establish a factory that would produce the same amount of work as was made under the merchant employer system. In view of the rapid changes of fashion and of the fluctuating nature of employment in the ribbon manufacture, it would be a poor business policy indeed to embark upon large investment in a factory.[8] The technical nature of the manufacture, too, was an obstacle to the introduction of the new system. "The Jacquard trade is, indeed, obviously the last stronghold of hand weaving against the encroachments of steam power." "It is the *plain* branch alone that, for a long time, the rivalry can be formidable, for it is that in which merely automatic labor bears the greater proportion, and can proceed with the least interruption."[9]

Merchant Employer System

With the exception of a few hand and power loom factories as above described, the ribbon manufacture was organized under the merchant employer system. The latter system varied in accordance with the branches of the manufacture such as engine loom and single hand loom. The older system—the "undertaking" system—prevailed in the single hand loom branch in the rural parishes; while the newer system—the "journeywork" system—prevailed in the engine loom branch in the city and suburb of Coventry. The engine loom and the single hand loom "are both a species of handloom weaving; the engine loom has several shuttles from 8 to 40, and makes as many ribbons as shuttles; the higher the number of the shuttles the narrower the width of the ribbons; but the single hand branch is a one shuttle loom only, making with one shuttle one ribbon at a time."[10] The 13,219 ribbon looms in 1838, including the 598 factory looms and 121 looms employed by master weavers on their own account[11] (i.e. under the master craftsman system in our terminology), were distributed, as to branch of manufacture and locality, in the following two tables.[12]

Table 153. Distribution of ribbon looms in Warwick according to branch of manufacture, 1838

Branch		No. looms
Engine		5,730
Plain	3,502	
Jacquard	2,228	
Single hand		7,489
Total		13,219

Table 154. Distribution of ribbon looms in Warwick according to locality of looms, 1838

Locality	No. Looms
Coventry, city and suburbs	4,859
Coventry, rural parishes	8,360
Total	13,219

In the following paragraphs, the two forms of the merchant employer system—"undertaking" system in the single hand branch and "journeywork" system in the engine branch—are analyzed.

The chief difference between the two forms of the merchant employer system—the "undertaking" and the "journeywork"—appeared in the presence or absence of middlemen known as "undertakers". "Up to 1812 the trade was conducted on the 'undertaking' system, the manufacturer [i.e. merchant employer] providing the silk, dyed in the hank, and the 'undertaker' providing the looms and other machinery himself. The operations were generally conducted by the 'undertaker' himself and the members of his family, assisted by apprentices and 'journey hands'. The finished article was returned to the manufacturer, who paid the undertaker a well understood price for its production, according to the time and skill required. Two thirds of the price thus received went, under this system, to the journeyhand weaver for his labor in filling the shoot, picking up the warp, and weaving and picking the ribbon, the remaining one third being retained by the undertaker as an equivalent for his labor, time and responsibility in procuring work, finding machinery and shop room, winding of warp and shoot, warping, assisting the journeyhands in putting in new patterns, and general superintendence."[13] Since 1812 a new system has arisen to challenge the persistence of the "undertaking" system. On the one hand, the large merchant employers, in view of the growing competition of small ones made possible by a change in the method of marketing, "soon discovered that their only course was to employ machinery, and to take upon themselves the duties of the 'undertaker'. Machinery was accordingly purchased; the 'throwing' was frequently done on the premise; hand-loom factories were built, improved engine looms procured, and the weaving performed by journeyhands under the superintendence of a foreman." On the other hand, "the best and steadiest journeyhands . . . , who did not possess looms, were enabled to purchase them, and supplied with work from the warehouses with the preparatory processes already in a great measure performed. This was called the 'journeyhand system' [i.e. 'journeywork system'] . . . As this system became more and more developed, the old 'undertaking system' declined, and has now [1866] long been almost obsolete, except in some minor departments of the trade."[14]

In 1838 the old "undertaking" system was called by the Handloom Weavers' Commission "this primitive industrial economy".[15] It was then

> "in full vigour only in the rural parishes. The whole of the single hand country trade is on this system, because of the want of some responsibility between the manufacturer and the dispersed journeyhands [chiefly women and children], and because the minutiae of its detail render this the best economy of labor. . . . Each undertaker has generally seven or eight looms; a few have a greater number up to sixteen; and one even fifty; but the man who has seven or eight looms finds himself fully occupied in managing the business of them, and when he has more than that number he requires the aid of an assistant, of nearly the same ability as himself, at wages . . . The number of single hand looms in the rural parishes being about 7,500, assuming that there are on an average about seven looms to each undertaker, the number of these master weavers, who are nearly all heads of families, will be upwards of 1,000."[16]

The "journeywork" system, unlike the "undertaking" system, employed no middlemen known as "undertakers", and the workmen themselves owned the looms. The "journeywork" system, in the words of Commissioner Fletcher, was

> "the system of the whole engine trade of the city and neighborhood of Coventry, with the exception of the factory looms, and the few in the Foleshill and Bedworth districts on the undertaking system. It presents a change from the undertaking system, by which the business of the undertaker (his intervention being no longer rendered necessary by either distance, economy of labor, or want of responsibility) is divided between the manufacturer and the journey-hands. The 'first-hand journeyman' has his own looms, which, though more valuable, are fewer in number than those of the single hand undertaker; and his house is generally so constructed, especially the modern ones, which are by far the best, as to have the household apartments on the ground floor, the sleeping apartments on the first floor, and the workshop at the top, capable of accommodating three or four looms. He obtains the work from the master, the warp prepared, and the shute in the hank; he finds loom-shop, looms, jacquard machinery (if he be in the figure-trade), and the wear and tear of all gearing; he takes in the work, he is paid usually at the rate of 1 d. per ounce, though sometimes less. In his own family, also, the filling for his looms is commonly done, for which he receives payment in the price of the weaving; and besides working in one loom himself, he superintends the

others, in which various members of his family, his journeymen, or his apprentices, are proceeding with work for which he is responsible. If he have not a family by which the winding and filling can be done, he has to pay one-fourth more to get the winding done than he receives for it, and the filling forms a considerable item of reduction from his gross earnings. The 'first-hand' is responsible to the manufacturer for all deficiency of weight beyond one ounce in twenty allowed for waste in the shute, and the same for thrums in the warp. The winding of the shute was, after its withdrawal from the undertaker, for some time, done by the manufacturer, but this is now again given to the first-hand journeyman, on terms as cheap to the master, while some employment is afforded to the younger members of the weaver's family." [17]

Under the "journeywork" system there were generally two as contrasted with three social classes under the "undertaking" system. The middleman, in the former system, was absent, and we find only the merchant employer on the one hand, and the workmen on the other. It is true that the journeyman sometimes employed other journeymen, or "journeymen's journeymen", as his assistants, but on the whole, such employment was uncommon. In the following table, the proportion between the "firsthand journeymen" and "journeymen's journeymen" is shown. [18]

Table 155. Proportion between "firsthand journeyman" and "journeymen's journeymen" under "journeywork" system in engine loom branch of ribbon manufacture, in city & suburbs of Coventry, Warwick, 1838

Type of journeymen	Number	Percent.
"Firsthand journeymen"	6,796	82
"Journeymen's journeymen"	1,520	18
Total	8,316	100

In the following table, a general notion of the variation in the size of merchant employers under the "journeywork" system is gained by an analysis of the number of looms employed by each merchant employer. [19]

Table 156. Merchant employers under "journeywork" system in engine loom branch of ribbon manufacture in city and suburbs of Coventry, Warwick, 1838, classified according to number of looms employed per merchant employer

Looms per merchant employer	No. merchant employers
10	30
10 to 50	29
50 to 100	16
100 to 200	6
200 to 300	5
400	1
Total	87

Abuses

In the words of Commissioner Fletcher, "all wages I believe to be faithfully paid in money throughout the trade. Truck is at present unknown in it, the demand for hands being remarkably brisk: although Messrs. Smith and Williams, one of the Bedworth firms, keep, contiguous to their public factory and partly under it, a public-house, a butcher's shop, and a grocer's shop. They say to their hands that they will never use any compulsion over them to purchase from them, but, that, while they sell as good and cheap as others (which they do), they might as well come to them. Most of their hands do go to their shops; though some do not."[20]

Embezzlement was probably carried on to a lesser extent in 1838 than in previous years, for the undertakers more than the workmen were generally the authors of this practice. "The undertakers", according to one account, "are many of them very dishonest to their employers. Some of them will put in more than a due share of shute, which enables them to subtract some of the warp, which is the more valuable; or will sell the first cuts of a pattern on their own account, and then buy silk to make for the master, who gave them it out. They will also make up not only the stolen silk immediately subtracted, but other silk which they purchase, knowing it to be stolen."[21] To prevent such a practice, the merchant employers in 1830 united to organize a "Silk Manufacturers' Association" "for the protection of the manufacturers of ribbons against all felony, embezzlements, misdemeanors, and penal offences connected with the trade". This association confined its exertions chiefly to paying the expenses of the few legal proceedings to which it was found necessary to resort.[22]

B. Hosiery

The hosiery manufacture of England in 1841 may be defined as including all persons who were fully employed in the winding, knitting, seaming, and finishing of such knit goods, of cotton, worsted and silk, as hoses, gloves, drawers, shirts, pantaloons, and caps. Under the Census definition which included knitters chiefly, a total of 44,183 persons were employed in the hosiery manufacture of England in 1841, and were distributed in the different counties as follows.[23]

Table 157. County distribution of persons employed in hosiery manufacture, England, 1841

County	Number	Percent.
Leicester	18,600	42.1
Nottingham	16,569	37.5
Derby	7,114	16.1
Other	1,900	4.3
England	44,183	100.0

As the Census statistics did not distinguish the branches of hosiery manufacture, we have to resort to the statistics of frames as given by the Frame-work Knitters' Commission in 1844. The frame statistics for 1844 distinguished the hosiery manufacture into three branches as follows.[24]

Table 158. Classification of frames employed in different counties of England according to branches of hosiery manufacture, 1844

County	Cotton	Worsted	Silk	Total
Leicester	6,933	11,457	104	18,494
Nottingham	12,440	61	2,094	14,595
Derby	4,380	2	1,454	5,836
Other	686	183	99	968
England	24,439	11,703	3,751	39,893

In the hosiery manufacture the prevalence of the merchant employer system was almost complete. Many obstacles, industrial as well as technical, tended to hinder the development of the factory system. From the workman's viewpoint the discipline of a factory was an infringement upon his "liberty and independence". He would not submit to it; it would be a great bone of contention.[25] Furthermore, the workmen in the hosiery manufacture were largely dispersed in the 240 parishes of the three Midland counties of Leicester, Nottingham and Derby.[26] If they were to be concentrated in the factories in the towns, "they would lose their time in coming from their homes to the factory, which at present they do not, except in some few cases."[27] From the merchant employer's viewpoint, several disadvantages would also have to be accounted for. A factory meant an additional outlay in fixed capital; new buildings would have to be erected to accomodate the frames that were now scattered in shops or homes. Fixed capital was a permanent instead of temporary investment. A merchant employer, in view of the constant fluctuations of business in the hosiery manufacture, could not stop the operation of his factory at the time of depression, and thereby free himself of the fixed charges. The concentration of labor in a factory, again, gave greater opportunity to the workmen for strikes against unsatisfactory conditions of employment. These factory employees, in the opinion of one large merchant employer, "are among his chief troubles. . . . They are hotbeds of discontent, and seldom produce work so good as that made by a single hand or a single family."[28] These, then, were the industrial obstacles to the introduction of factory system. The technical obstacles, in the next place, may be enumerated. Even in a date as late as 1902, the merchant employer system still maintained its stand in the hand knitting branch of hosiery manufacture, "for the steam power frames cannot produce the very finest work, and they also inevitably stretch the wool or merino in the process of manufacture, so that the garments have not the same elasticity of fit, and do not wear so well. Machinery cannot handle these materials with the same delicacy of touch as the human hand."[29] In 1844 the backwardness of mechanical inventions in the hosiery manufacture in itself would have been sufficient to bar the introduction of steam power. "Its [i.e. the hosiery manufacture's] peculiar characteristic of having escaped the propelling and invincible power of steam, to which even the winds and the waves may be said to have been constrained to submit, has been ascribed to the fact that the varied movements of the body, hands, and legs, which are each called into action in the working of a stocking frame, are all necessarily mainly regulated and guided by the eye."[30]

In conclusion, we may quote the accounts of one merchant employer concerning his experiment of and failure in introducing the factory system into the hosiery manufacture. His efforts to bring about the change were probably typical of other merchant employers.

> "I had the most anxious desire to carry on that [i.e. factory] system. I will state why I had that anxious desire to do so. Of course, one circumstance that suggested itself was that it could be done more advantageously to myself, and with better supervision, and without the chance of so much loss of material, and with a probability of keeping the machinery in better order and under my own observation. Another reason was, I hoped to establish regular habits and regular hours amongst the men that were employed. In order to do this, I put a lot of new machinery into a large factory, and I established rules to be observed by the men, who were to attend upon an average of 12 hours a-day, including hours for the two meals, allowing one hour and a half for the meals, and half an hour over at each meal, during which the gates were not closed. . . . I carried this on for I think nearly 12 months, and I did everything I could to conciliate and get the good opinion of the men. I employed their children as winders, or allowed them to do so on the premises; and, in order further to conciliate them, I gave them a handsome dinner altogether, and dined with them; but all was unavailing. I found the utmost distaste, on the part of the men, to any regular hours or regular habits, as a body. . . . The men themselves were considerably dissatisfied, because they could not go in and

out as they pleased, and have what holidays they pleased, and go on just as they had been used to do; and were subject, during after-hours, to the ill-natured observations of other workmen, to such an extent as completely to disgust them with the whole system, and I was obliged to break it up." [31]

Merchant Employer System

In the hosiery manufacture the merchant employer system had its fullest development. Because of its highly flexible character the system could be so modified as to meet the need for industrial organization under diverging circumstances. In the first place, there were the large and the small merchant employers, who engaged outworkers directly or indirectly. Directly, the work was given out to the workmen from the warehouse, or through the agent employed by the merchant employer (generally called "putter-out"). Indirectly, the work was given out to the workmen through the medium of middlemen known under various names such as "undertakers", "bagmen" etc. [32] The middlemen, again, varied in their extent of operation, employing different number of knitters, seamers and winders. They constituted what were known as the "sweaters". With this general introduction we may now examine in detail the elements and abuses of this system in the hosiery manufacture—the merchant employer system.

Social classes. Under the merchant employer system the social classes could be divided into three groups, namely, the merchant employer, the middleman and the workman.

In addition to his usual role, a merchant employer, large or small, also owned frames, or hired them from the independent owners who had no connection whatsoever with the hosiery manufacture, but rented frames for profit-making purposes. These frames were then let out by the merchant employer to his workmen or middlemen, whenever work was given out to them for execution.

There were two classes of merchant employers, namely, large and small. A good index for each class is found in the number of frames employed. [33] In the following table the largest merchant employers in 1844 are listed with the number of frames employed. [34]

Table 159. Largest merchant employers in hosiery manufacture, with number of frames employed by each, England, 1844

Name	Location	No. frames employed
Messrs. Ward	Belper	4,000
Messrs. Hurst, Sons and Ashwell	Nottingham	2,000
Messrs. Brettle & Co.	Belper	2,000
Messrs. Biggs	Leicester	900 to 1,000
Messrs. Stokes, Nephews	Earl Shilton	400 to 500 in Leicester Several hundred in country
Mr. Sills	Hinckley	800
Mr. Rogers	Nottingham	700
Mr. Carver	Nottingham	680
Messrs. Rawson	Leicester	500
Mr. Woodcock	Hinckley	400 to 450
Mr. Wileman	Earl Shilton	300
Mr. Spencer	Whitwick	300
Mr. J. Clarke	Hinckley	300

These thirteen merchant employers, of whose frames an enumeration is available, employed a total of about 14,000 frames in 1844. As the number of frames employed in the three Midland counties in that year was 38,925, there were still a balance of about 25,000 frames employed by other merchant employers. We know from the Census that there were in the three Midland counties in 1841 a total of 783 "hosiers", all of whom were, with the exception of a few middlemen, merchant employers in the hosiery manufacture. Supposing that the economic conditions in 1841 were largely similar to those in 1844, the remaining 25,000 frames would have been employed by 770 merchant employers. The average number of frames employed per merchant employer was therefore 32, which, in view of some merchant employers employing large number of frames, would probably be an over-estimate. At any rate, a general notion of the large and small merchant employers can be formed now that these statistics are at our disposal. Although a few merchant employers were unusually large, yet the majority of them could only be called small. This fact might be accounted for in different ways. The facility to become a merchant employer in the hosiery manufacture was great. Little capital was required, as frames could be readily hired from the "independent owners". The provision of workshop was not necessary. It was either delegated to the middlemen; or, when the work was given out directly from the warehouse, to the workmen. It is true that a merchant employer had to purchase the yarn, in addition to which a small office and warehouse were indispensable to conducting his business. The prices for work done, however, were not paid until after the goods were sold to some London merchant, generally a wholesaler or exporter, for it had not infrequently been the custom for the middlemen—especially the "bagmen"—to advance credit to his employer. When these "bagmen" come to the merchant employer's warehouse with their work done, "they take out bills.....The work is taken and sold in London by the time the bills come due; so that a man of small capital is able to be equal with the man who has a large capital."[35] At times, when these very middlemen had acquired sufficient capital from their own profits, chiefly by sweating their workmen, they established themselves as small merchant employers.[36]

The middlemen in the hosiery manufacture were highly diverse in character, and the very fact of their gradation defied classification. Two terms, as a rule, were used to designate the classes at the two extremes, namely, "bagmen" and "master men"; but these at times also overlapped. Other terms, such as "masters", "undertakers", "middlemen", "second masters", "bag hosiers", were used, and one can seldom reach at a right conclusion of their meaning unless the extent of the middleman's business is examined into and determined. The "bagman", again, was not always a middleman only; he could be a master craftman as well as a middleman. For instance, a "bagman" "keeps many frames upon his own premises, mostly in country villages, and makes for many warehouses as well as selling them goods that he makes up himself."[37]

This system of merchant employer giving out work through the medium of middlemen was probably of recent date. One witness remarked that, prior to its introduction into Nottingham, "we had a market in the town, and the men could be comfortable", and that "nothing would benefit their [the workmen's] condition more than to return to the old and honorable system of making the article direct to the warehouse."[38] Its rise was an outcome of several factors that became prominent with the growth of the hosiery manufacture under changed economic conditions. In the first place, "the system may have arisen partly from the workmen finding it inconvenient to carry in their work from the villages distant from Nottingham, to Nottingham; so that they avoid the loss of the day usually occupied in bringing in the work, and taking back the yarn."[39] Such function was performed by the agents of the merchant employer at first, but was gradually taken over by the middlemen.[40] This substitution of agents by middlemen leads us to the inquiry into a second factor responsible for the rise of the middleman. From the viewpoint of the merchant employer, a middleman was a more economic intermediary than an agent. The former could be dispensed with at any time when the trade became slack, while the latter drew a regular salary on the payroll. When the trade became brisk, a larger number of middlemen could be easily found to perform the work in a required period of time. In case of shipping orders where promptitude of execution was of prime importance, such middlemen filled a real need. "Bagmen are chiefly useful to shipping houses, and to those generally which make goods of indifferent quality. . . . They have a strong hold upon manufacturers [i.e. merchant employers] now, from the rapidity of movement which is necessitated by the modern system of business. In 14 days a large bagman will sweep a tract of country of a huge lot of goods. He gives out materials on all sides, and drives the men to get the work done in time, and picks up, besides, everything of the same kind which may be ready-made in the hands of other bagmen or masters."[41] Finally, one more factor that accounted for the rise of the middleman was the service performed by the middleman, which had been well described by a friendly critic in the following terms:

> "I think the journeymen generally do not sufficiently appreciate the pains taken, and the value of the services rendered to them, by the middlemen; as, for example, in the taking in and bringing out and preparing their work; and their general superintendence and responsibility. If every workman had a separate account with the warehouse as they would like to have, they would lose some time on Monday and some time on Saturday in bringing in and taking out their work; and they would necessarily lose other time in preparing it and superintending its finish, for all of which services they do not make proper allowances; in fact it would be relinquishing all the advantages of the division of labor. Beyond that, if it were to be adopted, and every workman were to come to the warehouse, the detail of it would be so irksome and infinite that no amount of business could be carried on. In giving out orders it would be excessively teasing and annoying to have to subdivide a large order among 100 or 150 hands, and to give a hundred directions, not half of which would be appreciated. They would inevitably miscarry, and you would have no protection for your manufacture, in fact."[42]

There were two classes of workmen in the hosiery manufacture, namely, those working "direct to the warehouse" at own homes, and those working at the shops of the middlemen. For both classes of workmen, "the framework knitting trade, like that of the handloom weavers, presents this anomaly. Vast numbers of women and children are working side by side with men. Often employed in the same description of frames, making the same fabrics, and at the same rate of wages; the only advantage over them which the man possesses being his superior strength, whereby he can undergo the fatigues of labor for longer hours than the weaker physical energies of the women and children enable them to bear; and therefore he earns more money, by turning off more work."[43] This competition for employment among workmen of different sexes and ages on an equal basis except that of physical strength arose from the very nature of the hosiery manufacture. The facility of acquiring a working knowledge and skill in the employment opened the way for child and woman labor; while the system of frame-renting greatly extended it. "In the hosiery trade, the rent derived from the workman for the use of the frame, forms an important element of profit; and this is obtained to the same amount and extent from the veriest novice, as from the most expert and best-skilled workman in the manufacture."[44] Besides, "there is also a constant competition going on between the several classes of frame-owners to obtain tenants for their frames, which greatly facilitates their acquisition by many, who, but for the system of frame-rents, would probably find it very difficult to obtain one."[45] In the following table, a general notion of the proportion between male and female labor can be obtained.[46]

Table 160. Proportion between male and female labor in hosiery manufacture, 1844

Place	No. frames employed	Male Labor	Female Labor
Swanwick	121	82	39
Shirland	116	69	47
Alfreton	144	104	40
Tibshelf	18	13	5
Enderby	40	26	14

In general, the men and women were employed in knitting, and the boys and girls in seaming, winding and finishing.

Outwork. Under the merchant employer system, the work was done outside the premises of the merchant employer. Outwork included either shop work or home work. A shop varied in size with the extent of the middleman's undertaking. In Nottingham, "some have a great quantity; some have as many as a score, some 12, some 14," and some four.[47] In the cotton glove branch, the average number of frames per shop varied from 9 to 14.[48] One witness preferred a shop of from 6 to 8 frames, for as he told the Commissioner, "we have them [i.e. shops] more under our control; we can point out faults better, and, in fact, a man can manage a few frames better than he can a great many."[49] Large shops that approached the size of factories were, however, found in a few cases, although prejudice against their introduction was great. In the glove branch in Leicester, the greatest shop had about 60 frames.[50]

Those workmen who worked "direct to the warehouse", or for the middlemen in some cases, carried on the employment at own homes. "The common custom is for each married man to hire two or more frames, according to the size of the family. One of these frames is worked by the husband, another by the wife, and the others by any of the children who are sufficiently old. The younger children are almost invariably employed in winding the cotton, worsted or silk for the frames, or in seaming the articles which have been made. Thus all the members of the family, except the very youngest, find employment."[51] In Bridgeford, nearly all work was carried on in the homes of the workmen, the number of frames in different houses varying from 2 to 6;[52] so also in Earl Shilton and other small towns and villages.[53]

Abuses

Sweating. The prices that the merchant employer paid to the middleman for the execution of the work were diminished in many ways before they were actually turned over to the workmen. These deductions constituted the middleman's chief sources of sweating, and could be classified under such items as shop charges, frame rents, and fraudulent exactions.

Under shop charges were deducted from the prices such items as seaming, winding, standing, taking in (or carriage),[54] needles, candles and coals for fire in the winter. These charges varied in many localities, and in various branches; but generally speaking, they were unreasonably high.[55] More than that, these charges were deducted from the prices even though the workmen did not derive any benefit, say for standing. "It is a custom of the trade, that if a journeyman takes a frame from his master to his own house, he must pay for the standing, though he pays rent for his own house to accomodate his frame; that is what he pays for the privilege."[56]

Deductions for frame-rents constituted another source of sweating for the middleman. The middleman hired the frames either from the merchant employer or from the "independent owners". Besides charging the workmen the frame-rent which the middleman had paid out in hiring the frame, he collected a commission from each of the owners for every frame employed. In this way the greater the number of frames employed in his shop, the larger the amount of commissions a middleman collected from his workmen. It was not strange that in many cases, especially during the period of slack trade, more frames were employed than the amount of work available could have justified. Two devices of sweating were therefore introduced by the middleman, namely, "stint" and "spreading the work". "Stint' was a practice whereby the middleman limited a workman to produce only a given pairs of stockings per week, or to work only on short time;[57] while "spreading the work" was just another form of it. It constituted "in the bagmen hiring frames of frame-smiths, and other independent frames, receiving full work for the hosiers' [i.e. merchant employers'] frames, and dividing that work with an equal number of independent frames, so that half-work is allowed for the whole, but full rent and charges demanded."[58] The frame-rents charged, again, were regulated by no fixed rule. They might vary from 8d. to 3s. 6d. a week for each frame, and no feature presented greater injustice than the disproportion between the capital value of a frame and the rent charged for its use. It was told the Commission that the frame-rent which one firm might receive from 400 frames if employed, "would, according to the custom of Leicester, range from 1s. a week upwards . . .; and therefore would . . . have realized a rental of £20 per week at the least, or upwards of £1,000 per annum, on a property the intrinsic value of which was thus proved to have been but £1,350. The expenses of keeping them in repair would have to be deducted, and of course the rental at any periods at which they were totally unemployed."[59]

Fraudulent exactions from the workmen were made in four ways, through "infringements", abatements, non-delivery of tickets, and truck payment. By means of infringements, the workmen were required to put in extra length to their work, without additional remuneration for the extra expenses incurred. Statement prices, generally agreed upon between the middleman and the workmen at certain times, were often unchanged for a long while, although lengths etc. had changed and thereby caused more expense to the workmen.[60] Abatements were deductions for work alleged to be defective or soiled, frequently made without sufficient cause. They were excessive in amount, beyond the actual damage done to the work. Frauds arising from the non-delivery of tickets were frequently complained of. A petition was signed by the workmen to the effect that "hosiers or manufacturers may be required to deliver out tickets with their work as to the price given—the quantity of work required, and making it penal for master stockingers or bagmen taking out hosiery to manufacture, not showing their journeymen such tickets when they deliver the material to them to work".[61] Truck payment, finally, was made directly or indirectly. Directly the prices were wholly or partly paid by the employer—merchant employer or his middleman—in bread, sugar, bacon, flour, or other articles. Indirectly they were paid in money in the first instance, with the understanding that the money be spent in the purchase of articles in a shop kept by the employer, his family or relatives.

Embezzlement. The practice of embezzlement prevailed among both classes—the workmen as well as the middlemen. The middleman embezzled the yarn as given out by the merchant employer,[62] while the workman embezzled the yarn as given out by the merchant employer or the middleman.[63] In some cases, such embezzlement gave rise to a new class of traders who dealt in embezzled yarn,[64] in addition to the bagman who purchased it as raw material for his own manufacture.[65]

C. Hat

The hat manufacture of England in 1841 employed a total of 16,665 persons; it was located chiefly in the northern and eastern counties.[66]

Table 161. County distribution of persons employed in hat manufacture, England, 1841

County	Number	Percent.
Lancaster	5,546	33.3
Chester	1,811	10.9
Surrey, Middlesex, Kent (a)	3,603	21.5
Gloucester	984	5.9
Stafford	710	4.3
Other	4,011	24.1
England	16,665	100.0

(a) London, which was located in a part of each of these three counties, had 3,506 persons.

In the hat manufacture, most processes were carried on in the factory, to be supplemented by outwork in a few other branches. According to Mr. Stopford, of the Messrs. Pole and Stopford's hat factory at Denton,

> "several branches of the trade are done by men at their own homes, in places attached to their cottages. 'Body-making' is generally done at home; that is the first process of forming the body, or frame-work, of the hat. 'Roughing', the process in which the fur or beaver is worked on to the body of the hat, is also done out. The branches we do in our establishment are 'dyeing', 'proofing', 'finishing', and 'trimming'."[67]

In Lancashire in 1841, Denton and Oldham were the centres of hat manufacture. In Oldham there were, according to Bamford, 3,000 hat makers employed in 57 factories, or an average of 53 hat makers per factory.[68] In Denton, the whole village was "composed almost entirely of hat makers of the different branches of the trade" [69] In Oldham, Messrs. Henshaw established a hat factory around 1780, which in a few years ranked as one of the leading hat establishments in the United Kingdom. The capital invested at the time of erection amounted to £2,000, but by 1807 it had increased to £154,000. In 1815 the only steam power employed was for the processes of fur blowing and wool carding.[70]

London, another centre of hat manufacture, either made its own hats or purchased them from the Lancashire manufacturers. "Manchester and its neighborhood," said one writer in 1834, "have also long been celebrated for an extensive manufacture of hats; and a very considerable proportion of what are called London hats are originally made in this vicinity, the finer sorts being finished in London."[71] In London Messrs. Christy had "a vast hat-factory now [1841] existing in Bermondsey, and respected to be the largest in the world".[72] The factory made beaver hats, silk hats, etc., and was selected by Dodd as an excellent example because, "from the completeness and systematic arrangement of the details, it well illustrates the economy of a large factory,—the concentration of many departments within the walls of one establishment, the division of labor, the exercise of delegated authority by foremen of each department, and a general supervision of the whole by the proprietors."[73] The factory, furthermore, was somewhat integrated. Besides manufacturing the hats from the raw materials into finished products by a connected series of processes in one establishment, "it may excite surprize to hear of saw-mills, and blacksmiths', turners,', and carpenters' shops, on the premises of a hat-maker."[74] With regard to the same factory of Messrs. Christy, the Children's Employment Commission remarked, "everything seems to be very well conducted. The premises are large and spacious. The welfare of the mechanics, especially of the younger ones, is promoted with solicitude by the principals."[75]

Besides London and other centres of hat manufacture, there was in 1833 "a considerable hat manufactory, the property of Mr. Hall, the son of the magistrate," in Newcastle-under-Lyne, Stafford.[76] In South Gloucester, "a large colony of hatters is located at Oldlands Common, and the neighborhood, principally, if not wholly, employed by London and Bristol houses".[77]

D. Glove

The glove manufacture of England in 1841 may be defined as including all persons who were fully employed in dressing, preparing, dyeing, cutting, sewing and finishing the raw leather or kid into the finished gloves. Under this definition the glove manufacture employed a total of 8,746 persons, and was located chiefly in the two counties of Somerset and Worcester,[78] Somerset employing 2,856 persons (or 33 percent) and Worcester employing 1,848 persons (or 21 percent).

In the glove manufacture, the system of industrial organization varied with the technical process employed. With the exception of sewing, and occasionally of preparing or paring, all processes were carried on in factories. A good notion of the extent of the factory system in the glove manufacture may be gained from an estimate of one experienced glove manufacturer, giving the proportion of different classes of glove workers employed in Yeovil, the Metropolis of the industry in Somerset.[79]

Table 162. Glove workers in Yeovil, Somerset, 1841

Kind of worker	Number of worker
Dresser	100
Preparer	125
Dyer	35
Cutter and finisher	500
Sewer	10,000
Total	10,760

The preponderance of sewers over other classes of workers in glove manufacture excited no surprize, as the sewers, unlike the dressers, preparers, dyers, cutters and finishers, were part-time outworkers. With regard to the sewers, one witness, who stated himself to have been employed as an operative, and afterwards as a manufacturer of gloves for 28 years, observed that

> "the whole number of gloves made in the town [Yeovil] and immediate neighborhood is about 5,000 dozen per week; and, although some persons [women and children] devote themselves entirely to this occupation, and make a great many pairs in the course of a week, yet, in the majority of cases, the gloves are sewn at leisure hours, and at odd times, by the wives and children of the laboring men and small tradesmen; so that the average may be reckoned to be half-a-dozen pairs per week for the whole number of sewers of all ages. This would give 10,000 women and children who are so occupied, besides others living at greater distances, and scattered over a considerable span."[80]

The sewers were organized under the merchant employer system, of which a clear account was given by one writer for the Penny Magazine in 1843. According to this writer,

> "the glove trade of Worcester, in its general arrangement, is very interesting. The master manufacturers [i.e. merchant employers] were estimated in 1832 at 120; and the operatives, including men, women and children, at 30,000. But it must not be supposed that this large number inhabit the city of Worcester: there are, on the contrary, only a small number of glove makers within the city; the bulk of them being scattered throughout the villages lying ten or a dozen miles on every side of it. The sewing of a pair of gloves requires so little preparation or arrangement, that an humble cottager can carry on the occupation in her own poor dwelling; while her

> husband, and perhaps her children, are at work in the fields or the farm yard. . . . If the operative glovers live in or near Worcester, they go to the house of the manufacturer, receive the leather in small parcels, and carry it home to work up into gloves. But if they live ten or twelve miles out, a curious system is followed. The manufacturer sends an agent, once a week, or as often as may be deemed necessary, who opens a temporary warehouse at a public house or at some hired room, and there meets all the humble workpeople who live within a few miles on every side. Each operative brings to the agent the gloves which he or she may have made since the last visit, receives the money payment for the labor, and takes home another supply of leather, to be worked up before the period of the next visit. Precision, certainty, and economy of time result from this plan. There is one great focus at Worcester; around which, at a distance of several miles, are minor foci. These smaller foci draw towards themselves the labor of a little circle each, and then yield them up to the central mart, whence the capital comes which sets all this productive industry into action." [81]

In Yeovil, Mr. Denman, a glove agent, was engaged in distributing the gloves of the merchant employer to the sewers, so as to get them made, and for this purpose had to go to "Lopen, South Petherton, Martock, Shippen Beauchamp, Barington, Seamington, Ilminster, Donyatt, Broadway, and many other towns and villages". [82]

The by-employment which sewing afforded to the women and children of the glove making districts had oftentimes been considered a benefit. In years of great agricultural depression and scarcity, and during the incendiary outbreaks of 1830, the glove making centres of Worcester and Somerset, said Hull, were comparatively happy and tranquil, owing to the women and children in the villages being employed in the glove manufacture, and mainly supporting the men. [83] In agreement with this view, we may quote the remarks of the Children's Employment Commission that "even where the gain of individuals is small in a family, yet, by their working together, and assisting with their contributions, they make out a living. [84] Besides, the independence and liberty which sewing gave to the outworkers was taken as a distinct advantage. One woman considered it "a very good thing for young woman to have this trade", as "it is always a resource, and makes them more independent." [85]

E. Wearing Apparel

The wearing apparel manufacture of England in 1841 employed a total of 209,891 persons, [86] including the following branches.

Table 163. Number of persons employed in wearing apparel manufacture, England, 1841

Branch [87]	Number of persons
Tailor	102,251
Dressmaker	89,588
Seamstress	18,052
Total	209,891

The wearing apparel manufacture was primarily a local industry, and was scattered in practically every town or village of England. This widespread distribution of the persons employed in the wearing apparel manufacture has been a subject of elaborate statistical analysis by Professor Day, from whose conclusions the following table is taken. [88]

Table 164. Distribution of persons employed in wearing apparel manufacture in selected districts, England, 1841 (in 10,000)

Branch	Hereford	Huntingdom	Rutland	York (ainsty)
Dressmaker, town	178	113	84	115
Dressmaker, country	34	34	21	36
Seamstress, town	33	14	7	8
Seamstress, country	6	3	4	—
Tailor, town	155	125	109	147
Tailor, country	48	60	69	73

The industrial organization of the wearing apparel manufacture of England in 1840 was not given the same amount of attention in contemporary literature as that of many other manufactures, not excepting the comprehensive Reports of the Children's Employment Commission of 1841. The latter restricted the inquiry into the wearing apparel manufacture chiefly to the London dressmakers and seamstresses, while the tailors were but casually reviewed in passing. In view of this lack of materials for a conclusive analysis, the present study of the industrial organization of the wearing apparel manufacture in 1840 is confined to the Metropolis only. The industrial organization of the wearing apparel manufacture in the Metropolis offered a tolerably typical example for that in other populous manufacturing and trading centres where a large local demand for wearing apparels existed. Besides, the Metropolis alone employed about 20 percent of the total persons employed in the wearing apparel manufacture, that is, 40,948 persons out of a total of 209,891.[89]

In the London wearing apparel manufacture the diversity of industrial systems was a notable feature. The merchant employer system was predominent among the tailors and seamstresses. The master craftsman system and the immature factory system were not uncommon among the tailors, but prevailed among the dressmakers.

Merchant Employer System (Among tailors)

Among the tailors the two forms of merchant employer system were found in parallel existence, namely, the merchant employer system and the middleman or "sweating" system. Under the merchant employer system, the work was given out directly by the merchant employer, oftentimes the proprietor of a shop or an immature factory, to his outworking tailors. This system could be traced back to the beginning of the nineteenth century, at which time the journeymen tailors must have been well organized. In 1810 a "Report submitted by the Master Tailors' Committee to the Trade at Large" complained that

> "if a master, in the busy time of the year, cannot be supplied with additional men, he dare not, however he may offend his customers by their work not being done to the time he has pledged himself, give out a single garment to be made without incurring the threat (which in many instances has been put into execution) of not being supplied for one calendar month for the first offence, two months for the second, and the Committee are informed, that for the third time, if hardy enough to offend a third time, he must retire from business altogether, as no supply is allowed him from the Combined Houses of Call. Thus they convert their union into an engine of oppression, which ruins the man on whom it lights. The same law, if it deserves the name, operated in a similar manner against any master who gives even to his own sister, the lightest garment to make, though she and her family may be starving for bread."[90]

In the following year, 1811, the same complaint of the master tailors against the journeymen tailors was repeated in the "Address from the Committee of Master

Tailors, submitted to the Trade at Large, etc."[91] The journeymen tailors, however, did not keep up this strong stand in their dealings with the master tailors, and in 1843, "home work, women's labor, and increase of cheap 'slop' clothing, combined with the absence of union among the men, had reduced them to a great and real distress."[92] In "The Address of the Journeymen Tailors of the Metropolis, May 1834", the merchant employer system and its unfavorable effects upon the tailors were clearly set forth. According to the Address,

> "there has been a large portion of the trade, for various reasons, under different circumstances, compelled to work at their homes, if so they may be called—the price of whose labor is necessarily reduced by the cost of extra firing for the irons, for extra candles, and also by the loss of time which they experience in going to and from their employers, for their work; moreover, they in general have to find their sewing trimmings, which are always supplied to men when they work in the shops of their masters; and the men working at home are scarcely ever able to earn more than 3/6 or 4/- per day, with the assistance of their wives and children, and then, as I have before stated, only when they can get work to do. Even these men, working in this manner at this great reduction, are very frequently days, nay months, without employment, and consequently without pay. If, however, an order comes in to be executed immediately, the journeyman must labor night and day to accomodate the customer and master, and make every sacrifice of health, and the only remaining domestic comfort on such occasions, or risk the chance of being discharged from his shop altogether. In the spring he is repeatedly called upon to make these sacrifices; but all the other parts of the year he is never certain of one week's constant employment, by which the pawnbroker's shop becomes the only daily refuge of the unfortunate man, ever needy, always wretched, always in debt, always in fear, and if he has the courage to remonstrate (for many of these masters are as haughty as they are the mean, crowling, bowing slaves of their customers of the aristocracy), the poor man is discharged with all the horrors of want, without a shadow of hope, to do as he can."[93]

Under the middleman or "sweating" system, the merchant employer gave out the raw materials to the middlemen known as "contractors" or "sweaters", who were to manufacture them into finished goods at a contract price. These "sweaters", in turn, employed varying number of workmen to manufacture the raw materials into finished goods, and made their profits from the difference between the price received from the merchant employer and the wages paid out to the workmen plus other manufacturing expenses. A London "sweater", according to the Assistant Handloom Weavers' Commission in 1838, "will fit a room as a workshop, and will employ women and boys, perhaps a dozen, to work for him. He takes home the clothes, and on the (sic) Saturday he draws the money for them, and he pays his workpeople at what rate he may have bargained, altogether without reference to the prices paid by the trade."[94] This system of sweating, which is said to have had its origin in the early thirties,[95] "rapidly increased, and reduced the circumstances of the men to the lowest point"[96] around the 'fifties. In 1850 the system was effectively described by Charles Kingsley. "It appears," wrote Kingsley anonymously under the pseudonym "Parson Lot",

> "that there are two distinct tailor trades—the 'honorable' trade, now almost confined to the West End,[97] and rapidly dying out, and the 'dishonorable' trade of the show shops and slop shops.....The honorable shops in the West End number only 60; the dishonorable, 400 or more.....At the honorable shops, the work is done, as it was universally thirty years ago, on the

premises and at good wages. In the dishonorable trade, the work is taken home by the men, to be done at the very lowest possible prices, which decrease year by year, almost month by month. At the honorable shops, from 36s. to 24s. is paid for a piece of work for which the dishonorable shop pays from 22s. to 9s. But not to the workmen; happy is he if he really gets two thirds, or one half of that. For at the honorable shops, the master deals directly with his workmen; while at the dishonorable ones, the greater part of the work, if not the whole, is let out to contractors, or middlemen—'*sweaters*', as their victims significantly call them—who, in their turn, let it out again, sometimes to the workmen, sometimes to fresh middlemen; so that out of the price paid for labor on each article, not only the workmen, but the sweater, and perhaps the sweater's sweater, and a third, and a fourth, and a fifth, have to draw their profit. And when the labor price has been already beaten down to the lowest possible, how much remains for the workmen after all these deductions, let the poor fellows themselves say!"[98]

The extraction of profit by "sweaters" was a means of direct sweating; the profit constituting the difference between the price received for the work and the price paid out to the workman plus other manufacturing expenses of a "sweater". The means of indirect sweating, which was first applied around 1835, was described in minute detail in 1850 by one tailor, probably himself a victim. According to this tailor,

> "many of the sweaters obtain considerable profits; others are very little raised in circumstances above the men they employ. Those men are always on the lookout for young hands, and sometimes employ their wives to watch at certain places and accost novices from the country. They ask any man whom they suppose to be a tailor, 'Do you want work?' If he replies in the affirmative, they at once say that work in the greatest abundance can be given. A woman will thus engage the man to come home with her. Work is given him, and they induce him to board and lodge with them. The result, in a few weeks, is generally that he runs into debt with the family. If he has any little valuables, they are pledged or sold, and he is bound down to the place where, by ill luck, he has been thrown. It is impossible for the poor man to get work on his own account. A security is required for its safe return. The operation of this system is most iniquitous and injurious. Sometimes a man will get some huckster or publican whom he has known for a length of time to become security for him. He is thus bound to the huckster, and must deal at his shop, or he must spend a part of his earnings at the publican's, or their [his?] security is withdrawn. Lodging house keepers do the same thing. This practice is the growth of the last fifteen years."[99]

The effects which this system of sweating produced upon the tailoring population of London were indeed deplorable, but perhaps they were exaggerated by contemporary writers such as Kingsley. "We were all bound to expect this," wrote Kingsley in his autobiography, "Alton Locke", about the period 1838-1842.

> "Every working tailor must come to this at last, on the present system; and we are only lucky in having been spared so long. You all know where this will end—in the same misery as 15,000 out of 20,000 of our class are enduring now. We shall become the slaves, often the bodily prisoners, of Jews, middlemen, and sweaters, who draw their livelihood out of our starvation. We shall have to face, as the rest have, ever decreasing prices of labor, ever increasing profits made out of that labor by the contractors

who will employ us—arbitrary fines, inflicted at the caprice of hirelings—the competition of women, and children, and starving Irish—our hours of work will increase one third, our actual pay decrease to less than one half; and in all this we shall have no hope, no chance of improvement in wages, but ever more penury, slavery, misery, as we are pressed on by those who are sucked by fifties—almost by hundreds—yearly, out of the honorable trade in which we were brought up,into the infernal system of contract works, which is devouring our trade and many others,body and soul. Our wives will be forced to sit up night and day to help us—our children must labor from the cradle without chance of going to school, hardly of breathing the fresh air of heaven,—our boys, as they grow up, must turn beggars or paupers—our daughters, as thousands do, must eke out their miserable earnings by prostitution. And after all, a whole family will not gain what one of us had been doing, as yet, single-handed."[100]

To this crying evil and misery of the tailoring population, Kingsley suggested that "the true method for the workingmen, if they wished to save their brothers and their brothers' wives and daughters from degradation, was to withdraw their custom from the slop sellers, and to deal, even at a temporary increase of price, with associate workmen."[101]

Merchant employer system (Among seamstresses)

The merchant employer system among the seamstresses, like that among the tailors, also took two forms, namely, merchant employer system and middleman system. The middleman or "sweating" system was the same among the seamstresses as among the tailors, with the exception that whereas among the tailors the work was generally done at home, it was done by the seamstresses or shirtmakers both at home and in workhouses and other charitable institutions.[102] In one of the London Union workhouses, for instance, "the whole of the female paupers were employed in shirtmaking."[103] These workhouses, in general, carried on the employment of shirtmaking under the superintendence of a class known as "contractors", and employed different numbers of paupers. In the following table a few of the workhouses are listed for which statistics on the number of employees were given by the Children's Employment Commission in 1841.[104]

Table 165. Number of persons employed per London workhouse, 1841

Name of workhouse	No. employees
City of London Union	130
Wapping	100
Bermondsey	100
St. Pancras	80 to 100
East	70
Islington	30
Stepney	30

When the work was not done in workhouses, it was done at the shirtmakers' houses. Mrs. Rowlandson, a "contractor", executed orders for Messrs. Silver and Co., an establishment employing in London and its neighborhood about 3,000 persons.[105] She "employs 50 women who make shirts, blouses, caps, collars etc. The women work at their own houses, and many of them two or three hands each, some of them are children."[106] It was generally not known how much such "contractors" charged for giving out work.[107]

In both cases above, the work, whether done at home or in workhouses, was given out by the merchant employer through the medium of "contractors". But "contractors" were not invariably employed. They were dispensed with when the work was given out directly to the workmen from the merchant employer's warehouse. Ann Foreman, for instance, "got her work first hand from the warehouse".[108] Messrs. Silver and Co., referred to above as an establishment employing about 3,000 persons in London and neighborhood, had a branch establishment for shirtmaking at Portsea employing from 1,200 to 1,400 persons—women. "With very few exceptions," said Mr. Hogg, the managing agent of the branch, "these females are engaged directly by witness [i.e. Mr. Hogg himself], not through the medium of small mistresses. The occupation is entirely domestic, the women employing no children but their own. .. There are no set hours of labor; sometimes they work early and late if an order requires it."[109] The branch establishment of Messrs. Silver and Co. was one example of a large merchant employer giving out work directly without the intermediation of "contractors"; but "there are at Portsea six or seven other establishments of a similar kind."[110]

Master Craftsman and Immature Factory Systems

In view of the general prevalence of the merchant employer system among the tailors and seamstresses, it is strange that, according to the contemporary writings studied, no mention whatsoever was made concerning the existence of the same system among the London dressmakers.[111] This omission of any mention of the merchant employer system leads us to believe, on the basis of the following evidences, that the London dressmakers were organized under the master craftsman and the immature factory systems.

In many respects the master craftsman system resembled the immature factory system. The pivotal figure under both systems, the master craftsman or the immature factory manufacturer, acted in the same capacity of capitalist, employer and foreman, only on a varying scale. Between the two systems the distinction consisted chiefly in the fact that under the immature factory system, we can trace the beginning of labor concentration under the employer's supervision. There were a larger number of employees concentrated in an immature factory than in a shop, and whereas the function of a master craftsman was partly manual and partly supervisory, that of an immature factory manufacturer was largely supervisory and organizing. An immature factory, on the other hand, differed from a mature factory, or a factory proper, insofar as the former had a smaller degree of labor concentration than the latter, and employed no machinery or other fixed instrument of production.

The size of a factory varied with the requirement of technical efficiency for each particular manufacture. The size of an iron factory, for instance, would be by far too large for that of a woolen factory. The same may be said of an immature factory or a shop. For our purpose of analyzing the industrial organization that prevailed among the London dressmakers, any establishment with a size of fifteen or more employees might be tentatively called an immature factory, while a shop would have fifteen or fewer employees. With regard to the dressmakers in 1841, Mr. James Grant, of the London Female Mission,

> "has endeavored to ascertain the number of establishments in London. There were in the Directory of 1838 about 900 names, and he has ascertained that there are many small employers whose names are not inserted. Calculates these at about 500; and it is probable the number in the Directory has with the increase of population increased, he conceives there may be altogether about 1,500 employers. These employ from 2 or 3 each, to 25 or 35; thinks the average is 10 in each establishment, making in the whole

> 15,000. This does not include journeywomen who work at their own houses." [112]

This estimate of 15,000 dressmakers in London was lower than the Census return of 20,780, even admitting the fact that Grant's estimate excluded the journeywomen working at own houses. Grant's estimate, however, is significant in the number of establishments given. An establishment of ten persons each was large enough for a shop, but a little too small for an immature factory. An average, nevertheless, leaves a good deal untold. Many of these establishments had a high degree of labor concentration. The Children's Employment Commission in 1841, for instance, mentioned three establishments that employed respectively 65, 50 and 41 persons. [113] On the other hand, a large number of the establishments, as Mr. Grant tells us in his estimate, employed not more than a dozen persons each.

In both classes of establishments, shops and immature factories, work was given out to be done by the workmen at own homes during the rush season, so as to lessen the necessity of overwork and late hours. [114] This practice, however, was limited in extent, and as a result, the evils of overwork and late hours in the shops were rampant. In the opinion of Commissioner Grainger for the London district, "the protracted labor .. is, I believe, quite unparalleled in the history of manufacturing processes. I have looked over a considerable portion of the Report of the Factory Commission [1831-34], and there is nothing in the accounts of the worst-conducted factories to be compared with the facts elicited in the present inquiry." [115]

Chapter IX. Clothing : II

A. Lace

The lace manufacture of England in 1840 may be divided into two branches—machinelace and handlace. Each of the two branches may again be divided into two sub-branches, according to the source of power employed for the machine lace and the kind of implement employed for the hand lace. In the following diagram a classification of the lace manufacture is given.

Branch	Sub-branch
1. Machine (or "bobbin net")	a. "power" (i.e. steam power)
	b. "hand" (i.e. human power)
2. Hand (or "pillow")	a. Bobbin
	b. Needle point

The Census definition of the lace manufacture in 1841 agreed with the one given above, except that it excluded a large number of persons who were employed as part-time outworkers in both branches, viz. children and women. These excluded persons were employed either in the hand lace branch, or in the technical processes of the machine lace branch such as preparing, embroidering (or lace-running), and finishing. [1] In the following table, the county distribution of the two branches of lace manufacture is shown. [2]

Table 166. County distribution of persons employed in lace manufacture, England, 1841

Branch	County	Number		Percent	
Machine lace	Nottingham	9,711		29.8	
	Derby	2,411		7.4	
	Leicester	1,890		5.8	
	Devon	2,248		6.9	
	Somerset	685		2.1	
			16,945		52.0
Hand lace	Buckingham	5,474		16.8	
	Northampton	3,356		10.3	
	Bedford	3,193		9.8	
	Oxford	684		2.1	
	Other	2,933		9.0	
			15,640		48.0
	England		32,585		100.0

In view of the late application of the stocking frame to the production of machine lace, which began with the efforts of men like Crane and others in 1760, the preponderance of machine lace over hand lace manufacture in the number of persons employed in 1841 was a striking triumph for the machine lace manufacture. [3] In that year the percentage of persons employed was 52 for the machine lace manufacture, as against 48 for the hand lace manufacture.

In the lace manufacture of England in 1840 the flexible character of industrial organization resulted in a variety of systems. These systems, in brief, are classified as follows, with reference to the branch of lace manufacture and the technical process employed.

System	Branch	Process
1. Craftsman		
a. Family	Hand	All
b. Master	Machine	Preparing, knitting
2. Merchant Employer	Machine	Preparing, embroidering, and finishing
3. Factory	Machine	Preparing, knitting and finishing

In other words, the hand lace branch was organized under the family craftsman system; while the machine lace branch was organized under the master craftsman, the merchant employer, and the factory systems.

Family Craftsman System

The family craftsman system in the hand lace branch was largely similar to the system that prevailed in the straw plait manufacture. Its chief features included home (or school) work and sale of product to the dealers.

Home work. The family craftsmen, including the wife or mother and children of a family, carried on the hand lace manufacture as a by-employment in practically every town or village of the Midland hand lace districts.[4] The tools—the bobbin or needle point and the pillow—were simple and inexpensive; the raw materials—flax, silk, woolen, worsted or cotton threads—could be purchased in small amount.[5] The home constituted the workshop, for which no additional outlay needed be expended. With all these, the element of capital was absent, and any family could carry on the hand lace manufacture under the family craftsman system. In the poetic lines of William Cowper, the conditions of work were indeed ideal for the hand lace maker: [6]

> "Yon cottager who weaves at her own door,
> Pillow and bobbins all her little store,
> Content though mean, and cheerful if not gay,
> Shuffling her threads about the livelong day,
> Just earns a scanty pittance, and at night
> Lies down secure, her heart and pocket light;
> She, for her humble sphere by nature fit,
> Has little understanding, and no wit,
> Receives no praise, but (tho her lot be such,
> Toilsome and indigent) she renders much."

The children, however, were generally sent to the "lace schools" to learn and to work under the supervision of an elderly woman; similar conditions that existed in the straw plaiting schools were also found in the lace schools. The lace schools were practically the only schools in the Midland hand lace districts, and the teaching there largely superseded the ordinary school teaching. In the words of Mrs. Palliser,

> "the girls left the day school at the age of 8 years and joined the lace school, and here the hours were from 6 A.M. to 6. P.M. in the Summer, and from 8 A.M. to 8 P.M. in the Winter. Half an hour was allowed for breakfast and for tea, and one hour for dinner, so that there were ten hours for actual work. The girls had to stick ten pins a minute, or 600 an hour; and if at the end of the day they were five pins behind they had to work for another hour. On Saturdays, however, they had a half-holiday.

> They paid 2d. a week (3d. in winter) for lights, and in return they received the money realized from the sale of the lace they made, and they could earn about 6d. a day. Pay day was a great event; it came once a month."[7]

These schools varied in the number of children attending, generally from 20 to 30. The age of the children varied from 5 to 15, and their works were purchased by the dealers who made arrangements with the instructresses.[8] The price received, deducting the school charges, went to enhance the family earnings of the children.

Sale to dealers. The works of the family craftsmen were sold in different ways to the dealers, "from whom they were often obliged to purchase their thread at exorbitant prices"[9] In the first place, "these dealers would travel from village to village buying up the completed lengths, and in many cases making large profits on their transactions."[10] Secondly, the dealer "appoints a day when his 'workers', as they are called, met him at an inn, and he buys the lace, and perhaps gives them new patterns and parchment to work on, and sees no more of them until the next journey, which may be a month or six weeks afterwards". [11] Finally, the family craftsman brought the finished lace to the shop of the dealer in town, and sold it to him at a certain price.[12]

Abuses

Truck. In Devon, where the famous Honiton lace was made, truck was alleged to have existed to a large extent. The system was complained of by many witnesses [13] to the Children's Employment Commission in 1841, of whom the following is a good example. This witness, Maria Lewis,

> "says that the custom of paying in goods, instead of money, at the shop, is universal, and states that it is generally considered they do not get the full value of their work, so they often do not know the real quality of the goods, and are often induced to take articles which they do not want."[14]

Theft. Hand lace was made according to different patterns, and each dealer, in giving out his order to the family craftsman, always required that the lace be made according to his own pattern. The theft of a pattern, in such cases, was complained of by the dealers as an evil inherent to the system. One dealer was asked, for instance, by the Committee on Arts and Manufactures in 1835,

> "Is not that [theft of patterns] an evil that must necessarily attach to a manufacture that is purely domestic, and when the labor is not immediately under the control of the master?"

The answer to the question was,

> "It seems attributable in great part to that; if conducted on the premises, I should take care they should not carry the lace [pattern] away of course, but they might then have other means of copying and carrying the pattern away."[15]

Master Craftsman System

Under this system the master craftsman had the ownership of the machine and the raw materials employed, and sold the lace to the "finishing houses". The character of his workshop varied under circumstance. If the machines were moved by hand power, they might be worked in his own home or shop; if they were moved by steam power, they were generally placed in a "tenement factory". Such "tenement factory", which was also typical in the Sheffield cutlery manufacture, was rented by a varying number of master craftsmen with a view to the economical utilization of steam power, and owned and managed by some enterprising outsider. In Mr. Taylor's "tenement factory", for instance, "there are 23 masters, owning from one to ten machines each. The steam power is hired from the proprietor of the mill; the engine going 20 hours a day, except Saturday, when it stops earlier, to make up for which they work all Friday night."[16] The master craftsman carried on his work with the assistance of his family members,[17] sometimes employing a few journeymen. If he intended to carry on the work by himself, and yet owned more machines than he needed, he would let off the extra ones to the journeymen.[18]

Merchant Employer System

Social classes. The merchant employer system in the machine lace manufacture illustrated excellently the difficulty of generalization in an analysis of industrial organization. In the process called embroidering, for instance, the merchant employer system at times shaded into the factory system, and it would indeed be hard to distinguish the one from the other. Meantime, the merchant employer system itself possessed a high degree of flexibility in organization. In an attempt to make the system intelligible, the following diagram may prove suggestive.

1. Merchant employer ——————————————→ Outworker
2. Merchant employer ——————→ First hand agent ——————→ Outworker
3. Merchant employer ——→ First hand agent ——→ Second hand agent ——→ Outworker

According to the above diagram the merchant employer system is subject to variations and gradations under different circumstances, by the introduction of middlewoman and sub-middlewoman known as "first-hand agent" and "second-hand agent". Under system (1) the merchant employer might or might not execute part of the work at his own warehouse, by giving out a part or all of the work to the outworkers. In case he executed part of the work at own warehouse, such warehouse oftentimes resembled an immature factory, with 10, 15, or more employees. The system of giving out a part or all of the work directly to the outworkers was the simplest, and was employed by the small and large merchant employers alike,[19] the small merchant employer invariably and the large merchant employer occasionally. The large merchant employer had resort to systems (2) and (3). Under the last two systems the merchant employer gave out work through the middlewoman, or the middlewoman and sub-middlewoman. The middlewoman, or "first-hand agent", executed part of the work at her own shop, and gave out the remainder to the sub-middlewoman or outworkers. The sub-middlewoman, again, executed part of the work at her own shop, and gave out the remainder to the outworkers.[20] These two systems of employing middlewoman and sub-middlewoman were adopted by the large merchant employers whose outworkers were dispersed throughout the rural villages. The barrier of distance was one factor for their growth; while the simplicity of administration was another.

Home work. In the lace-running (or embroidering) branch of the machine lace manufacture, home work presented a rather dark picture despite certain of its over-rated advantages. To quote from the Children's Employment Commission in 1833,

> "lace running is a wholly domestic employment. It is remarkable that the lace runner, at the present time, is the hardest worked, and the worst paid of all the operatives connected with the lace trade. .. They begin early, and work late, and though they rejoice in the liberty of their own homes, and the power of leaving off when they please, yet the hours during which they actually work from the voluntary motives of gain may be called not only long, but excessive, when the nature of the work, the posture, and the confinement are considered."[21]

This description of the life and work of a lace embroiderer agreed closely with a popular poem of the time:[22]

"She plies her needle till the lamp
Is waxing pale and dim:
She hears the watchman's heavy tramp,
And she must watch like him—
Her hands are dry; her forehead damp,
Her dark eyes faintly swim.

"Once, maiden, thou wast fresh and fair
As those sweet flowers of thine;
Now shut from sunny light and air;
How canst thou choose but pine?
Neglected flows thy raven hair,
Like the uncultured vine.

"Poor maiden! if the fair ones who
Thy graceful 'broidery buy,
Only one half thy struggles knew,
And filial piety,
Methinks some drops of pity's dew
Would gem the proudest eye!"

Abuses

Sweating. Under the merchant employer system in the machine lace manufacture two abuses arose from the presence of midlewoman and sub-middlewoman, namely, sweating and truck. Sweating was primarily an abuse attributed to the practice of the middlewoman and sub-middlewoman, while truck in some cases was forced upon the worker by both the merchant employer and the middlewoman and sub-middlewoman. In the following circular sent out by a committee of the lace-runners to their fellow workers, and intended to incite a unanimous turnout against the system of middlewoman and sub-middlewoman, an insight may be gained with regard to the oppressions of sweating:[23]

"Sisters:

On Monday last, a meeting was held, in the Democratic Chapel, Rice's place, Nottingham, to consider the best means of putting an end to the present injustice practiced towards you, by those who call themselves mistresses [middlewomen and sub-middlewomen]; you are well aware, that they are in the constant habit of going round to the warehouses taking out nearly all the work, and dealing it out to you, at what price they think proper to give; and you are not only obliged to take work from what are called 'second-handed mistresses', but also from 'third-handed ones'; that in consequence of this method of giving out work, the lace-runners in many cases receive for their work not more than one-half the original price, the mistresses reserving to themselves the remainder for that most useful of all purposes, walking with your work to the warehouse! Are you to be thus robbed of your hard-earned pittance to maintain these cormorants in idleness, and many of their husbands in drunkenness and profligacy;—no wonder that misery enters our dwelling—that we are in the depth of poverty, that our children are crying for bread, while there is a swarm of locusts hovering between us and the manufacturers, ready to devour one-half of our hire."

Meantime, another letter was addressed to the merchant employers of Nottingham with the plea that,[24]

> "if the manufacturers [merchant employers] were to cause to be stamped on each lace-piece, the price to be given for running; or to append to each piece a ticket stating the price, it would effectually put an end to the system of which we complain. .. We are willing (and we think it is reasonable) to allow the mistresses a penny in the shilling from our earrings, for the trouble they take in going with our work to the warehouse."

Sweating, indeed, was a common woe to the lace-runners; its persistence had not been broken down until the government took steps to legislate on all forms of home work at the end of the nineteenth century.[25] The outcry against this evil was perhaps justified, but we must not overlook the obvious fact that every industrial system, however oppressive, did have its need to meet. A recent writer took a different view of the role that the "agents" or "mistresses" in the machine lace manufacture had played, and such view would probably be not unreasonable. An agent, he said, "is the scapegoat of trade reformers. But he or she (for in the industries under consideration it is almost invariably a 'middlewoman') has useful functions to perform. Usually she is some trusted, elderly woman, often a former employee of the firm from which she gets the work. Perhaps she has left the warehouse to get married, lost her husband, and does not like the idea of going back among a lot of girls. She is therefore entrusted with a certain amount of home work, part of which she often does herself and part of which she lets out to others. An ordinary first hander has a certain number of women working for her, as a rule from three to a dozen. In addition to this form of middlewomen, there are a certain number of little private warehouses which perform the same functions. It is impossible to say to what extent the workers are cheated and sweated by these sub-employers. Probably this evil is greatly exaggerated. The fetching and carrying of the lace, responsibility for damage, etc., is not overpaid by one third of the profits: This may be considered as a fair average."[26]

Truck. Another abuse under the merchant employer system was the payment of wages in truck. One witness reported that "many of the runners [embroiderers] are paid by the agents in clothes, bacon, and bread; has heard that the bacon has been charged 2d. a pound more than the market price; many young persons under 18 have been paid in this manner."[27]

Factory System

The factory system in the machine lace manufacture, like that in the textile manufactures, contained two features, namely, labor concentration under employer's supervision and employment of hand or power machinery. In the following paragraphs each feature is discussed in brief.

In 1840 there were 42 lace factories in the five counties of machine lace manufacture, namely, Nottingham, Derby, Leicester, Devon and Somerset, 30 in the first three counties and 12 in the last two.[28] As these factories were enumerated under the factory returns for the textile factories, it is hard to approximate the number of persons employed.[29] Furthermore, many factories moved by hand power and not by mechanical power were excluded. The excluded factories, according to Mr. Felkin, were larger in number than the ones included by the Factory Inspectors. His estimate gave a total of 70 factories for the year 1836, only 30 factories being moved by mechanical power.[30]

In view of the scanty statistics, it is hardly possible to give the size of an average factory as measured by the number of persons employed. We shall have to resort to a brief mention of the size of a few factories for which data can be obtained. At Chard, Somerset, there were in 1833 seven lace factories operated "on a small scale", according to the report of the Children's Employment Commission.[31] Of these seven factories, one factory employed 150 to 160, and another employed about 500 persons.[32] Some of them, furthermore, employed varying number of outworkers at their own homes. These factories, as compared with those at Derby,[33] were fairly good-sized, and one wonders why they should have been called "small" by the Commission. Perhaps after having visited the extensive factory of Mr. Heathcote at Tiverton, Devon,—a factory employing 800 to 850 persons inside the factory and about 1,500 outside the factory—the Commission was justified in its conclusions. Mr. Heathcote's factory, it was generally agreed, was the largest throughout the United Kingdom in the machine lace manufacture.[34] Other large factories belonged to Messrs. Boden & Morley at Derby and Mr. Sewell at Nottingham. The former employed 500 persons,[35] and the latter 420 persons.[36]

Power. In 1833 the hand power drove more machines than the steam power, as may be seen in the following table.[37]

Table 167. Number of hand and power machines in machine lace manufacture, England, 1831-33

Year	Power	Hand	Total
1831	1,000	3,500	4,500
1833	1,100	3,900	5,000

From the above table it is clear that in 1833 "the introduction of power factories in the bobbin net trade can scarcely be considered as yet complete. . . The hand worker competes successfully with the power machine; the article he produces is confessedly not inferior, and the chief advantage enjoyed by the owner of the power machine is, that he works it with no labor but that of superintendance."[38] In addition, the work performed by the power machine was different from that performed by the hand machine. The power machine turned out plain nets, while the hand machine produced fancy nets. Mr. Felkin estimated that the proportion of plain nets made was only 15 percent. of the whole machine lace output. This proportion was perhaps underrated.[39]

Machinery. The number of both hand and power machines for 1831 and 1833 in the machine lace manufacture is given in the above paragraph. In order to get some notion of the factory size as measured by the number of machines employed, we may take the figures of machine ownership for 1831 and 1836 as an index. It is true that the owner might have his machines scattered in different places, but on the whole such instances were not numerous.[40]

Table 168. Statistics on machine (both hand and power) ownership in machine lace manufacture, England, 1831-36

No. machines owned	No. owners	
	1831	1836
12 to 20	17	10
20 to 30	9	9
30 to 40	5	6
40 to 50	1	5
50 to 60	1	2
60 to 70	2	1
70 to 80	1	1
80 to 100	1	1
100 to 120	1	1
Upwards of 120	0	1
Upwards of 200	1	1
Total	39	38

The medium size of machine ownership, according to the above table, was 20 to 30 in 1831, and 25 to 35 in 1836. Mr. Heathcote, whose factory was the largest in England, owned 206 machines in 1833,[41] while Messrs. Boden and Morley owned 150 machines in 1829.[42]

Among the effects arising from the employment of machinery may first be mentioned the overhead cost. The overhead cost for a factory of 40 machines was calculated at £2,300 by Mr. Johnson in 1840. "I have made an estimate," said Mr. Johnson, "of what I believe would be the present cost of building a warehouse, engine, gearing, etc., for forty 12-quarter machines, and the capital to work them. I take the buildings, steam engine, and gearing at £3,000; I take forty 12-quarter machines at £21 per quarter, which is £10,080; 40 double or extra sets of bobbins and carriages at £50 each, £2,000, making a total of £15,080; I have taken the capital to carry on the business at £2,920, which brings it in round numbers to £18,000."[43] Taking 5 percent. as the interest upon the fixed and floating capital of £18,000, and 7½ percent. as depreciation charges upon the fixed capital of £15,080, and adding the expenses of £269 for "wages of clerks, warehousemen and other helpers", we got the total overhead cost of £2,300.

Five percent. interest on £18,000	£900
Seven and half percent. depreciation charge on £15,080	£1,131
Wages of clerks, warehousemen and other helpers	£269
Total overhead cost	£2,300

The depreciation charges in machine lace manufacture were unusually high, as invention superseded invention almost every year. One observer remarked in 1835 that "an acquaintance of his had two machines, for which he paid his only *thousand* pounds, a few years ago, that would not now sell for *thirty* pounds; and that another had three machines, then worth *four thousand* pounds, which would not now bring *fifteen* pounds, not because they are worn out, but because they are superseded."[44]

A second effect is the increased productivity. We were told by Mr. Felkin in 1866 that "while a pillow lace maker can form 5 meshes per minute by her skillful and pliable fingers, Heathcoat, in his first essay upon his bobbin net machine, made 1,000, and before the expiration of his patent, 10,000 of those meshes per minute; a man sitting to overlook his machine now, will watch its movements producing 50,000 meshes per minute—an increase of 10,000 fold on the cushion of laborer's arduous and painstaking task."[45]

B. Straw Plait

The straw plait manufacture of England in 1841 employed a total of 9,775 persons, and was located chiefly in the Eastern Counties.

Table 169. County distribution of persons employed in straw plait manufacture, England, 1841

County	Number	Percent.
Hertford	4,751	48.6
Bedford	1,753	18.0
Buckingham	1,332	13.6
Essex	431	4.4
Other	1,508	15.4
England	9,775	100.0

The localization of the straw plait manufacture around the Eastern counties of Hertford, Bedford and Buckingham went back to the seventeenth century, when for various reasons the Scotch workmen—the descendents of the Lorrainers who migrated across the Atlantic at the invitation of the ill-facted Queen Mary of Scots—transferred the industry to the Luton Districts. Since then the straw plait manufacture had grown up under the favorable climate of the Eastern Chiltern Hills, an area of chalky subsoil, and had spread to the Eastern counties.[46]

Family Craftsman System

In the straw plait manufacture the industrial organization bore a close analogy to that in the pillow lace manufacture. In both manufactures, the family craftsman system prevailed. Under this system, the craftsman was generally a wife or mother carrying on the occupation as a by-employment, with the assistance of her own children. She worked at home, owned the tools and raw materials, and sold the finished product to the dealers or at the public markets.

Home work. In general, the husbands—chiefly agricultural laborers of the straw plaiting district—purchased the raw straw from the dealers who had collected it from the farmers for sale. The wives and children of the family would then cut the straw into lengths at each joint, and split it down with cunning little "engines" which divided the straw into three, four or more strips according to the fineness of the plait to be made. The strips were then steeped in water, to make them less brittle and more pliable.[47] As a rule, the children learnt plaiting not from their parents, but at a 'plaiting school' in the neighorhood. At first, they paid fees for technical instruction, but after a short while they began to work on their own account, and turned over their finished product for sale by the mistress (i.e. director) of the school or by their own parents. In both cases, the price received, deducting the school charge for the use of working facilities, went to enhance the earnings of the family. The mistress of the school, in addition to her capacity as a teacher, had to drive the children to work, at the request of the parents. Thus, one mistress told the Children's Employment Commission in 1841: "In my judgment I think the mothers task the children too much; the mistress is obliged to make them perform it; otherwise they would put them to other schools."[48] The following description of the plaiting school was typical of other schools.[49]

> "In most villages there is a Plaiting School, which is generally conducted by an elderly dame, who receives from each scholar, 2d. or 3d. per week. The children are there some time before they can plait so as to earn anything, but after a year or two they contrive to obtain 6d. to 1s. 6d. per week, after the plait is disposed of by their parents.
>
> "They remain at school the usual school-hours; afterwards, during the time they do not play, they plait a little till sent to rest. They continue working at school till they can plait sufficiently well, and when they are above 8 or 9 years old, they earn 2s., or 3s., per week. On leaving school, they earn 4s. to 5s., if expert plaiters, and after they become skilful, they may obtain as much as 7s."

The school also furnished board to the children at a charge. If the parents did not undertake to sell the plaits done at the school by the children, the mistress or her husband would sell them to the dealers and turned over the amount received to the parents.[50] In the following poem entitled "The Straw-plaiter", the work and life of a young straw-plaiter were vividly described.

> "Here I sit and plait the straw,
> All the live long day, ma'am;
> Neater never lady saw,
> So I'm sure you'll say, ma'am.

"'T is on and on concern
  Nothing very sprightly;
Yet our daily bread I earn,
  Giv'n my mother nightly.

"Much I thank the lady fair
  Who would have taught it:
Once I used to romp and tear,
  'T was not I who sought it.

"Then in rags I used to go—
  I've a sickly mother:
I could neither knit nor sew,
  Only nursed my brother.

"Now my brother runs alone,
  Able just to totter:-
Long my mother had to groan,
  Till her meals I got her.

"How it cheer'd her languid eyes,
  When my gains I brought her!
Oft I hear her sigh and cry
  'Bless thee, my good daughter!"[51]

Sale to dealers or at public markets. The important role that the dealers of various sorts played in the straw manufacture could not be over-emphasized. From the field where the raw straw was grown and cut, to the factory or warehouse of the straw bonnet and hat manufacturer, the dealers were ever the connecting link between the different classes of producers. The dealer purchased the raw straw from the farmers throughout the country, and sold it at a public market to the plaiters. The plaiters, in turn, sold their plait to the plait dealers at the public market, who finally sold the plaits to the bonnet and hat manufacturers at Luton, Dunstable and other centres.[52] This method of disposing the product at the public market was subscribed to by the persons employed in the straw manufacture; and attempts were early made to taboo officially any other method of purchase or sale. In 1802, for instance, resolutions were passed and submitted to the leading Luton straw bonnet and hat manufacturers to the following effect:[53]

> "Resolved that as there are several markets conveniently situated and fully adequate to the buying and selling of every article in the trade, the undersigned hereby agree and pledge themselves not to buy any platt, trimming, or other article in the trade, before the market shall be assembled, which shall be done by the ringing of a bell at each market, from Lady-day to Michaelmas, at 8 o'clock in the morning; and from Michaelmas to Lady-day at 9 o'clock; and that a second bell that the market is fully met, shall be rung which shall be a signal to begin buying.
>
> "Resolved that it having been a practice of many persons to ride about the country to buy platt, etc., and by such means to forestall the markets, tending very materially to injure the essential interests of the trade; the undersigned hereby declare their intention immediately to discontinue such practice, and as much as lays in their power to discourage the same."

In the following quotation from Charles Knight in his British Almanack and Companion for 1861, much that was said of the plait market at that time could have been applied to the conditions throughout the early part of the nineteenth century.[54] The description reminds one of the cloth markets at Leeds and elsewhere around the Yorkshire woolen districts.

> "Let me attempt to describe Luton straw plait market. . . . At 9 o'clock the market bell rings, and the traffic begins. My attention is first attracted by the dealers in straw prepared for plaiting; these come from the neighboring hamlets, in which they are employed in the selection of straws from the farmer's barns; in sorting it into different degrees of fineness; in cutting it into a regulated length; in bleaching it by exposure to sulphur fumes; and in making it up for sale in little bundles. The straw plaitters come to the market to buy this straw; as they also come to sell their plait. Those women whose goods have not been collected by a middleman, stand in rank, their small dealings being principally confined to the private makers of bonnets at their own homes, who chaffer with the plaitters for a score or two of the plait. Carts have come in from distant places with loads of plait. The dealers are opening their bags upon the stalls. The commodity will sustain no material damage from the rain; and so the trade goes forward, as if all were sunshine. The buyers here are the agents of the great houses. They rapidly decide upon quality and price; enter the bargain in their note books; the bags are carried to the warehouses; the loaded trestles are soon relieved of their burthens; and in an hour or two the street is empty."

These plait markets, however, proved unfit to the expanding manufacture, and finally in 1869 two plait halls were built and officially opened for regular business transactions.[55]

CHAPTER I.

ENGLISH INDUSTRIAL ORGANIZATION ABOUT 1840.

1. Day, **Distribution of Industrial Occupations in England,** 1841-61, New Haven, 1927, p. 89.

2. Lloyd, **Cutlery Trades,** London, 1913, pp. 398-9.

3. This unevenness of materials available accounts for the varying degree of completeness in treating the different manufactures included in the present study.

4. U. S. A.

5. Day, **op. cit.,** p. 91. For a summary of the various meanings of the term 'manufacture', see Cooke-Taylor, **Introduction to a History of the Factory System,** London, 1886, pp. 12-16.

6. Day, **op. cit.,** p. 94. As regards the total number of persons employed in English manufactures in 1841 here treated, the only difference between Professor Day's and my own lies in my inclusion of 13,118 persons employed in the "clock and watch" manufacture and 946 persons employed as "founders".

7. Each group of manufactures is given one or more chapters, and each subgroup, i.e., each manufacture, a chapter or a separate section thereof. For this reason, complete (full) references are cited for each chapter.

8. For an authoritative criticism of the Occupation Census of 1841, see Day, **op. cit.,** p. 90.

9. In this and the following references to facts as presented in this chapter, the reader is advised to consult the different sections in the following chapters devoted to each of the twenty-four manufactures under our analysis.

10. For a brief critical summary of important writers including Schaffle, Schwarz, Marx, Thun, Held, Bücher, Schmoller, see Sombart, Die gewerbliche Arbeit und ihre Organisation, in **Archiv für soziale Gesetzgebung und Statistik,** 1899, XIV, pp. 316-20.

11. See his article on "Gewerbe", in Conrad (ed), **Handwörterbuch der Staatswissenschaften,** 1st. ed., 1890-94, 6 vols., III, pp. 922-50; or 4th ed., 1923-27, 8 vols., IV, pp. 966-99; Id., **Die Entstehung des Volkswirtschaft,** Tübingen, 1893 (tr. fr. 3rd German ed., 1900, by S. M. Wickett, as **"Industrial Evolution"**, New York, 1901, and used in this Study), ch. IV, pp. 150-184.

12. Schmoller, Die Geschichtliche Entwicklung der Unternehmung, in 12 pts., pts. 4-5, in **Jahrbuch für Gesetzgebung,** 1890, XIV, pp. 1043-76; **Grundriss der allgemeinen Volkswirtschaftslehre,** 1st ed., Leipzig, 1900, 2 vols., I, pp. 417-34.

13. Ashley, Early History of the English Woolen Industry, **in American Economic Association Publication,** 1887, pp. 72-3; An **Introduction to English Economic History and Theory,** I, pt. II, The End of the Middle Ages, 1892, p. 219; **Economic Organization of England,** 1914, pp. 35-6.

14. Thun, **Industrie am Niederrhein,** Leipzig, 1879, 2 vols., II, p. 242 et. seq.

15. These terms are translated from the original in German by Wickett, and will be modified by the present writer in a later paragraph, on p. 7.

16. Usher, Generalizations of Economic History, in **American Journal of Sociology**, 1916, XXI, p. 474.

17. Below, **Territorium und Stadt**, München und Leipzig, 1900, pp. 321ff.

18. Bücher, Industrial Evolution, **op. cit.**, pp. 151, 170.

19. Gerlich, **Geschichte und Theorie des Kapitalismus**, München und Leipzig, 1913, pp. 319ff.

20. **Infra**, pp. 182ff.

21. Sombart, Die gewerbliche Arbeit, **op. cit.**; **Der Moderne Kapitalismus**, 1st ed., Leipzig, 1902, 2 vols., I, pp. 3-49.

22. **Id.**, Der Moderne Kapitalismus, I, p. 26.

23. The classification of industrial organization into large-scale (Grossbetrieb) and small-scale (Kleinbetrieb) is a hobby with many German writers. See Thun, **op. cit.**, II, p. 242; Hasbach, Zur Characteristik der englishen Industrie, in **Jahrbuch für Gesetzgebung**, 1902, XXVI, p. 1018.

24. Marx was the first writer to distinguish the kind of technique by such terms as the "manufacture" and the "machine technique". See Marx, **Capital**, in 3 vols., American edition (Kerr), Chicago, 1912-13 (original date of publication in German under the title **"Das Kapital"**, I, 1867, II, 1885, III, 1894), I, pp. 369—70. Later writers, such as Riekes, have retained this classification by the incorporation of a third kind of technique,—the handicraft. See Riekes, Der Fabrikbegriff und die Handwerksorganisation, in **Jahrbuch für Nationalökonomie**, 1902, LXXIX, p. 204.

25. Hasbach, **op. cit.**, p. 1018.

26. **Growth of English Industry and Commerce**, 3 vols., 1882, II, pp. 497, 616; **Western Civilization in its Economic Aspects**, Cambridge, 1900, 2 vols., II, p. 236 note.

27. Other terms include "capitalist" (Cunningham, Growth, **op. cit.**, II, p. 616), "undertaking" (Marshall, **Principles of Economics**, 1st ed., 1890, p. 357), "factor" (Sprague, **English Woolen Industries in the Seventeenth and Eighteenth Centuries**, Cambridge, 1897, unpublished doctoral dissertation, p. 9), "little master" (Lloyd, **op. cit.**, passim), "clothier" (Heaton, **Yorkshire Woolen and Worsted Industries**, Oxford, 1920, p. 89), "manufacturer" (contemporary usage in 1840), etc.

28. **Infra**, p. 71.

29. **Op. cit.**, II, p. 497. According to Cunningham, the term "domestic system" "is used in the sense in which it was current in Yorkshire at the beginning of the nineteenth century (**Reports**, 1806, III, 1058, printed pag. 444). Professor Schmoller opposes the gild system to the domestic system, as separate and successive phase of development, but this arrangement does not seem to me to apply in English history. I prefer to say that the domestic system existed from the earliest time till it was superseded by capitalism; that craft gilds were a form of industrial organization which was appropriate to the domestic, rather than to the capitalist system; and that these gilds were convenient instruments for enforcing civic, as contrasted with national policy."

30. Liefmann, **Uber Wesen und Formen des Verlags,** Freiburg, i.B., 1899, pp. 5-6, 7, 94-5.

31. Usher, **Industrial History of England,** Boston, 1920, pp. 13-14. This term is also used by Miss Hazard, **Organization of the Boot and Shoe Industry in Massachusetts before 1875,** Cambridge, 1921 (Harvard Economic Studies, Vol. XXIII).

32. For an account of the origin and development of the term "factory" see Cooke-Taylor, **op. cit.,** pp. 1-12.

33. Cunningham, Growth, **op. cit.,** III, pp. 668 et seq.

34. Here the irregularity was primarily attributed to the unrestrained habits of the craftsmen themselves. But under the merchant employer system, it was attributed to the actions of the merchant employer and the outworkers alike. See **infra,** pp. 15, 16.

35. Industrial Evolution, **op. cit.,** ch. V.

36. An excellent summary of reasons for the persistence of the craftsman and the merchant employer systems is given in Gonner, The Survival of Domestic Industries, in **Economic Journal,** 1893, III, pp. 23-32. For other studies, cf. Hasbach, **op. cit.,** pp. 1032-52; Brodnitz, Betriebskonzentration und Kleinbetrieb in der englishchen Industrie, in **Jahrbuch für Nationalökonomie,** 1909, XC, pp. 51-86, 145-84; Bücher, Industrial Evolution, **op. cit.,** pp. 189-197.

37. This argument is often advanced by the gild socialists as a raison d'être for the resuscitation of the craftsman system. See writings by Morris, Ruskin, etc.

38. The best critical study is by Liefmann, **op. cit.,** passim.

39. Cf. **supra,** note 9.

40. **Infra,** Chapter III, note 133.

41. Burrows, Machinery, in **The Advocate,** or, **Artizans' and Laborers' Friend,** 1833, no. 7, p. 55.

42. According to Ure, "the term 'factory' designates the combined operation of many orders of work-people, adult and young, in tending with assiduous skill a system of productive machinery continually impelled by a central power. This definition includes such organizations as cotton-mills, flax-mills, silk-mills, woolen-mills, and certain engineering works, but it excludes those in which the mechanisms do not form a connected series, nor are dependent on one prime-mover. Of the latter class, examples occur in iron-works, dye-works, soap-works, brass foundries, etc. Some authors, indeed, have comprehended under the title factory, all extensive establishments wherein a number of people cooperate towards a common purpose of art; and would therefore rank breweries, distilleries, as well as the workshops of carpenters, turners, coopers, etc. under the factory system, but I conceive that this title, in its strictest sense, involves the idea of a vast automaton, composed of various mechanical and intellectual organs, acting in uninterrupted concert for the production of a common object, all of them being subordinate to a self-regulated moving forces." (in his **"Philosophy of Manufactures",** 1835, pp. 13-14). Ure's emphasis upon the presence of power machinery as the prerequiste to a factory is also typical of the Government's definition in the Factory Acts of 1833; it represents one extreme. To Ure the factory system was a product of the Industrial Revolution, and was therefore unknown

there being many large as well as small mills. In Ashton, Stalybridge, in any former period of human history. His failure to comprehend the element of labor concentration and supervision constituted a serious omission. On the other hand, Cooke-Taylor, who would perhaps be included as one of "some authors" referred to by Ure, defined the factory as "an establishment where several workmen are collected together for the purpose of obtaining greater and cheaper conveniences for labor than they could procure individually at their homes; for producing results, by their combined efforts, which they could not accomplish separately: and for saving the loss of time which the carrying of an article from place to place, during the several processes necessary to complete its manufacture, would occasion. The principle of a factory is, that each laborer, working separately, is controlled by some associating principle, which directs his producing powers to effecting a common result, which is the object of all collectively to attain." (in the **"Factories and the Factory System"**, 1844, p. 1) Cooke-Taylor's definition, unlike Ure's, is too broad, and represents the other extreme. According to him, "there is scarcely any trade or occupation in the world which is not conducted, more or less, on the principle of the factory system,," such as a printing office, a shipbuilding yard, and a farm (**Ibid.**, pp. 1-2). A factory, in other words, "is indeed nothing more than systematized industry" (**Ibid.**, p. 118). Cooke-Taylor's definition, which was accepted *verbatim* by the first historian of the American factory system, Carroll D. Wright (in his **"Factory System of the United States"**, 1882), and regrettably unknown to the first historian of the English factory system, the younger Cooke-Taylor (in his **"History of the Factory System"**, 1886, pp. 28 et seq.; **"Factory System and the Factory Acts"**, 1894, pp. 25-30), failed to include in the definition of a factory the element of machinery and other fixed instrument of production in addition to that of labor concentration. Secondly, Cooke-Taylor failed to point out the element of supervision—the keynote to labor concentration, and instead, noticed only the effects of cooperation arising from labor concentration. Finally, Cooke-Taylor's factory may be a shop in our definition, for in a shop what Cooke-Taylor calls "the associating principle" is also present. He did not distinguish between the extent of labor concentration in a factory and that in a shop. Our definition lies between the two, namely, Ure's and Cooke-Taylor's, and meantime embraces the element of supervision in addition to the two elements of labor concentration and employment of machinery or other fixed instrument of production.

43. For an excellent analysis of the causes for the variation of sizes of industrial unit in textile industries, see Chapman and Ashton, "The sizes of businesses, mainly in the textile industries", in **Royal Statistical Society Journal, 1913-14,** LXXVII, p. 519.
44. **Infra,** p. 28.
45. **Infra,** Chapter IX, note 29.
46. **Infra,** Chapter II, note 18.
47. Reasons for such changes will be explained in treating each of the five textile manufactures in the following chapters.
48. Factories and the Factory System, **op. cit.,** Preface.
49. Wright, Factory System as an Elment of Civilization, in **Journal of Social Science,** 1882, XVI, p. 107.
50. For a brief, impartial summary, see Knowles, **Industrial and Commercial Revolution in Great Britain in the 19th century,** 1921, pp. 84, 103.
51. Factories and the Factory System, **op. cit.,** p. 111.

CHAPTER II. COTTON

1. Baines, **History of the Cotton Manufacture in Great Britain,** London, 1835, pp. 396-98, 412. His estimate of the employees in the cotton manufacture of Great Britain was as follows:—

Spinning and weaving (1835 Factory Returns)	237,000
Handloom weaving (Baines)	250,000
Lace-making and embroidery (Felkin)	159,000
Hosiery (Felkin)	33,000
Calico printing (Baines)	45,000
Total	724,000

We exclude in our (i.e. Census) definition lace-making and embroidery, and hosiery, for two reasons: (1) These two manufactures employed raw materials besides cotton; (2) outworkers working part time such as women and children formed the majority of the employees. The exclusion of calico printing is a matter of expediency. Calico printing in 1840 was generally carried on independently of spinning and weaving, in establishments that were not subject to the legislative interference, and were therefore not enumerated in the Factory Returns. These establishments, however, were operated under the factory system. See **Committee on Copyright of Designs, Report,** in P.P., 1840, VI, p. 494.

2. **Occupation Abstract,** in P.P. (H.L.), 1844, XXXIII.

3. Qualifications to this statement may be made in the case of fustian manufacture. In 1841, fustians were mostly woven by power looms in the factories, and were subsequently 'cut' by the outworkers under the merchant employer system. (See **Asst. H.L. Weavers' Com'n., Report,** in P.P., 1840, XXIV, p. 601; Moore, Industrial Evolution of a Manufacturing Village, in **Economic Journal,** 1911, XXI, pp. 613-24). Fustian manufacture, however, employed only 3,553 persons in England according to the Census, and in view of its small extent, it is excluded in our analysis of the cotton manufacture in order to facilitate statistical computation.

4. Cotton manufacture proper employed 213,565 persons. The rest, 49,589 persons, are apportioned to the cotton manufacture from the "weavers", "spinners", and "factory workers"—whose branch of manufacture was not specified in the Census—according to the principle adopted by the Census; but modified by the present writer. "In order to give a general view of the relative importance of the different staple manufactures [i.e. cotton, woolen, worsted, silk, flax and linen, lace, and hosiery]," said the Census, "we have prepared the following table, in which we have allotted to each fabric its fair proportion of the total number of persons here returned under the general appelation of weaver, spinner, and factory worker (fabric not specified)." (Occupation Abstract, **op. cit.,** p. 15) The Census, however, treated the fair proportion in a very summary fashion, giving the proportions without explaining the procedure of obtaining them. It may even be doubled if the proportion given each fabric was a *fair* one. In view of our definitions of the cotton, woolen, worsted, silk, flax and hemp, lace and hosiery manufactures, (see Chapters II, III, IV, and IX of this study), the 109,710 persons enumerated in the Summary Returns for England under the headings "weaver", "spinner", and "factory worker" are apportioned among the cotton, woolen, worsted, silk, flax and hemp, lace and hosiery manufactures according to the relative proportion of the number of persons employed in each manufacture.

The following table gives the details of such an apportionment.

Manufacture	Total persons employed	Add Proportion No.	%	Revised total persons employed
Cotton	213,565	49,589	45.2	263,154
Woolen	86,779	20,187	18.4	106,966
Worsted	30,430	7,131	6.5	37,561
Silk	54,123	12,507	11.4	66,630
Flax and hemp	24,949	5,815	5.3	30,764
Lace	26,441	6,144	5.6	32,585
Hosiery	35,846	8,337	7.6	44,183
Total	472,133	109,710	100.0	581,843

5. Ellison, **Cotton Trade of Great Britain**, London, 1886, p. 118.
6. The apportionment among counties of the "weavers", "spinners" and "factory workers" is based on the relative proportion of the number of persons employed in each county. Cf. note 4.
7. Shaw, **Manchester Old and New**, London, (date?), pp. 30-33.
8. Chapman, **Lancashire Cotton Industry**, Manchester, 1904, pp. 152-7.
9. We take 1838 and 1841 as the nearest years for which comparative statistics are available. Other factory returns were given in 1835; and Census returns in 1851.
10. Number of handloom cotton weavers in Great Britain, 1820-40.

Year	Number
1820-30	240,000 (a)
1833	240,000 (b)
1835	250,000 (c)
1840	"almost as great as at any former period" (d)

(a) **Committee on Fluctuation of Employments in Manufacturing Districts, Report**, in P.P., 1830, X, pp. 3-4.
(b) **Committee on Manufactures, Commerce and Shipping, Report**, in P.P., 1833, VI, Q. 10171 (James Grimshaw, a cotton manufacturer from Lancashire). Another witness, George Smith, gave 200,000 (Q. 9449).
(c) Baines, **op. cit.**, pp. 238, 383.
(d) Hickson, **Report on the Condition of the Handloom Weavers**, in P.P., 1840, XXIV, p. 12.

11. Occupation abstract, **op. cit.**, p. 15.
12. Cooke-Taylor, **Introduction to the History of Factory System**, London, 1886, Chap. IX.
13. Chapman, **Cotton Industry and Trade**, London, 1905, pp. 37-8.
14. **Children's Employment Commission, Supplementary Report**, in P.P., 1834, XIX, Pt. I, p. 181; Ure, **Philosophy of Manufactures**, 2d ed., 1835, pp. 330, 332.
15. Ure, **Cotton Manufacture of Great Britain**, London, 1835-6, 2 vols., II, p. 448.
16. In P. P., 1840, X, Q. 4857-8.
17. **Committee on Allotments of Land, Report**, in P.P., 1843, VII, Q. 1109.
18. "There are several small mills with regard to which it has been impossible to enforce some of the minor details required by the statute. This arises from the fact that the presence of a water wheel brings within the operation of the statute any building in which any process enumerated in the first section is carried on, however few in number may be the persons employed.

Thus it happens that the occupier of a building, containing a single carding machine, impelled by water, is amenable to the law, although the persons employed in the building, including adults as well as children, do not amount to half a dozen." Inspector Howell, **Half-Yearly Report ending August, 1835,** in P.P., 1835, XL, p. 5.

19. **Factory Returns of 1838,** in P.P., 1839, XLII.
20. Inspector Horner, **Half-Yearly Report ending December, 1841,** in P.P., 1842, XXII, pp. 20-21. A year ago in 1840, Horner reported that in his district there were 1,819 mills belonging to 1,536 firms. See **Report of Committee on Mills and Factories, 1840,** in P.P., 1840, X, Q. 8,648.
21. Committee on Mills and Factories, 1840, **op. cit.,** Q. 2911.
22. See Chapters V and VI.
23. These small factories that were grouped together in one building were strictly speaking too small to qualify as factories, some of which carried on spinning under the commission system. (see Chapman, Lancashire Cotton Industry, **op. cit.,** pp. 62-4) Butterworth told us that "Messrs. Jones had become possessed of Fog Lane or King Street mill, previous to 1796, and let it to several individuals, who commenced the cotton spinning business on a small scale, forming about one of the first instances of a system since become extremely common, particularly at Oldham, the appropriation of mills to various concerns of moderate extent." (**Historical Sketches of Oldham,** 1847-8, p. 139).
24. In our analysis the Factory Returns for 1838 and later years were chosen, although there had been an earlier Return published in 1835. "The Return for 1835," the factory inspectors pointed out in their Returns for 1856, "is evidently incomplete as a return of all the factories in the country. The total number of factories is given in that Return as 3,164, whereas in the succeeding Return of but three years later date, viz. 1838, the number is reckoned at 4,217, an increase so great that it must have arisen either from very many occupiers having neglected or refused to comply with the requisitions made to them, or from the inspectors at that early period not being fully acquainted with all the factories in their districts." See **Factory Inspector's Report, October, 1856,** in P.P., 1857, (Sess. I), III, p. 11.
25. References to the **Factory Returns** of various years are given below, and will not be repeated hereafter.

For the	year	1835	see	P.P.,	1836, XLV;
„	„	1838	„	„	1839, XLII;
„	„	1847	„	„	1847, XLVI;
„	„	1850	„	„	1850, XLII;
„	„	1856	„	„	1857 (Sess. I), XIV.

1847 Returns are used in this study in connection with the child and woman labor statistics only, as they give no statistics in other respect.

26. Butterworth wrote in 1847 that "although Oldham is still remarkable above all other places in the cotton district for a great number of small concerns, yet there are now several extensive manufacturing establishments," employing respectively 944, 900, 550 500, 500, 470, 470, 450, 404, 400, 400, 400, 400 hands. "There are numerous mills here employing 300 hands each, and a still greater number which provide work for 200 and 100 hands each, but the largest member of all are various small concerns employing under 100 hands each. In 1833 three of the principal mills in Manchester employed about 1,400 hands each, 8 from 500 to 900 each, 8 from 300 to 500 each, and 17 from 100 to 300 each. In Bolton there are numerous large mills, few concerns being under 200 hands each. In Stockport the manufactories vary greatly in extent,

Dukinfield, and Hyde, most of the concerns employ from 300 to 800 hands each; there are some few above and below these numbers, but they are not many. In Rochdale, Bury, and Heywood, with the exception of a trifling number of large establishments, the mills are principally of moderate size." (**op. cit.**, pp. 118-9).

27. Chapman, Lancashire Cotton Industry, **op. cit.**, pp. 57-8.
28. Inspector Horner, Half Yearly Report ending December, 1841, **op. cit.**, pp. 34-56, Appendix no. 1. The Inspector gave a total of 887 factories, for in the parishes of Saddleworth and York, there were 10 instead of 11 factories employed in spinning and power loom weaving (see Report, **op. cit.**, p. 51). The correct number of factories, therefore, would be 886.
29. Clapham, Some factory statistics of 1815-16, **Economic Journal**, 1915, XXV, p. 477; Baines, **op. cit.**, p. 382. Except in the case of Factory Returns for 1835, 1838, 1850 and 1856, a factory is not used in this study to designate a separate building only. A factory is equivalent to an occupancy, which may comprise a whole or a part of the building or several buildings. Besides, when the same occupier had several factories (i.e. plants in modern usage) in one locality or several localities, such plants were collectively called a factory. During the period under our analysis, viz. 1840, this practice was not exceptional for the fact that in many cases statistics for several plants belonging to the same occupier were merged together. To call these plants a factory gives us a uniform standard for comparing the extent of factory operation for the different occupiers.
30. In P.P., 1816, III, p. 374.
31. Supplementary Report, **op. cit.**, pp. 119-23; Ure, Cotton Manufactures of Great Britain, **op. cit.**, I, pp. 334-6. The returns gave 43 factories in 1816, and 151 factories in 1833. I have deducted from the returns for respective years those factories employing 184 or fewer persons in order to get the first-rate factories. I have chosen the factory employing 184 persons as the smallest first-rate factory, which is twice the size of a median factory. This definition of a smallest first-rate factory as being twice the size of a median factory is arbitrary but useful, and has been applied to the cotton factory as well as woolen, worsted, silk, or flax-and-hemp factory throughout this study.
32. Children's Employment Com'n., 1834, **op. cit.**, XIX, pp. 119-23.
33. Granville, **The Spas of England**, London, 1841, 2 vols., II, pp. 19-20.
34. Daniels, Early Records of a Manchester Cotton Spinning Firm, in **Economic Journal**, 1915, XXV, pp. 178-88. See also his latest work on **The Early English Cotton Industry**, 1920, which is primarily a source study of the present firm's records and Crompton's unpublished letters.
35. See for description of the welfare work, Hickson's Report, 1840, **op. cit.**, p. 681; Children's Employment Commission, 1834, **op. cit.**, p. 207; Baines, **op. cit.**, pp. 447-51; Faucher, **Manchester in 1844** (tr. by J.P. Culverwell), London, 1844, pp. 105-6.
36. **Memoirs of the Right Honorable Sir Robert Peel**, by the author of "The Life of the Duke of Wellington", 2 vols., London, 1842, I, pp. 9-10, 14.
37. Faucher, **op. cit.**, p. 96.
38. **Ibid.**, p. 105.
39. Butterworth, **op. cit.**, p. 146.
40. **Children's Employment Commission, Report,** in P.P., 1843, XIV, p. 63.
41. Bevan, **Industrial Classes and Industrial Statistics**, 1876-77, Vol. on "Textiles", pp. 10-11. In the Report of the Committee on Children's Employment in Manufactories, 1816, **op. cit.**, pp. 116-7, a good account was given of the

arrangements made by the cotton factories dependent upon water for motive power. "There is a small reservoir generally attached to a mill, and if the stream is small during the droughty season, the first mill would retain the stream for some time, and the other mills likewise would do the same, so that upon a small stream where there are 12, 14, or 16 mills, the water would be 6 or 8 hours travelling from the upper mill to the lower, and in that case the mill at the bottom would not be able to commence operation till the water came from the top."

42. Committee on Mills and Factories, 1840, **op. cit.,** Q. 4031.
43. Ure, Cotton Manufacture of Great Britain, **op. cit.,** II, p. 401.
44. (a) Humphrey, **Great Britain.... in 1835,** New York, 1838, I, p. 213. The small number of spindles reported despite the inventions of Hargreaves' and Arkwright's may be justified when we recall that after having driven poor Hargreaves from his home about 1780, the Blackburn spinners and weavers renewed attacks on his machine. "Every jenny for many miles round Blackburn was destroyed, excepting such only as had less than 20 spindles." See French, **Life and Times of Samuel Crompton,** 2nd ed., Manchester, 1860, p. 54.
    (b) Allen, **History of the County of York,** London, 1829-31, 6 vols., I, p. 308.
    (c) **Committee on Crompton's Mule,** in P.P., 1812, II, pp. 2-3. Estimates were given before the Committee by leading manufacturers such as James Watt, Thomas Ainsworth, and George Lee. See also Crompton, in French, **op. cit.,** pp. 148-50.
    (d) Burns in **Commercial Glances** gave 8,000,000 spindles for England, from which Baines concluded that there were 9,333,000 spindles in Great Britain. See Baines, **op. cit.,** pp. 367-8.
    (e) Ure, Cotton Manufacture of Great Britain, **op. cit.,** II, p. 405.
    (f) **Factory Returns of 1850 and 1856.**
45. The **Factory Returns of 1850 and 1856** for the first and second time gave returns on the number of spindles and power looms classified according to the different types of factories, namely, "spinning", "weaving", "spinning-and-weaving", and "not included in either". Statistics for 1831-32 are taken from the **Report of Committee on Regulation of Factories,** in P.P., 1831-2, XV, Q. 9409.
46. Committee on Regulation of Factories, 1831-2, **op. cit.,** Q. 9409; Ure, Philosophy of Manufactures, **op. cit.,** p. 341; Bevan, **op. cit.,** p. 17; Cooke-Taylor, **Modern Factory System,** London, 1891, p. 338.
47. For an excellent brief survey on the history and extent of power looms, see Asst. H.L. Weavers' Com'n., 1840, **op. cit.,** p. 591.
48. **Committee on Cartwright's Loom,** in P.P., 1808, II, pp. 3-6.
49. Chapman, Lancashire Cotton Industry, **op. cit.,** pp. 29-32.
50. Hardwick, **History of the Borough of Preston,** Preston, 1857, p. 374.
51. **Committee on Handloom Weavers,** in P.P., 1834, X, Q. 5204.
52. For 1813 to 1833 see Asst. H.L. Weavers' Com'n., 1840, **op. cit.,** p. 591. For 1835 to 1856 see Factory Returns for the same years. (a) Judging from the 1820 estimate by Baines in which it was set down that 86% of all power looms in Great Britain belonged to England, we may reasonably conclude that about the same proportion of 2,400 power looms in 1813 in Great Britain, or 2,064, belonged to England. See Baines, **op. cit.,** p. 235.
53. **Factory Returns of 1835.**
54. **Ibid.**
55. Bücher, **Industrial Evolution (Die Entstehung der Volkswirtschaft),** tr. fr. 3rd German ed. (1900) by S. M. Wickett, New York, 1901, p. 292.

56. Supplementary Report, 1834, **op. cit.**, p. 33.
57. Usher, **Industrial History of England**, Boston, 1920, pp. 358-9.
58. Child is here defined as being of 18 years of age or below. For **1850-6,** statistics of this nature are not available for England (but "England **and** Wales"), nor for each county.
59. Excellent statistics on the overhead cost of a factory in Manchester, Bolton, etc. were given in Inspector Horner's Half Yearly Report ending December, 1841, **op. cit.**, pp. 79-82.
60. This well-known phrase comes from the famous economist, Senior. See for detailed discussion his **Letters on the Factory Act, as it Affects the Cotton Manufacture**, London, 1837.
61. **Inspector Howell's Report**, April, 1839, in P.P., 1839, XIX.
62. Henry Ashworth (one of the directors of the Manchester Chamber of Commerce), "Statistics of the present depression of trade at Bolton; showing the mode in which it affects the different classes of a manufacturing population", in **Royal Statistical Society Journal**, 1842, V. pp. 74-81.
63. Inspector Horner's Half Yearly Report, December, 1841, **op. cit.**, p. 24.
64. Quoted by M'Culloch, **Statistical Account of the British Empire**, 2nd ed., 2 Vols., London, 1839, I, p. 648 n.
65. Baines, **op. cit.**, p. 240.
66. Children's Employment Commission, Supplementary Report, 1834, **op. cit.**, p. 187, Table V. Schulze-Gavernitz, **Cotton Trade in England and on the Continent**, London, 1895 (tr. by Oscar S. Hall), p. 89.
67. Baines, **op. cit.**, Chapter XV.
68. See **Supra**, p. 34.
69. **Children's Employment Commission**, in P.P., 1833, XX, pp. 784-6. It is surprising that such valuable statistics were not quoted at all by the leading authorities of the subject, the only exception being Mr. Wheeler in his **Manchester, its political, social and commercial history**, Manchester, 1836, pp. 172-3. The latter author, however, made a few misquotations.
70. Symons, **Arts and Artizans at Home and Abroad**, Edinburgh, 1839, pp. 234-5.
71. Children's Employment Commission, 1833, **op. cit.**, pp. 846-7; Ure, Cotton Manufacture of Great Britain, **op. cit.**, II, pp. 441-3.
72. Symons, **op. cit.**, p. 233.
73. Ure, Cotton Manufacture of Great Britain, **op. cit.**, I, pp. 312-14.
74. Baines, **op. cit.**, p. 415.
75. Dodd, **British Manufactures**, London, 1844-46, 2 vols., I., p. 14.
76. The total for England did not agree with that obtained by adding up the county totals in a few cases. In the present computation, the latter is adopted in preference to the former.
77. Cunningham, **Growth of English Industry and Commerce**, 1903, 3 vols., **II**, p. 794.
78. Hickson's Report, 1840, **op. cit.**, pp. 9-10, art. 23.
79. **Ibid.**; also quoted in Cunningham, **op. cit.**, II, pp. 792-3.
80. Cf. Census return of 44,977 persons and discussion on p. 26 of this Chapter.
81. Asst. H.L. Weavers' Com'n., 1840, **op. cit.**, p. 578.
82. Committee on Handloom Weavers, 1834, **op. cit.**, Q. 5058; 1835, XIII, Q. 2929.
83. Committee on Manufactures, Commerce and Shipping, 1833, **op. cit.**, Q. 11822.
84. Committee on Handloom Weavers, 1834, **op. cit.**, Q. 5862-65, 6244.

85. **Ibid.**, Q. 5670. Quoted also by Chapman, Lancashire Cotton Industry, **op. cit.**, p. 25.
86. Committee on Handloom Weavers, 1834, **op. cit.**, Q. 5645. Also see Asst. H. L. Weavers' Com'n., 1840, **op. cit.**, p. 609.
87. **Manchester Historical Recorder**, Manchester, 1874, p. 37.
88. Chapman, Lancashire Cotton Industry, **op. cit.**, pp. 15-6.
89. V. C. H., **Lancashire**, London, 1908, II, p. 383.
90. Committee on Handloom Weavers, 1835, **op. cit.**, Q. 1700.
91. Committee on Manufactures, Commerce and Shipping, 1833, **op. cit.**, Q. 9032-33, 9282-84.
92. Asst. H. L. Weavers' Com'n., 1840, **op. cit.**, p. 586.
93. Chapman, Lancashire Cotton Industry, **op. cit.**, pp. 44-5. See also **Committee on Emigration, 2nd Report**, in P.P., 1826-27, V, p. 5.
94. Hickson's Report, 1840, **op. cit.**, p. 7, arts. 19-20. He continues the above description in saying, "This description is not of universal application. In some instances I met with streets of small modern built houses, in which the cellars occupied by the cotton weavers were light and convenient apartments, and in many towns the cotton weavers work not in cellars, but in an unboarded room on the ground floor."
95. Kohl, **England and Wales**, Philadelphia, 1846, published in "The Foreign Library" Series, pt. 4, p. 63.
96. Asst. H. L. Weavers' Com'n., 1840, **op. cit.**, p. 594.
97. **Ibid.**, pp. 592-3.
98. Committee on Handloom Weavers, 1834, **op. cit.**, Q. 4367, 5032, 5854-9, 6636-7.
99. Asst. H. L. Weavers' Com'n., 1840, **op. cit.**, pp. 607, 608-9.
100. Committee on Handloom Weavers, 1834, **op. cit.**, Q. 5030.
101. **Ibid.**, Q. 5127.
102. **Ibid.**, Q. 5645-7. Also see Asst. H. L. Weavers' Com'n., 1840, **op. cit.**, p. 609.
103. **Ibid.**, p. 610.
104. **Ibid.**, p. 601.
105. Hickson's Report, 1840, **op. cit.**, p. 12, art. 29.
106. Committee on Handloom Weavers, 1834, **op. cit.**, Q. 5734.
107. Hickson's Report, 1840, **op. cit.**, p. 13, art. 29.
108. **Ibid.**

CHAPTER III. WOOLEN AND WORSTED

1. **Supra**, Chapter II, note 4.
2. Lipson, **History of the English Woolen and Worsted Industries**, London, 1921, pp. 1-6.
3. **Ibid.**, p. 6.
4. Baines, On the Woolen Manufacture of England, in **Royal Statistical Society Journal**, 1859, XXII, p. 15.
5. Bischoff, **Comprehensive History of the Woolen and Worsted Manufactures**, London, 1842, 2 vols., II, Table IV.
6. **Asst. H. L. Weavers' Com'n., Report**, in P.P., 1840, XXIII, pp. 147, 175, 399,417.
7. Dodd, **British Manufactures**, London, 1844-6, 2 vols., I, p. 87.
8. "Walk mill" and "fulling mill" are synonymous terms.
9. James, **History of the Worsted Manufacture in England**, London, 1857, pp. 585-6.
10. **Report**, in P.P., 1806, III, p. 9.
11. Asst. H. L. Weavers' Com'n., 1840, **op. cit.**, XXIII, p. 528.

12. **Children's Employement Commission, Supplementary Report**, in P.P., 1834, XX, C. I: North-Eastern Districts: Yorkshire, pp. 793-1080. Wherever possible, the year of erection is preferred to the year of occupation as an index of growth. Where the year of erection is not available, the year of occupation is given. The year of erection, however, means in some cases the year of rebuilding. When an occupier has several factories (i.e. plants) erected or occupied or rebuilt at different dates, the earliest is chosen. This volume of valuable statistics, to my knowledge, has not been analyzed elsewhere.

13. "In the 1780's," said Wadsworth, "the carding machines and slubbing billies began to be put in little mills, and for nearly 50 years these were the only power driven mills in the Rochdale woolen industry. Even as late as 1833 it could be said that in general the carding process alone is done by power." Wadsworth, History of the Rochdale Woolen Trade, in **Rochdale Lit. and Sci. Soc. Trans.**, 1925, XV, p. 107. Also Heaton, **Yorkshire Woolen and Worsted Industries**, Oxford, 1920, pp. 357-8.

14. The Act of 3 and 4 will. IV, (1833) C. 103, stated that "nothing in this Act shall apply or extend to the working of any steam or other engine, water wheel or other power in or belonging to any mill or building, or machinery, when used in that part of the process or work commonly called fulling, roughing or boiling of woolens, nor to any apprentices or other persons employed therein." (In P.P., 1833, II) This omission of fulling mills from legislative interference might have influenced the factory inspectors in omitting to enumerate the fulling mills in the Factory Returns. In that case, the Factory Returns would have been under-estimated with respect to the number of factories, employees, motive power, spindles and power looms to the extent that the fulling mills were thus omitted.

15. Baines, An account of the Woolen Manufacture to 1870, in his **Yorkshire, Past and Present**, 1870, 2 vols., II, p. 646. This article was first printed in **Royal Statistical Society Journal**, 1859, XXII, in which the above quotation could be found on page 19.

16. **Inspector Horner's Half Yearly Report ending December, 1841**, in P.P., 1842, XXII, pp. 64-69.

17. **Supra**, note 12.

18. See **infra**, pp. 75-76.

19. Clapham, Industrial Organization in the Worsted and Woolen Industries of Yorkshire, in **Economic Journal**, 1906, XVI, p. 520.

20. "The manufacturer's success depends so largely on the skillful composition of his yarn, that it must as a rule be made under his own eye and not bought from a neighbor. He is perhaps scribbling new wool and rag wool, or wool and cotton together, in proportions nicely calculated so as to produce the maximum of effect in the finished cloth at the minimum cost. Or maybe he is twisting a tender coating of shoddy about a core of cotton yarn that will stand the tension of the loom and of subsequent wear. These are tasks which a Yorkshire man, last of all, would care to trust as a regular thing to an outside spinner." Clapham, **Woolen and Worsted Industries**, London, 1907, p. 148.

21. Clapham, Industrial organization, **op. cit.**, pp. 521-2.

22. The procedure adopted in computing the first-rate factories in woolen manufacture is the same as that in cotton manufacture. Cf. **supra**, Chapter II, note 31. The median used here is an average of the two medians, viz. 15 and 19, which is 17.

23. **Children's Employment Commission, Reports,** in P.P., 1833, XX, p. 960.
24. **Ibid.,** p. 958. In 1832, an "Association for the Promotion of Order" was formed among his work people "to ensure a better observance of Sundays, and to check the great evils of idleness and drunkenness, which the number of beer-houses lately opened has greatly contributed to aggravate and encourage". In Children's Employment Commission, 1834, XX, **op. cit.,** p. 304.
25. In P. P., 1833, VI, Q. 1859, 1867, 1876, 1917, 1925.
26. Children's Employment Commission, 1833, **op. cit.,** p. 927.
27. Asst. H. L. Weavers' Com'n., 1840, **op. cit.,** XXIV, p. 362.
28. **Ibid.,** p. 363.
29. Clapham, Woolen and Worsted Industries, **op. cit.,** p. 135.
30. Again, the unsatisfactory character of the factory statistics must not be overlooked. In referring to the Factory Returns of 1856, Baines remarked that "according to Mr. Baker, one of the most laborious of the factory inspectors in the collection of statistics, the returns of the woolen spindles are not in the least trustworthy, as some of the inspectors have returned only the billy spindles, which are used in the first stage of spinning, whilst others have returned the mule spindles, used in the second stage." (Baines, Yorkshire Past and Present, **op. cit.,** II, p. 647).
31. Fletcher, **The Making of Modern Yorkshire,** 1750-1914, London, 1918, pp. 100-102.
32. Clapham, Industrial Organization, **op. cit.,** p. 522.
33. Children's Employment Commission, 1833, **op. cit.,** p. 502.
34. Asst. H. L. Weavers' Com'n., 1840, **op. cit.,** XXIV, p. 434.
35. Chapman, **Lancashire Cotton Industry,** Manchester, 1904, p. 47.
36. This total includes all power looms returned under "woolen" and "woolen and worsted", while Lipson's includes the following power looms in addition to those returned under "woolen" (Lipson, **op. cit.,** p. 188 and footnote): Leicester, 89 "woolen and worsted" looms; Somerset, 3 "woolen and worsted" looms; Montgomery, 4 "woolen" looms. I can hardly see the reason why Professor Lipson in his total of "woolen" power looms includes 89 "woolen and worsted" looms in Leichester and 3 in Somerset; while in Yorkshire there are 226 "woolen and worsted" looms which he mentions but does not include. My total is computed from 1835 Factory Returns, in P.P., 1836, XLV, pp. 1-9.
37. Hickson, **Report on the Condition of the Handloom Weavers,** in P.P., 1840, XXIV, p. 19.
38. Asst. H. L. Weavers' Com'n., 1840, **op. cit.,** XXIII, pp. 439-41. Quoted also in Lipson, **op. cit.,** pp. 258-60.
39. Cf. Heaton, Yorkshire Woolen and Worsted Industries, **op. cit.,** pp. 90-91.
40. Report, 1806, **op. cit.,** pp. 9, 11.
41. There was evidence of the existence of a few handloom factories in Leeds and vicinity as reported by Commissioner Chapman. See Asst. H. L. Weavers' Com'n., 1840, **op. cit.,** XXIII, pp. 533, 552.
42. **Ibid.,** XXIV, pp. 368,459.
43. **Ibid.,** pp. 376-8, 386.
44. **Ibid.,** p. 378.
45. **Ibid.,** p. 437.
46. **Ibid.,** XXIII, p. 460.
47. Report, 1806, **op. cit.,** pp. 12-13.

48. Asst. H. L. Weavers' Com'n., 1840, **op. cit.**, XXIII, pp. 529, 540-1, 543, 546-7, 549-50.
49. Ashley, **Economic Organization of England**, 1914, p. 142.
50. Heaton, in commenting on the unsatisfactory character of the term "domestic system," concluded that "confining our attention to the textile industry, we might use the phrase 'clothier system', but usage has established the claims of the older term." (Heaton, **op. cit.**, p. 89).
51. Cunningham, **Growth of English Industry and Commerce**, 1903, 3 vols., **II**, p. 497. Professor Sprague, in his unpublished dissertation on the **English Woolen Industry in the 17th and 18th Centuries**, Cambridge, 1897, p. 24, told us that "the Yorkshire industries did not grow up out of a gild industry, nor were the masters organized in crafts or companies, except for a short time for the specific purpose of carrying out the provisions of certain statutes. From the first it seems to have been a domestic industry. It has, however, all the characteristics of an industry carried on under the gild system, with the additional element of production for a wide market, just enough to make it a domestic industry."
52. Report, 1806, **op. cit.**, pp. 8-9. Cf. Heaton's account in his "Yorkshire Woolen and Worsted Industries", **op. cit.**, p. 92 et seq.
53. Cf. **infra**, pp. 75-76. The extension of the joint stock mills in 1840 constitutes perhaps the most notable change, although its introduction can be dated several decades before 1806.
54. Whenever accounts applicable to an earlier period than 1840 are drawn for our analysis, I have done so from my own inference, suported by evidences available to my knowledge, that with a few exceptions, the facts have remained unchanged in 1840 as in the eighteenth century.
55. Heaton, **op. cit.**, pp. 298-9. Heaton's account of the two classes during the 18th century was also true of a later period, although a great deal of the social stratifications among the clothiers had in all probablity disappeared. Some of the comparatively larger clothiers had now become factory manufacturers, and the small clothiers had thus become a more homogenous body than ever before.
56. Lipson, **op. cit.**, pp. 79-85.
57. In West Riding, Yorkshire, Commissioner Chapman remarked that "the manufacture of cloth is carried on by small manufacturers, who are themselves both weavers and spinners. They are completely ignorant of the circumstances affecting distant markets; and in adapting their supply to the demand, they are guided solely by the state of the market on the last market day, and not by any enlarged prospective views. In the event of a slack market, they will refuse a warp perhaps to one weaver; a second dull market, and a second weaver will be refused work, and so they go on until their production is confined to their own families. In the event of a brisk demand, they, being ignorant, are apt to be over-sanguine, and they work on, until the consequence is a great aggravation of subsequent depression. The only check upon this is the intelligence of the Leeds merchants; but they, again, are less cognizant of the circumstances affecting supply and demand than the merchants of great shipping markets, and they are consequently guided by their local market to a greater extent than a more enlarged view would induce." (Asst. H. L. Weavers' Com'n., 1840, **op. cit.**, XXIII, p. 563).
58. Quoted in Heaton, **op. cit.**, pp. 344-6, from whom Lipson quoted in his "History", **op. cit.**, pp. 72-3. Another poem, "The Brothers", by the poet Wordsworth, Described family labor as follows (Quoted by Morris, **The Golden Fleece**, Oxford, 1922, p. 105):

"It was a July evening; and he sate
Upon the long stone-seat beneath the caves
Of his old cottage,—as it chanced, that day,
Employed in winter's work. Upon the stone
His wife sate near him, teasing matted wool,
While, from the twin cards toothed with glittering wire,
He fed the spindle of his youngest child,
Who, in the open air, with due accord
Of busy hands and back-and-forward steps,
Her large round wheel was turning".

59. Asst. H. L. Weavers' Com'n., 1840, **op. cit.**, XXIII, pp. 540-1, 543.
60. Lipson, **op. cit.**, pp. 73-4; Heaton, **op. cit.**, pp. 291-2, 293.
61. **Committee on Handloom Weavers**, in P.P., 1835, XIII, p. 204.
62. Cloth halls were established in Leeds, Bredford, Halifax, Huddersfield, Dewsbury in Yorkshire, and Rochdale in Lancashire for woolen and worsted manufactures. See Asst. H. L. Weavers' Com'n., 1840, **op. cit.**, XXIII, p. 527.
63. Parsons, **History of Leeds**, Leeds, 1834, 2 vols., II, pp. 209-11. In 1711 the earliest cloth hall was established in Leeds, but was soon abandoned because of inadequacy in serving the needs of an expanding trade. In Halifax, the first cloth hall was erected in 1700. As regards Leeds white cloth hall in 1711 see the contemporary account by Thoresby, in Baines, Yorkshire Past and Present, **op. cit.**, II, p. 134. Heaton has made a study of the same cloth hall and its subsequent development from MSS obtained from the trustees of the Cloth Hall, in **Thoresby Soc.**, 1913, XXII, pp. 131-171.
64. **Committee on Woolen Manufacture of England**, 1806, op. cit., pp. 13-14.
65. Children's Employment Commission, 1833, **op. cit.**, p. 502.
66. Committee on Woolen Manufacture, 1806, **op. cit.**, p. 13.
67. Parsons, **op. cit.**, II, pp. 210-11.
68. For an excellent summary of the merits and demerits of the system see Heaton, Yorkshire Woolen and Worsted Industries, **op. cit.**, pp. 347-353.
69. Committee on Woolen Manufacture, 1806, **op. cit.**, p. 10.
70. **Ibid.**, p. 11.
71. Clapham, Woolen and Worsted Industries, **op. cit.**, pp. 57-8.
72. **Committee on Joint Stock Companies, First Report**, in P.P., 1844, VII, pp. 349-351, Appendix 5. The articles of partnership of one joint stock mill, dated May 1825, are quoted in full, and illustrate excellently the chief features of its organization. Our quotation above is also to be found in Lipson, **op cit.**, pp. 177-9. For other accounts see Asst. H. L. Weavers' Com'n., 1840, **op. cit.**, XXIII, pp. 549-550; Baines, **op. cit.**, in Royal Statistical Society Journal, 1859, XXII, pp. 29-30.
73. Clapham, Woolen and Worsted Industries, **op. cit.**, pp. 37-8. See also Kropotkin, **Fields, Factories and Workshops**, Boston and New York, 1899, p. 166; Id., "Industrial Villages of the Future", in **Nineteenth Century**, 1888, XXIV, pp. 516-7.
74. Lipson, **op. cit.**, pp. 43-51. Sprague, however, told us that in the West of England the woolen manufacture "was very largely established in the 16th century by the manufacturing class of wealthy merchants who desired to escape the restrictions to which the gild subjected industry, and it grew up without much origin, either among clothiers or workmen. In other instances it appears to have grown up of itself without any outside influence." (**op. cit.**, pp. 25-6)

75. The merchant employer system could be found chiefly in West Country in 1840, although Commissioner Chapman, by devoting disproportionate attention in his Report to the borough of Leeds where only 541 woolen cloth weavers were employed, gave the misleading impression that the merchant employer system was in extensive operation in West Riding. (Asst. H. L. Weavers' Com'n., 1840, **op. cit.**, XXIII, p. 529, et seq.)
76. Asst. H. L. Weavers' Com'n., 1840, **op. cit.**, XXIV, p. 370.
77. Report, 1806, **op. cit.**, p. 8. See also Asst. H. L. Weavers' Com'n., 1840, **op. cit.**, XXIII, p. 529.
78. Asst. H. L. Weavers' Com'n., 1840, **op. cit.**, XXIV, p. 523. See also pp. 414-5, 521-6.
79. Asst. H. L. Weavers' Com'n., 1840, **op. cit.**, XXIII, pp. 546-7.
80. **Ibid.**, p. 473.
81. **Ibid.**, pp. 473-4.
82. **Ibid.**, p. 458.
83. **Ibid.**, p. 461.
84. **Ibid.**, p. 460.
85. **Ibid.**, p. 462.
86. **Ibid.**, pp. 458-9. Letter to the Commission from Edward C. Ross, grocer, draper, and druggist at Woodchester, June 14th, 1838.
87. "The word 'worsted' has been derived from the Dutch Ostade. Camden and other authorities derive it from the Norfolk village of Worstead, or Worstede, called Woreste-dosta in the Conqueror's day. It is on the Thym, an affluent of the Yare, though twenty-one miles from Yarmough."—Bonwick, **Romance of the Wool Trade**, London, 1887, p. 365.
88. Combing, instead of carding, constitutes the main difference between worsted and woolen manufacture. The French and German equivalents of woolen and worsted are 'carded' and 'combed'—terms descriptive of the destinctive preliminary processes through which the wool goes.
89. This definition of Dr. Playfair's for the use of the jurors in the Great Exhibition of 1851 is modified by the present writer. Cf. Forbes, History of the Worsted Manufacture in England, in **1851 Exhibition Lectures**, II, pp. 320-21.
90. For the apportionment of "weavers" "spinners" and "factory workers" (branches not specified), cf. **supra**, note 1.
91. The subdivision of worsted manufacture into the three branches of worsted, stuff and carpet was given by Baines, **op. cit.**, in Royal Statistical Society Journal, 1859, XXII, p. 21. It is here chosen with a view to conform to the Census emmeration of 1841.
92. The apportionment among the counties of 7,131 persons assigned to the worsted manufacture (cf. note 90 of this Chapter) cannot be carried out upon the same principle as that adopted for the cotton, woolen, silk, flax and hemp manufactures—the principle of giving to each county its proportion according to the relative percentage of persons employed in that county. In the worsted manufacture in 1841 a peculiar condition arose. The Census returned only 183 persons under "worsted manufacture" for Norfolk, the ertswhile centre and birthplace of the worsted manufacture, but 2,846 persons under "weavers" for the same county. According to various authorities on the subject, (Clapham, Transference of the Worsted Industry from Norfolk to the West Riding, in **Economic Journal**, 1910, XX, pp. 196-7; Assistant Handloom Weavers' Commission, 1840, **op. cit.**, XXIII, pp. 309, 328; James, **op. cit.**, pp. 483-4) these 2,846 weavers, although not specified as to the branch of manufacture, were "worsted" weavers, and must be classified under the worsted manufacture. For this reason, I include these 2,846 "weavers"

under "worsted manufacture" for Norfolk, and thus get a total of 3,029 persons employed in "worsted manufacture" in Norfolk and 23,681 persons employed in "worsted manufacture" in England. I then deduct 2,846 persons from 7,131 persons apportioned from "weavers", "spinners", and "factory workers" (branches not specified), leaving a balance of 4,285 persons. These 4,285 persons, instead of 7,131, are then apportioned among the counties on the same principle as that adopted for the cotton, woolen, silk, flax and hemp manufactures.

93. For a detailed discussion of the subject, see an excellent article by Clapham, Transference, **op. cit.**, passim. Also Asst. H. L. Weavers' Com'n., 1840, **op. cit.**, XXIII, pp. 306-7, 341, 342-4.
94. Baines, **op. cit.**, in Royal Statistical Society Journal, 1859, XXII, p. 16.
95. Lipson, **op. cit.**, p. 184 footnote.
96. Clapham, Transference of the Worsted Industry, **op. cit.**, p. 203.
97. Id., Industrial organization, **op. cit.**, pp. 517-18.
98. **Ibid.**
99. James, **op. cit.**, pp. 605-6.
100. Lipson, **op. cit.**, p. 185.
101. The number of persons employed under the factory system needs a word of explanation. After an examination of the Factory Returns of 1838 for the worsted factories, it is found that in several counties the factory return exceeded the census return in the number of persons employed. Thus, in the four leading counties of lace manufacture—Leicester, Nottingham, Derby and Northampton—the factory return of 1838 exceeded the census return by a total number of 1,780 persons. The excess in the number of persons would of course be larger, as in our computation it is assumed that by deducting the census return from the factory return and thus obtaining the excess, every person included under the census return would *ipso facto* be a factory employee. The excess of factory return over census return is as follows:—

County	Factory return (I)	Census return (II)	Excess of (I) Over (II)
Leicester	1,734	412	1,322
Nottingham	363	22	341
Derby	71	—	71
Northampton	57	11	46
Total	2,225	445	1,780

This excess of more than 1,780 persons in the lace counties may be explained by the includsion of the lace "factories" under the "worsted" factories. (For reasons, see Chapter IX). Meantime, the excess in counties other than the above mentioned—Leicester, Nottingham, Derby and Northampton—may be explained by the difference in the years at which the census and the factory returns were taken respectively, viz. 1838 and 1841.

102. The median average for 1841, which we have compiled for the cotton and woolen factories, cannot be similarly compiled for the worsted factories because of insufficient statistics. See Inspector Horner's Half Yearly Report ending December, 1841, **op. cit.**, pp. 71-2. Instead, we take the statistics for 1833 as the source materials for the compilation of a median average. See **supra**, note 12.
103. From **Yorkshire form of Factory Returns**, in P.P., 1834, XX.

104. The procedure adopted in computing the first-rate factories in worsted manufacture is the same as that in cotton or woolen manufacture. Cf. **supra**, Chapter II, note 31. The median used here is an average of the two medians, viz. 56 and 57, which is 56.5, and the smallest first-rate factory, which, in our definition, is twice the median size of an average factory, is, therefore, 113. The actual sizes of the 21 first-rate factories are as follows: 125, 129, 130, 130, 155, 155, 165, 166, 176, 176, 207, 221, 250, 278, 301, 321, 387, 438, 527, 550, 610.

105. These quotations were taken from an account by a contemporary in **Penny Magazine**, Nov. 16, 1833, II, pp. 445-6.

106. Clapham, Woolen and Worsted Industries, **op. cit.**, pp. 141-2. See also pp. 144-5. Commissioner Chapman reported in 1838 that, in West Riding, "in the worsted manufacture the business of spinning is more completely separated from that of manufacturing than in the case of the woolen manufacture. Though in some few cases houses will be found who are both spinners and manufacturers, it generally happens that the worsted spinner sells, and the stuff-manufacturer buys, the yarn in the market." (In Asst. H.L. Weavers' Com'n., 1840, **op. cit.**, XXIII, p. 557.

107. **Supra**, note 12.

108. Parsons, **op. cit.**, II, p. 215. Also James, **op. cit.**, p. 414; or his other work, **History of Bradford**, 1841, p. 168.

109. James, History of the Worsted Manufacture, **op. cit.**, pp. 597-8.

110. Power loom statistics are taken from the Factory Returns for the respective years mentioned in the table.

111. **Inspector Saunder's Report dated January, 1844**, in P.P., 1844, XXVIII, pp. 558-9; James, History of the Worsted Manufacture, **op. cit.**, p. 511.

112. Asst. H. L. Weavers' Com'n., 1840, **op. cit.**, XXIII, p. 586.

113. **Ibid.** p. 566.

114. **Ibid.** p.558.

115. Sprague, **op. cit.**, p. 15.

116. **Ibid.** pp. 17, 18.

117. Turner, Bradford Piece Halls, in **Bradford Antiquary**, 1888, I, pp. 135-9.

118. Asst. H. L. Weavers' Com'n., 1840, **op. cit.**, XXIII, p. 309; James, History of the Worsted Manufacture, **op. cit.**, pp. 483-4; Clapham, Transference of the Worsted Industry, **op. cit.**, pp. 196-7.

119. Asst. H. L. Weavers' Com'n., 1840, **op. cit.**, XXIII, pp. 309, 333.

120. **Ibid.**, p. 329. See also pp. 331-4.

121. **Ibid.**, p. 331.

122. **Ibid.**, p. 332.

123. **Ibid.**, p. 312.

124. **Ibid.**, p. 314.

125. **Ibid.**, p. 559.

126. **Ibid.**, p. 582. See also p. 316.

127. **Ibid.**, p. 312.

128. **Ibid.**, p. 552.

129. **Ibid.**, p. 571.

130. **Ibid.**, p. 571.

131. These 3,409 handloom weavers in carpet manufacture were enumerated by the Census for the whole of England. The actual number, including part-time outworkers such as women and children, would probably be much larger than this return by the Census.

132. Asst. H. L. Weavers' Com'n., 1840, **op. cit.**, XXIV, p. 530.
133. **Children's Employment Commission, Report,** in P.P., 1833, XX, p. 909; Cunningham, **op. cit.**, II, pp. 781-2 footnote.
134. Children's Employment Commission, 1833, **op. cit.**, p. 913.
135. **Children's Employment Commission, Report,** in P.P., 1843, XIV, C24 art. 35.
136. **Ibid.**, c107.
137. **Ibid.**, C21 art. 15.
138. **Ibid.**, c113, c116.
139. Asst. H. L. Weavers' Com'n., 1840, **op. cit.**, XXIV, p. 530.
140. The original total is 2,021; but the correct total, according to actual addition, is 2,011.
141. Children's Employment Commission, 1843, **op. cit.**, XIV, c107.

CHAPTER IV. SILK, FLAX AND HEMP

1. Weaving includes winding, warping and weaving. Technically speaking, we may use the term "manufacturing" as distinguished from either "throwing" or "spinning".
2. Plush manufacture is not included in our definition, as it was separately enumerated in the Census, and employed only 260 persons in England in 1841. It also employed materials other than silk. As to crape, the Census of 1841 returned a total of 323 persons. It is hard to justify an independent enumeration of this nature, as silk goods under the name "crape" constituted a far more extensive branch of silk manufacture. (See **Asst. H.L. Weavers' Com'n., Report,** in P.P., 1840, XXIII, pp. 286, 287, 293, 378, 398, 448.)
3. For the apportionment of "weavers", "spinners", and "factory workers" (branch not specified), Cf. **supra**, Ch. II, note 4.
4. Rawlley, **Economics of Silk Industry**, London, 1919, p. 215.
5. The number of "factory" employees as reported by the Factory Returns of 1838 exceeded the number of total employees as reported by the Census of 1841 for a few silk counties, perhaps due to the fact that factories making silk goods of mixed materials such as worsted were classified under silk factories by the Factory Inspectors in their Returns. The following list is an example:

County	No. employees (Census, 1841)	No. employees (Factory Returns, 1838)	Excess of Factory Returns over Census
Derby	2,865	3,226	361
Norfolk	2,132	2,274	142

6. We exclude Norfolk and Essex on the ground that the factories in these two counties were few and unusually large—they were not typical.
7. **Children's Employment Commission, Supplementary Report,** in P.P., 1834, XIX, p. 145. **Supra,** Ch. III, note 12.
8. **Committee on Silk Trade, Report,** in P.P., 1831-2, XIX, pp. 691-4.
9. Asst. H. L. Weavers' Com'n., 1840, **op. cit.**, XXIII, p. 317.
10. The term "throwing" is used here to include both throwing and spinning for the sake of brevity.
11. Porter, **Silk Manufacture,** in Lardner's Cabinet Cyclopaedia Series, London, 1831, p. 275.
12. Committee on Silk Trade, 1831-2, **op. cit.**, Q. 10307.
13. **Ibid.**, Q. 11494-99.
14. **Ibid.**, Q. 3106.
15. Report, 1840, **op. cit.**, XXIII, p. 299.

16. **Ibid.**
17. For references other than those here given, see **Committee on Ribbon Weavers and Silk Manufacturers, Report,** in **P.P.**, 1818, IX, pp. 63, 66, 76.
18. Asst. H. L. Weavers' Com'n., 1840, **op. cit.**, XXIII, p. 347.
19. **Ibid.**, p. 432.
20. **Ibid.**, p. 294.
21. **Children's Employment Commission, Report**, in P.P., 1843, XIV, C17, art. 49; Asst. H. L. Weavers' Com'n., 1840, **op. cit.**, XXIII, p. 301; Hertz, English Silk Industry in the 18th Century, in **English Historical Review**, 1909, XXIV, pp. 710-27.
22. Before the 2nd quarter of the 19th century "the manufacturers' only customers were shopkeepers and the leading drapery houses; in those days there was no middleman [i.e. warehouseman], and those who wanted silks had to go direct to the producer. The trade had just passed through a time of great prosperity, and, as in all times of affluence, the reaction is sooner or later sure to set in; and so it was with the silk trade, for it had begun to ebb when there sprang up a class of middlemen, or warehousemen. Some of them were drapers, who laudably anxious to extend their business, were allowed by the foolish manufacturers of those time to get all the buying into their own hands. They went to the retail silk mercer, and told him they would save him all the trouble and expense of collecting the goods himself; then to the manufacturer, and in turn told him they could save him all the expense and trouble of disposing of the goods for himself; and in short, made themselves as desirable as possible to both. It was in this way the manufacturers were entrapped to their ruin. These warehousemen or middlemen became at once masters of both the manufacturer and the buyer; they had secured a giant's strength, and with a giant's power they used it, which crushed the manufacturer with such oppressive effect that many who could afford to do so left the trade." (Whiter, **Silk Industry of Great Britain and its Revival**, London, 1882, pp. 33-4).
23. Article by George Dodd on "Spitalfields", in Knight, **London**, London, 1842, II, pp. 395-6.
24. **Report**, in P.P., 1833, VI, Q. 5138.
25. Committee on Silk Trade, 1831-2, **op. cit**,. Q. 7639-41.
26. In P. P., 1840, XXIV, p. 19 and passim.
27. Fay, **Life and Labor in the Nineteenth Century**, Cambridge, 1920, pp. 175-6.
28. Quoted in Warner, **Silk Industry of the United Kingdom**, London, 1921, pp. 87-8. Cf. Hickson's Report, 1840, **op. cit.**, XXIV, p. 51, art. 99; **Report of the Tariff Commission**, 1905, II, pt. VI.
29. Hickson's Report, 1840, **op. cit.**, XXIV, p. 51, art. 99. Quoted also in Knowles, **Industrial and Commercial Revolutions in Great Britain during the 19th Century**, London, 1921, p. 68.
30. Asst. H. L. Weavers' Com'n., 1840, **op. cit.**, XXIII, pp. 258-9.
31. **Ibid.**, p. 466.
32. Crankshaw, **Weaving**, London, 1924, p. 59.
33. For the apportionment of "weavers", "spinners", "factory workers" (branches not specified), **cf. supra**, Chapter II, note 4.
34. In our definition the following branches of the flax and hemp manufactures are excluded for various reasons:

Branch	No. employees	Reason for exclusion
Mat	782	Employs jute or mixed materials
Sack & bag	777	,, ,, ,, ,, ,,
Sail cloth	620	,, ,, ,, ,, ,,
Canvas	322	,, ,, ,, ,, ,,
Girth	77	,, ,, ,, ,, ,,
Net	219	,, different technical process (knitting)
Hemp	352	Faulty enumeration by Census.

One reason applies to each of the above 7 branches, namely, to facilitate statistical computation.

35. In the hemp (i.e. rope and cord) manufacture, "the plant and technique were much simpler than in the regular textile industries in which raw material could still be obtained from local sources." (Day, **Distribution of Industrial Occupations in England,** 1841-1861, New Haven, 1927, p. 129). Unlike other textile manufactures, the hemp manufacture of England in 1841 had a provincial instead of national distrubution (**Ibid.**, p. 129) with regard to the extent of its market. A manufacture was national in its distribution "if more than half the persons engaged in it were concentrated in districts comprising less than one quarter of the total population". (**Ibid.**, p. 95).

36. The wide dispersion of a small number of flax dressers in the different counties suggests that this preparing process in 1841 was still carried on under the master craftsman system. Sometimes, a flax dresser was also a spinner, weaver, or "worker". (e.g. Leicester, 9 "flax dressers and workers"; Oxford, 5 "flax dressers and weavers"; Suffolk, 5 "flax dressers and spinners"; Warwick, 9 "flax dressers and spinners".) In the following table, all counties having ten or fewer flax dressers in 1841 are listed with their respective numbers:

County	No. flax dressers
Berks	9
Buckingham	2
Hertford	3
Huntington	1
Kent	7
Leicester	9
Monmouth	2
Norfolk	4
Oxford	5
Rutland	1
Suffolk	5
Warwick	9
Worcester	10
Total	67

37. **Committee on the Linen Trade of Ireland, Report,** in P.P., 1825, V. p. 41.
38. **Ibid.,** pp. 35-6, 46, 48, 54, 100.
39. Dodd, **British Manufactures**, Vol. on Textiles, London, 1844, p. 167.
40. Asst. H. L. Weavers' Com'n., Report, 1840, **op. cit.**, XXIII, p. 478.
41. Timmins, **Industrial History of Birmingham and the Midland Hardware District,** London, 1866, p. 576.

42. **Report,** in P. P., 1843, XV, M40-45, M52-55, M58. Some of these roperies may be called immature factories, as the extent to which the concentration of labor was carried on in these roperies was already beyond that of a shop. **Infra,** pp. 193-4.
43. Armstrong, **Industrial Resources of Northern Districts,** 2nd ed., London, 1864, pp. 254-55.
44. Children's Employment Commission, 1843, **op. cit.,** XV, M42-3.
45. In this table the figures for various headings are derived as follows:—

a. Flax weaving	Census of 1841, under the heading "linen"
b. Flax and hemp spinning under factory system	Factory Returns of 1838, under the heading "flax factories"
c. Hemp spinning under master craftsman system	Total employees in flax and hemp manufactures minus employees under (a) and (b)

This table is not intended to be strictly accurate, but gives a rough approximation.

46. In 1835 there were only 41 power looms for flax weaving in the whole of England. This unusually small number may be explained by the want of elasticity in the flax yarn that hindered the application of power looms to flax weaving. The Marshalls of Leeds, the first and largest flax spinner by power machinery, did not introduce power looms into their factory until 1843. See Combe, **A Summary of the Origin and Processes of the Linen Manufacture,** Belfast, 1868, pp. 20, 23.
47. **Factory Returns of 1838; Inspector Horner's Half Yearly Report dated December, 1841,** in P.P., 1842, XXII, pp. 72-4. For definition of a "first-rate" factory cf. **supra,** chapter II, note 31.
48. **Report, op. cit.,** pp. 41-44.
49. **Committee on Export of Machinery, First Report,** in P.P., 1841, VII, Q. 2697-2701, 2721, 2747, 2802-3.
50. This view is further confirmed by the fact that in Yorkshire, there were, according to the Factory Returns, 38 spinning factories with 25,496 spindles in 1850, and 37 spinning factories with 149,201 spindles in 1856. The error lies again in reporting 25,496 spindles instead of 125,496.
51. In this compilation, the total number of spindles in England in 1850 is not the one as given by the Factory Returns, but the one as corrected by the present writer in agreement with the view expressed in the preceding footnote.
52. In Newark, Nottingham, Commissioner Mitchell in 1838 reported that "there is a great temptation to the manufacturers to try to introduce the power loom. At present, when an order comes requiring to be executed in a short time, there are not additional hands to be got to do it." (Asst. H. L. Weavers' Com'n., 1840, **op. cit.,** XXIII, p. 351).
53. The number of power looms as given by Gill, in his **Irish Linen Industry,** Oxford, 1925, p. 329, is incorrect for the years 1850 and 1856. For 1856 he misquoted 1,987 looms as 1,787 from Warden, **Linen Trade,** London, 1864, p. 387. For 1850 his number of power looms is 1,131, while mine is 1,083, as given by the **Factory Returns,** in P.P. 1857 (Sess. I), XIV.
54. Horner, **Linen Trade of Europe,** Belfast, 1920, p. 252.
55. **Committee on Linen Trade of Ireland,** 1825, **op. cit.,** p. 43.
56. Clapham, **Economic History of Modern Britain,** Cambridge, 1926, I, p. 162.
57. Asst. H. L. Weavers' Com'n., 1840, **op. cit.,** XXXIII, p. 352.
58. **Ibid.,** p. 488.
59. **Ibid.,** pp. 353, 355, 357.

60. **Ibid.**, pp. 350-51. See also p. 352.
61. **Ibid.**, pp. 352, 353, 356.
62. **Ibid.**, p. 483.
63. **Ibid.**, p. 483.
64. **Ibid.**, p. 356.

CHAPTER V. PAPER, POTTERY AND GLASS

1. That is, paper makers, not including paper stainers and paper hangers.
2. Day, **Distribution of Industrial Occupations in England, 1841-1861,** New Haven, 1927, pp. 100-103, 124.
3. Maddox, **Paper, Its History, Sources and Manufacture,** London, 1916, p. 6.
4. Spicer, **Paper Trade,** London, 1907, p. 173.
5. **Ibid.,** pp. 181-2.
6. "Although the Fourdrinier machine had been in common use for some time, and there were some considerable factories operated by steam power, the old hand mills persisted along the streams throughout England; there was not a county absolutely devoid of paper makers, and nowhere had the industry attained a very high degree of concentration." (Day, **op. cit.**, p. 124)
7. Spicer, **op. cit.,** pp. 258-260.
8. **Report,** in P.P., 1843, XV, r12 line 22. The mill referred to was located in Kent, and owned by Mr. Joynson.
9. Children's Employment Commission, 1843, **op. cit.**, XV, M66 art. 353.
10. **Bradshaw's Manchester Journal,** 1842, III, p. 35. See also p. 55.
11. Report, 1843, **op. cit.,** XV, 132 no. 33; 135 no. 51; M66 art. 353; r8 no. 22; r11 nos. 25,28,29.
12. Spicer, **op. cit.,** pp. 54-63.
13. Maddox, **op. cit.,** pp. 8-9.
14. Spicer, **op. cit.,** pp. 54-63.
15. **Ibid.,** pp. 65-66.
16. **Select Committee on Fourdrinier's Patent,** in P.P., 1837, XX, Q. 484.
17. Issue for September 30, 1833, II, p. 384.
18. Children's Employment Commission, 1843, **op. cit.,** XV, r8 no. 22.
19. Fourdrinier's Patent, 1837, **op. cit.,** Appx. A.
20. **Ibid.**
21. **Ibid.**
22. The 17, 822 persons enumerated in the Census returns for Stafford County in 1841 included: 15,158 under "potter and pottery manufacture", 349 under "pottery burnisher and enameller", 196 under "pottery engraver", and 2,119 under "pottery painter, printer, and gilder". These different classes of workers were often employed together in the same factory. See **Children's Employment Commission, Report,** in P.P., 1843, XIV, c1 no. 1.
23. Cunningham, Localization of Industry, in **Economic Journal,** 1902, XII, p. 503.
24. Shaw, **Staffordshire Potteries,** London, 1900 (First published in Hanley, 1829), p. 27.
25. **Ibid.,** pp. 3-4.
26. Kohl, **England and Wales,** Philadelphia, 1846, p. 12.
27. Children's Employment Commission, 1843, **op. cit.,** XIV, c12 no. 49.
28. Bladen, the Potteries in the Industrial Revolution, in **Economic Journal Supplement (Economic History no. 1),** January, 1926, p. 129.
29. Children's Employment Commission, 1843, **op. cit.,** XIV, C4 art. 14.
30. **Ibid.**, c22 no. 81.
31. **Ibid.**, c12 no. 49.

32. Quoted in Bladen, **op. cit.**, p. 125. Mr. Bladen, accepting this average of 47 persons per pottery factory, criticized Arthur Young as having given an underestimate of it in the following terms. "An account of Arthur Young in 1769 gives, I believe, an overestimate of the number of works and an understimate of their average size: I had the pleasure of viewing the Staffordshire Potteries at Burslem and the neighboring villages, which have of late been carried on with amazing success. There are 300 ***houses***, which are calculated to employ upon an average 20 hands.'" Bladen, **op cit.**, pp. 125-126 (Italics mine). The average, according to Young, was 20 hands per '*house*', but not necessarily per pottery factory, or '*works*'. Mr. Bladen, in my opinion, has overlooked the fact that 'house', in many cases, was a popular term for 'workroom' in the 'Staffordshire Potteries'. (See Children's Employment Commission, 1843, **op. cit.**, XIV, passim.) According to a little known Directory compiled by Parson and Bradshaw in 1818, there were 145 pottery factories in the "Staffordshire Potteries". (Wedgwood, **Staffordshire Potteries and Its History**, London, 1913, pp. 152-56). This total number of pottery factories in 1818, namely, 145, did not prove that the same for 1762, namely, 150, was correct; but, despite the spread of dates between 1762 and 1818, their closeness to each other was suggestive of our view that 300 houses did not mean 300 pottery factories.

33. Boyle, An Account of Strikes in the Potteries in the years 1834 and 1836, in **Royal Statistical Society Journal**, 1839, I, p. 43. Mr. Boyle divided these 130 pottery factories into two classes as follows:

Class of factories	No. factories	No. employees		Employees per factory
a. At strike	78	19,160	(95%)	246
b. Not at strike	52	940	(5%)	18
Total	130	20,100	(100%)	155

In class (b), many factories undoubtedly employed less than the average of 18 persons per factory, and were shops rather than factories. These shops, however, were insignificant in view of their relatively small proportion of persons employed, and may be dismissed as exceptions or survivals from the old craftsman system. The average of 155 persons per factory is preferred, however, for two reasons. First, this average, which applied to all factories in the district, could be used as a comparable unit for factory size in different years without violating the criterion of uniformity. Second, it seems that Mr. Boyle had somewhat distorted the statistics on the number of employees in favor of class (a) factories, namely, those at strike; as his chief purpose in writing the article was to show the unfavorable economic effects of strikes. A stronger impression could be made when it was shown that as many as 95 percent. of the total number of employees working at 78 of the 130 factories were thrown out of employement on account of the strike of 1836. This suspicion on my part of Mr. Boyle's distortion of employment statistics is partly confirmed by statistics from other sources. The Children's Employment Commission of 1841 reported, for instance, that out of the 127 factories in the Potteries in 1841, 79 ***principal*** factories employed a total of 12,407 persons, or an average of 157 persons per ***principal*** factory. (see **Report**, in P.P., 1843, XIII, pp. 16, 20; XIV, C2-3 art. 7.)

34. Children's Employment Commission, 1843, **op. cit.**, XIV, c12 no. 49; c14 no. 55; c18 no. 64; c20 no. 74; c24 no. 87; c26 no. 93; c29 nos. 108-109; c44 line 1; c45 no. 167; c54 no. 197; c73 no. 250; c75 no. 256; c77 no. 261.

35. Shaw, **op. cit.**, pp. 2, 35; Solon, **History of Old English Porcelain and Its Manufactories**, London and Derby, 1903, p. 25.
36. Wedgwood, **op. cit.**, p. 147. He quoted statistics here from Brongmart, **Traité des Arts Ceramiques**, II, p. 453.
37. Children's Employment Commission, 1843, **op. cit.**, XIV, c12 no. 49.
38. **Ibid.**
39. **Ibid.**, c15 line 44.
40. **Ibid.**, c15 line 38.
41. **Ibid.**, C1 art. 2.
42. **Ibid.**
43. Kohl, **op. cit.**, p. 14.
44. **Children's Employment Commission, Report**, in P.P., 1833, XX, p. 1088.
45. Children's Employment Commission, 1843, **op. cit.**, XIV, c53 line 60.
46. Children's Employment Commission, 1833, **op. cit.**, pp. 1089, 1090.
47. **An Account of the quantities of the several articles charged with duties of Excise in the four years ended 5th April, 1844**, in P.P., (H. L), 1844, XV, pp. 221-2.
48. The crippling effects of the excise duties upon the glass manufacture were felt by the glass manufacturers from all parts of the country as well as by the Commissioners on the Excise inquiry in 1835. See Commissioners of Excise Inquiry, 13th Report, **Glass**, in P. P., 1835, XXXI, passim.
49. **Ibid.**, p. 32.
50. **Ibid.**, p. 106.
51. **Ibid.**, pp. 112-3.
52. **Ibid.**
53. **Ibid.**, p. 68.
54. **Ibid.**, p. 107.
55. Excise Inquiry, 1835, **op. cit.**, p. 112.
56. **Ibid.**, p. 109.
57. **Ibid.**, pp. 74-75.
58. **Ibid.**, pp. 74-75. As different classes of glass products paid a different rate of excise duty, the amount of duty paid was not an ideal index for measuring the size of a glass manufacturer, but was the only one available. Generally, each manufacturer specialized in one branch, although it was not uncommon for a manufacturer to take up two branches at one time, such as flint and bottle glass. A brief summary of the Excise statistics for different manufacturers here quoted is given in Clapham, **Economic History of Modern Britain**, Cambridge, 1926, I, pp. 189-190.
59. Excise Inquiry, 1835, **op. cit.**, pp. 74-75.
60. **Children's Employment Commission, Report**, in P.P., 1843, XV, l14 nos. 40, 41; l16 no. 46.
61 Excise Inquiry, 1835, **op. cit.**, p. 28.
62. McCulloch, **Statistical Account of the British Empire**, 2d ed., London, 1839, I, p. 714.
63. **Report**, in P.P., 1843, XIII, p. 3 no. 2.
64. **Ibid.**, XV, l3 line 40.
65. **Ibid.**, l3 line 7.
66. Excise Inquiry, 1835, **op. cit.**, p. 66.
67. **Home Tour Through the Manufacturing Districts of England**, London, 1836, p. 84. Head did not mention the name of "this extensive establishment", but

we know from the Report of the Excise Inquiry Commissioners that there was only one plate glass factory in Lancashire in 1835, viz. that of Thomas Cockburn & Co.

68. Excise Inquiry, 1835, **op. cit.**, pp. 126-31.
69. Children's Employment Commission, 1843, **op. cit.**, XIV, fl50 no. 411.
70. Excise Inquiry, 1835, **op. cit.**, p. 120.
71. **Ibid.**
72. Children's Employment Commission, 1843, **op. cit.**, XIV, f290 no. 837.
73. Dodd, **Days at the Factories**, London, 1843, pp. 258-9.
74. Children's Employment Commission, 1843, **op. cit.**, XIV, fl47 lines 46, 53.
75. **Ibid.**, XV, l11 no. 25.
76. **Ibid.**, XIV, b54 no. 170.
77. **Ibid.**, XIV, c77 line 49.

CHAPTER VI. METAL: I

1. In the summary return for England no mention was given of "steel manufacture"; while in the county returns for the West Riding of Yorkshire and Salop, a total of 329 persons were enumerated under "steel manufacture". In Stafford, the most important county of iron manufacture, there was no separate enumeration of "iron manufacture" or steel manufacture", in the county return; instead, there was the composite enumeration of both manufactures under "iron and steel manufacture". In view of these facts, it may be concluded that "steel manufacture" was enumerated under "iron manufacture" in the summary return for England.
2. In 1841 it was not uncommon to find an "ironworks" founding such products as engine and machine, which, in Census definition as well as in popular usage, constituted a separate manufacture.
3. According to Professor Day, "internal evidence furnished by the Census figures shows that the term 'iron manufacture' was construed broadly, and included not only those engaged in the primary processes, but also any who worked on iron as a material". (In his **Distribution of Industrial Occupations in England, 1841-1861**, New Haven, 1927, p. 121).
   (a) The number of persons employed under this heading "iron manufacture" must be discounted, as it included the persons employed in the steel manufacture. See **Supra**, note 1. This discounting, however, did not affect materially the total number of persons employed in iron manufacture, for the number of persons employed in the steel manufacture was not large. West Riding, York, where the steel metropolis, Sheffield, was located, had only 312 persons employed in the "steel manufacture" according to the county returns of the Census of 1841. This small total was due to the fact that the steel manufacture was carried on in connection with the cutlery manufacture, and the persons employed in the steel manufacture might have been included under the cutlery manufacture.
   (b) "Moulders" and "founders" were enumerated independently, and the Census did not specify the manufacture to which they properly belonged. I include them under the "iron manufacture" without apportioning them between the "iron manufacture" and the "copper and brass manufacture", as the Census had a separate heading enumerating the "brass founders and moulders". "Smelters", whose branch of manufacture was also not specified, were omitted. A summation of "smelters" for the county returns gave a total of 170 persons, while the summary returns for England gave a total of 613 persons. The statistics in the county

returns, therefore, were incomplete, as they differed considerably from those in the summary return. Besides, for the leading iron counties such as Stafford, Monmouth, Lancaster, West Riding of York, etc. there was not a single smelter enumerated.

(c) The grouping of "iron products" under the iron manufacture is largely based on expediency in analysing the industrial organization. Many "ironworks" cast products other than those here listed, but it would not serve our purpose to include, for instance, such articles as nails, chains, etc. The latter products demand separate analysis in our study of industrial organization.

(d) "The water pipes of the period [1841] were made of lead, iron, baked clay or wood, and the occupation of pipe maker included probably workers in all those materials, but the Census gives reason to believe that the use of clay was unimportant, and that metal pipe was driving the old water pipe from the market." (Day, **op. cit.**, p. 131). A slight discounting in our total above is therefore necessary, so long as there were in 1841 pipes not made of iron.

4. Ashton, **Iron and Steel in the Industrial Revolution**, Manchester, 1924, ch. I.
5. **Ibid.**, p. 1.
6. Lloyd, **Cutlery Trades**, London, 1913, p. 400.
7. Ashton, **op. cit.**, p. 195.
8. For Table 137 see **Midland Mining Commission**, in P.P. (H.L.), 1843, XXXV, pp. 129-30. The statistics on output were originally given for the six months ending June 30, 1842, but for our purpose of quoting annual instead of semi-annual output, they were multiplied by two in the Table. For Table 138 see P.P. (H.L.), 1849, XXII, pp. 26-27.
9. The total annual output of pig iron for England in 1848 was 648,440 tons according to the Report of the Commissioner of Mining Population for 1849, **op. cit.**, pp. 26-27. This total, however, included the estimated output of 3 "ironworks" in Derbyshire that were "out of blast." This estimated output, which amounted to 29,120 tons, was deducted from the total annual output of 648,440 tons in our computation.
10. Clapham, **Economic History of Modern Britain**, Cambridge, 1926, I, p. 428.
11. Midland Mining Commission, 1843, **op. cit.**, pp. cxvii, 55,65. Another witness, Mr. Boydell, estimated that each blast furnace would employ about 300 persons, while Commissioner Tancred, compromising between the two estimates 157 and 300, concluded with his own estimate of 200. As neither Mr. Boydell's nor Commissioner Tancred's estimate gave detailed analysis, we prefer Mr. Best's. It is possible that the difference in technical conditions under which different blast furnaces were operated might account for the difference in estimating the number of employees per blast furnace. The detailed analysis for Mr. Best's estimate is quoted below.

Number of employees per blast furnace, 1843

Smelting:		Mining:	
Furnace men	30	Boat steers, carters &c. for raw materials	7
Moulders	4	Colliers, etc.	27
Refiners	6	Ironstone miners etc.	73
Overlooker	1	Limestone getters etc.	9
Total	41	Total	116

12. Commissioner of Mining Population, **Reports for 1846, 1847, 1849, 1850**, in PP. (H.L.): 1846, XXIV, pp. 28-29; 1847, XVI, pp. 20-21; 1849, XXII, pp. 6-8; 1850, XXIII, p. 56.

13. **Children's Employment Commission, Report**, in P.P., 1843, XV, Q87 art. 836.
14. **Commissioner of Mining Population, Report**, in P.P., (H.L.), 1844, XXVII, pp. 60-64; 1846, XXXVI, pp. 29-30; 1847, XVI, p. 26.
15. Ashton, **op. cit.**, p. 100.
16. The proportion between miners and other workers of an "ironworks" varied, although the one here adopted is conservative. See **Supra**, note 11. Mr. Glover, in his **History of Derby** published in 1829 (I, p. 232), said that the Alfreton Ironworks, which consisted of mines, blast furnaces and foundries, employed 500 persons, of whom "350 are connected with the ironstone works and colliers; and the remainder with the blast furnaces and foundries".
17. Ashton, **op. cit.**, pp. 40-43, 46-8. The Coalbrookedale Ironworks of the Darbys employed £100,000 in 1785; while the ironworks of the Walkers in South Yorkshire employed £213,393 in 1793.
18. Evidence against Orders in Council, 13, quoted in Ashton, **op. cit.**, p. 163.
19. In P.P., 1833, VI, Q9658.
20. **Ibid.**, Q. 9671.
21. Ashton, **op. cit.**, p. 164.
22. **Ibid.**, p. 40.
23. Meade, **The Coal and Iron Industries of the United Kingdom**, London, 1882, pp. 829-36; Scrivenor, **History of the Iron Trade**, London, 1854, New Edition (first published 1841), pp. 57, 87-8, 95-9, 132-36, 256. For the years 1796, 1806, 1823, 1830 and 1839, these authorities disagreed on the statistics given, although they employed the same sources. In our table Meade's is preferred as being the more recent compilation for all years except 1806.
24. Cf. Armstrong (ed.), **Industrial Resources of the Northern District**, London, 1864, 2nd ed. (1st ed., 1833), p. 292.
25. (a) The summary return of the Occupation Census of 1841 for England gave 20,349 persons under "engineer and engine-worker"; the county returns gave 20,347 persons under "engineer and engine-worker" and 828 persons under "civil engineer". "Civil engineer", therefore, was not included under "engineer and engine-worker"; while "military engineer" would more likely be included under "army" and "navy". In other words, engineer under "engineer and engine-worker" included mechanical engineer, who, in a restricted sense, was synonymous with engine and machine maker (see Pole, **Life of Sir William Fairbairn**, London, 1877, pp. 20, 23). Mrs. Knowles said that during the Industrial Revolution "engineers were required to make and repair the steam engines, to make machinery for the textiles, to make machinery for lifting coal out of the pits, to make machine tools and locomotive." (in her **Industrial and Commercial Revolutions in Great Britain during the 19th century**, London, 1921, p. 20). Engine worker under "engineer and engine-worker", however, probably meant engine tender. Professor Day's view that engineer under "engineer and engine-worker" "seems to have been what the Americans now understand by the term, an engine tender rather than metal worker" (in Day, **op. cit.**, p. 142) is in my opinion more applicable to engine-worker under "engineer and engine-worker" rather than to engineer. "Engineer and engine-worker", therefore, constituted a composite class for the two different occupations, namely, engine and machine maker and engine tender. As the 1841 Census gave no clue to an exact determination, we are forced to adopt an arbitrary procedure in assigning one half of the composite total to each occupation, or 10,175 persons. In our definition, engine and machine makers, not engine tenders, are included.
(b) "Engineer and machine maker" was a term commonly used for "machine maker" in the period of our investigation, 1840 or earlier; so also

was "iron founder and machine maker". At other times we meet with establishments such as "iron founder, millwright and engineer", "millwright machinery maker", etc. In other words, the three occupations as listed by the Census under "engine and machine maker", "engineer" (adapted from "engineer and engine-worker" according to the reasons given in the preceding note) and "millwright" were closely related, and in contemporary usage, represented the different branches of the same occupation, viz. engine and machine manufacture. See **Committee on Artizans and Machinery, Report,** in P.P., 1824, V, pp. 340, 347, 566; **Committee on Export of Machinery, First Report,** in P.P., 1841, VII, Q. 1266-7, 1299, 1533; Children's Employment Commission, 1843, op. cit., XV, M56-7, 02 art. 17; Pole, **op. cit.**, pp. 4, 20, 23-4, 26-7, 47; Clapham, op. cit., I, pp. 151, 152, 154-5, 446-9.

(c) Including machine tools. A machine maker, however, frequently made his own tools, and repaired his own machines. In the textile manufacture, it was not uncommon for a manufacturer, a cotton spinner for instance, to make his own machines as well as to repair them. See Artizans and Machinery, 1824, **op. cit.**, p. 303; Export of Machinery, First Report, 1841, **op. cit.**, Q. 232, 234, 1329, 1482; Children's Employment Commission, 1843, **op. cit.**, XV, M64 art. 328.

26. Day, **op. cit.**, p. 95. See also pp. 224-5, 231.
27. Rutland had no "engineers" but 5 "millwrights". On the number of "engineers" in each county, the Census figure is revised in accordance with the qualifications made in **Supra**, note 25a.
28. Day, **op.cit.**, p. 121.
29. Children's Employment Commission, 1843, **op. cit.**, XIV, B45 art. 315; b61, c121-4, e58. The card setting shops, however, were immature factories employing an average number of 15 to 20 persons each, chiefly children. The occupation was manual, generally supervised by some middleman executing orders for the merchant employer. One Joseph Knowles near Dewsbury told the Commission, "I employ children to come here to my house to set cards. All I do is to take care that they do their work, and I furnish them with light. I receive about one penny a week for this from each; and I receive the work from the manufactures, and take a small profit." **Ibid.**, e58 no. 78; also e58 no. 80.
30. Export of Machinery, First Report, 1841, **op. cit.**, Appendix No. 2, p. 230.
31. **Ibid.**, Q. 1309.
32. If we discount 17,382 employees by two fifths, we get 10,429 employees, which is 2,073 mployees in excess of the Census return of 8,356 employees for the county of Lancaster. Furthermore, Mr. Jenkinson, after having given the above estimates, remarked that "in addition...... there are a great number of small establishments, from which no sufficient returns have been obtained." (**Ibid.**, Appendix no. 2) The excess, in all probability, was due to the difference in classification on the part of the Census authority and Mr. Jenkinson. Some of the textile manufactures were also machine makers, and it is not unlikely that whereas the Census enumerated them under the textile manufacture, Mr. Jenkinson would enumerate them under the engine and machine manufacture.
33. Export of Machinery, Second Report, 1841, **op. cit.**, Appendix, p. 105. Mr. Fairbairn remarked that "this return is limited to Leeds; other towns of Yorkshire have considerable machine making establishments. It is estimated that Bradford, Bingley and Keigley employ in this way nearly 1,000 hands, and that the capital invested exceeds 60,000 pounds." Mr. Jenkinson, however,

estimated that the same localities, namely, Leeds, Bradford, Bingley and Keighley, employed 442 horse power, 5,000 hands, and £407,000 capital. See **Ibid.**, First Report, 1841, Q. 1301-2, 1305.

34. a. Children's Employment Commission, 1843, **op. cit.**, XV, 02 art. 17.
b, c, g. Butterworth, **Historical Sketches of Oldham**, 1847-8, pp. 186-7; Kohl, **England and Wales**, 1846, p. 59.
d, h. Children's Employement Commission, 1842, **op. cit.**, XVI, p. 686, no. 527; Armstrong, **op. cit.**, p. 308.
e. Export of Machinery, First Report, 1841, **op. cit.**, Q. 3064.
f. Nowrojee and Merwanjee, **Journal of a Residence..... in Great Britain**, London, 1841, pp. 358-60.
i. V. C. H., **Buckingham**, London, 1908, II, p. 127.
35. See section A of this Chapter.
36. See **Supra**, note 34.
37. Children's Employment Commission, 1843, **op. cit.**, XV, M45-51 arts. 252-69; M38 art. 242; M56-7 art. 296. The famous Soho Works, the full story of which was told by Smiles (**Lives of Boulton and Watt**, Philadelphia, 1865, pp. 169-77), made all kinds of small metal wares in addition to engines, such as toys and trinkets, buttons, buckles, clasps, watch chains etc. In 1834 it was said to have employed 600 persons (Lardner, Cabinet Cyclopaedia, series on **Manufactures in Metal**, London, 1834, III, p. 317). Butterworth, in 1846, said that the Soho Works at Birmingham "used to employ about 1,100 hands". See **supra**, note 34, b, c, g. To take another instance, Fawcett & Co. in Liverpool, from which "have proceeded many of the engines employed for propelling our largest and most celebrated steam vessels" (**Stranger in Liverpool**, 12th ed., Liverpool, 1838, p. 109), manufactured cannons as well as marine engines. "Some idea may be formed of the importance of the concern", said Kohl, "when we are told that the house undertakes the execution of orders such as 300 pieces of heavy ordnance for the King of Holland. Steam engines are made here of more than 500 horse power. Their largest was 520, the engines of the steamers on the Boston line of 440, and the largest employed in Manchester of 300 horse power." (In his "England and Wales", **op. cit.**, p. 24).
38. Children's Employment Commission, 1843, **op. cit.**, XIV, B41 art. 281-4.
39. Artisans and Machinery, 1824, **op. cit.**, p. 566.
40. Export of Machinery, 1841, First Report, **op. cit.**, Q. 3064-76.
41. **Chambers' Edinburgh Journal**, 1850, XII, p. 215.
42. **Ibid.**
43. See **Supra**, note 34 f.
44. Of a total of 2,913 employees in button manufacture in the county of Warwick, Birmingham had 2,888.
45. **Report**, in P.P., 1843, XIV, fl26-141.
46. **Ibid.**, fl29 no. 348.
47. **Ibid.**, fl29 no. 349.
48. **Ibid.**, fl35 no. 362. A visitor to one button factory in Birmingham in 1844 remarked that " in the large establishment I visited, nearly 500 persons were employed, the whole distributed through a number of brick buildings of 2 or 3 storeys in height, and surrounding interior courtyards—the rooms generally spacious and well ventilated. Each species of button had its own department, and as may be supposed, there was a minute division of labor." (**Chambers' Edinburgh Journal**, 1844, I, p. 370).
49. **Ibid.**, fl33 no. 356.
50. **Ibid.**, fl39 line 10, fl40 line 5.

51. **Ibid.,** fl41 no. 381.
52. In Sheffield there was one bone button factory belonging to Samuel Meggitt, which employed 65 persons. See **Ibid.,** e39 no. 6.
53. Special Assistant Poor Law Commission, **Reports on the Employment of Women and Children in Agriculture,** in P.P., 1843, XII, pp. 86-87.
54. **Ibid.,** p. 16.
55. **Ibid.,** p. 87.
56. **Ibid.**
57. Bevan, **Industrial Classes and Industrial Statistics,** in British Manufacturing Industries Series, 1876, XIII, p. 105.
58. **Royal Statistical Society Journal,** 1839, II, pp. 436-7.
59. Children's Employment Commission, 1843, **op cit.,** XIV, E8 art. 55., E11 art. 91; XV, Q6 art. 67.
60. Children's Employment Commission, Report, in P.P., 1833, XX, p. 1091.
61. Children's Employment Commission, 1843, **op. cit.,** XV, Q6 art. 67, Q60-61 art. 650, Q75 art. 741, q45 no. 208, q51 no. 244, q60 no. 285.
62. **Ibid.,** q51 no. 244.
63. **Ibid.,** Q75 art. 741, q60 no. 285.
64. **Ibid.,** XIV, fl53 no. 423.
65. **Ibid.,** fl54 line 26.
66. **Ibid.,** fl55 line 23, fl56 line 67.
67. **Ibid.,** F17 art. 158.
68. **Chambers' Edinburgh Journal,** 1844, I, p. 371.
69. Glover, **op. cit.,** I, p. 234.
70. Bore, **The Story of the Invention of Steel Pens,** New York, 1890, pp. 3, 24, 53.
71. Bume, **Josiah Mason: a Biography (1795-1881),** London, 1890, pp. 53-59; Bore, **op. cit.,** p. 4.
72. Edwards, **Personal Recollections of Birmingham and Birmingham Men,** Birmingham, 1877, p. 92. "By some fortunate accident, Gillot and Miss Mitchell met and after a brief courtship they entered into an engagement to marry. She spoke to her intended husband of the nature of her occupation and Gillot at once conceived the idea that the press, the useful implement then used principally in the button trade, might, if proper tools would be made to suit, produce pens in large numbers very rapidly. With his own hands, in a garret of his own house, he secretly worked until he had succeeded in making pens of a far better quality than had yet been seen. His process was one in which, unassisted, he could produce as many pens as twenty pairs of hands, working under the old system, could turn out. There was an enormous demand for his goods, and as he wanted help, and secrecy seemed needful, the young couple married, and Mrs. Gillot used to tell how, on the very morning of his marriage, he, before going to the church, made with his own hands a gross of pens, and sold them at one shilling each, realizing thereby a sum of seven pounds and four shillings." Also quoted in Dent, **The Making of Birmingham,** Birmingham, 1894, p. 454.
73. Bume, **op. cit.,** pp. 63-66.
74. **Ibid.,** p. 60. For further details on Perry's enterprize see Bore, **op. cit.,** pp. 27-8, 61; Dent, **op. cit.,** p. 453.
75. Edwards, **op. cit.,** p. 93.
76. **Penny Cyclopaedia,** London, 1835, IV, p. 445. McCulloch, **Statistical Account of the British Empire,** 2nd edition, London, 1839, 2 vols., I, p. 698.
77. Children's Employment Commission, 1843, **op. cit.,** XIV, fl65 no. 463.
78. **Ibid.,** fl65 no. 464.
79. **Ibid.,** fl65 line 17.

80. The 306 persons employed in the "fish hook" and "hook and eye" manufactures in 1841 are included under the needle manufacture, because of the fact that those two manufactures, in their localization, industrial organization, and technique resemble each other closely. The counties of Worcester and Warwick had 258 persons out of a total of 306 persons employed in these two manufactures, or 84 percent. The only hook and eye factory mentioned by the Children's Employment Commission of 1841, however, was that of Messrs. Wood and Sons, at Southwark Bridge Road, London. It employed 30 boys, and perhaps a large number of adults besides. (Children's Employment Commission, 1843, **op. cit.**, XIV, f279 no. 793).
81. Buffet, Making Needles 150 Years Ago, In **American Machinist**, XXX, pp. 740-2, 1907. Best summary of the manufacture was written by Morrall, **Short Description of Needle Making**, 1886.
82. Dodd, **British Manufactures**, London, 1844-6, volume on "Metals", p. 111.
83. **Penny Magazine**, 1843, XII, pp. 34-5.
84. Glover, op. cit., I, p. 234.
85. Children's Employment Commission, 1843, **op. cit.**, XIV, E18 art. 133. This factory of Henry Cocker and Sons at Hathersage, Derby, must not be confused with the factory of Samuel Cocker and Sons at Sheffield. In 1841, the latter was manufacturing "steel, files, wire, needles, etc.", with 134 employees and 63 steam and water horse power. See **infra**, p. 161.
86. Recently, writers such as Ashton and Clapham deplored this disproportionate prominence given to the pin manufacture as the illustration of an industry carried on under the principle of the division of labor. "In textbooks and examination scripts," writes Mr. Ashton, ("The Records of a Pin Manufactory, 1814-21", in **Economica**, 1925, V. p. 281), "the pin trade of a hundred or more years ago had been given a prominence which is far from justified by its true rank among economic activities. Babbage notwithstanding, 'the manufacture of pins does not afford the ideal illustration of the division of labor'; and one may echo Dr. Clapham's regret 'that Adam Smith did not go a few miles from Kirkaldy to the Carron Works to see them turning and boring their cannonades instead of to his silly pin factory' ". (See Babbage, **Economy of Manufacture**, Ch. XVIII; **Economic Journal**, XXIII, p. 401).
87. Smith, **Wealth of Nations**, 2, vols., I, chap. I, 6 (Cannan's edition, 1904). According to Cannan, "in Adam Smith's 'Lectures' p. 164, the business is, as here, divided into 18 operations. This number is doubtless taken from the 'Encyclopédie', tom. V, (published in 1755), s.v. Epingle. The article is ascribed to Mr. Delaire, 'qui décrivait la fabrication de l'épingle dans les ateliers même des ouvriers', p. 807. In some factories the division was carried out further. E. Chambers, 'Cyclopaedia', vol. ii, 2nd ed., 1738, and 4th ed., 1741, s.v. Pin, makes the number of separate operations 25."
88. The pin manufacture offers an interesting contrast to the other two manufactures, namely, the clock and watch manufacture and the gun manufacture. These two manufactures had a more minute division of labor, and required greater skill for each process than the pin manufacture. Because of this difference in division of labor and in skill required, merchant employer system rather than factory system was the prevalent form of industrial organization in these two manufactures. See **infra**, Chapter VII, sections on "Clock and Watch Manufacture" and "Gun Manufacture".
89. Lardner (ed.), **op. cit.**, II, p. 320. This total probably included outworkers too.
90. Taylor, **Scenes of Wealth**, London, 1826, pp. 36-37.
91. Children's Employment Commission, 1833, **op. cit.**, XX, pp. 925-6.

92. Children's Employment Commission, 1843, **op. cit.**, XV, 08 no. 6.
93. **Ibid.**, 010 no. 9.
94. **Ibid.**, XV, M2-3, art. 9; M5 art. 21; XIV, b49 lines 7, 39; b51 no. 164.
95. One reason for the excess lies in the difference in classification. The wire-drawers employed by these factories were not unlikely enumerated by the Census under the "wire-drawers", not "pin-makers". Some factories drew their wires, while others purchased them from the wire-drawers.
96. **Ibid.**, XV, m3 no. 6. See also M9, art. 50.
97. **Ibid.**, XV, m3 line 53.
98. Ibid.
99. **Ibid.**, XIV, b51 no. 164.
100. **Ibid.**, f119 line 49.
101. **Ibid.**, f122 line 37; f120 line 64.
102. **Ibid.**, f124 nos. 337, 338.
103. **Ibid.**, XV, M3-6 arts. 11-23.
104. **Ibid.**, M3-4 arts. 13-16; M5 art. 22; M9-10 arts. 54-55; 09 line 1.
105. In almost every pin factory, one half of the persons employed were headers each header employing one heading machine.

CHAPTER VII. METAL: II

1. The three centres of manufacture in these counties were Liverpool, employing 1,232 persons; Clerkenwell, employing 887; and Coventry, employing 831.
2. Lardner (ed.), Cabinet Cyclopaedia, Series on **Manufactures in Metal**, London, 1834, III, pp. 284-5.
3. Clapham, **Economic History of Modern Britain**, Cambridge, 1926, I p. 166.
4. Petty, **Growth of London**, 1682, reprinted in Hull (ed), **Economic Writings of Sir William Petty**, Cambridge, 1899, in 2 vols., II, p. 473. The manufacture of clocks, however, was not so subdivided as that of watches. (Dodd, Clerkenwell, in Knight, **London**, 1841-4, 6 vols., III, p. 144)
5. **Committee on the Petitions of the Watchmakers of Coventry etc.**, in P.P., 1817, VI, p. 17. In this and the following quotations from the 1817 Report or other early sources, it is understood that, in view of the persistence of industrial organization in the clock and watch manufacture as well as in other manufactures similarly organized, and of the limited contemporary (i.e. 1840) evidence now available to my knowledge, the early (i.e. 1817 and around) descriptions of industrial systems are used to picture the conditions in a later period, at least in 1840. Cf. Bevan, in **infra**, note 23.
6. Marx, **Capital**, Chicago, 1912-13, Kerr edition, I, p. 377.
7. Dodd, writing on Clerkenwell around 1840, said, "the 'watch-maker', or 'watch manufacturer', is a tradesman who understands the relative positions and the combined action of all the parts which have been thus made. They are generally persons possessing some considerable capital, as occupying the channel through which the purchaser deals with the actual makers. The watch manufacturers of Clerkenwell are the class to which we here more particularly allude; for many of the retail dealers in watches in other parts of London merely purchase the articles in a finished state to sell again at a profit." (Dodd, Clerkenwell, **op. cit.**, III, p. 143)
8. "If we wish to be introduced to the workman," wrote one contributor to the **Ree's Cyclopaedia** (Vol. IX), "who has had the greatest share in the construction of our best clocks, we must often submit to be conducted up some narrow passages of our Metropolis, and to mount into some dirty attic, when we find illiterate ingenuity closely employed in earning a mere pittance, compared with the price which is put on the finished machine by the vendor."

Also quoted in Dodd, Clerkenwell, **op. cit.**, III, p. 143. In watch making at Prescott, Lancashire, in the 18th century, "the system of manufacture in vogue was domestic. The manufacturer gave out his orders on the Monday morning and received the work from the job masters on the Saturday. Having assembled the parts, he dispatched them to customers in other towns." V.C.H., **Lancaster**, London, 1908, II, p. 366.

9. Dodd, **op. cit.**, III, p. 143; Id., **British Manufactures**, London, 1844-6, Series IV, p. 173.
10. "The principal London watch makers order the movements... of the movement makers of Prescott, who make them according to the callipers they receive from each maker with their orders. But the ordinary description of movements may be purchased at most of the watch tool shops in London." (**Ree's Cyclopaedia**, first American Edition, Philadelphia, 1810-42, XXXIX, under 'watchmaker'). These London watch makers then assembled the movements made in Prescott.
11. Watchmakers of Coventry, 1817, **op. cit.**, p. 84.
12. **Ibid.**, p. 84.
13. **Ibid.**, p. 80. For a list of names of these early factory manufacturers, see Poole, Coventry: **Its History and Antiquities**, London, 1870, pp. 364-5.
14. Watchmakers of Coventry, 1817, **op. cit.**, p. 43.
15. **Ibid.**, pp. 90-91.
16. **Ibid.**, p. 17.
17. **Ibid.**, p. 89.
18. **Ibid.**, p. 24.
19. **Ibid.**
20. Dodd, Clerkenwell, **op. cit.**, III, p. 143.
21. **Children's Employment Commission, Report**, in P.P., 1843, XV, r5: line 55.
22. **Ibid.**, r7: line 48.
23. Bevan, **Industrial Classes and Industrial Statistics**, 1876, p. 181.
24. The detailed enumeration is as follows:

Branch	England	Warwick	Stafford	Middlesex	Other
Gun & pistol	4,832	1,839	729	1,109	1,155
Gun barrel	307	285	—	—	22
Cock	162	156	—	—	6
Gun implement	39	39	—	—	—
Bayonet	28	13	10	—	5
Stock	17	—	—	—	17
Total	5,385	2,332	739	1,109	1,205

25. Lardner (ed.), **op. cit.**, II, p. 115.
26. Even in a period as late as 1866, the Birmingham gun workers were highly subdivided into none less than fifty classes. For detailed list see Goodman, Birmingham Gun Trade, in Timmins (ed.), **Industrial History of Birmingham and the Midland Hardware District**, London, 1866, pp. 392-3.
27. Dodd, British Manufactures, **op. cit.**, p. 193.
28. Artifex and Opifex, **Causes of Decay in a British Industry**, London, 1907, p. 69.
29. Goodman, **op. cit.**, p. 391; also Id., On the Progress of the Small Arms Manufacture, in **Royal Statistical Society Journal**, 1865, XXVIII, p. 497.
30. Goodman, in Royal Statistical Society Journal, **op. cit.**, p. 497.
31. Taylor, **A Brief History of the Westley Richards Firm, 1812-1913**, 1913, pp. 83-4.
32. **Report**, in P.P., 1843, XIV, fl51, line 55.
33. Lloyd, **Cutlery Trades**, London, 1913, pp. 175-8.
34. "The word cutlery is used in Sheffield in a sense narrower than its popular

and original one, being taken there to mean only the handling, making up, and finishing of knives, razors, scissors, etc., not the forging or grinding of the blades." **Children's Employment Commission, Fourth Report,** in P.P., 1865, XX, p. 2. See also **Children's Employment Commission, Report,** in P.P., 1843, XIV, E5 art. 25.

35. Poole, **Statistics of British Commerce,** London, 1852, p. 134.
36. Dodd, British Manufactures, **op. cit.,** p. 48.
37. Lloyd, **op. cit.,** p. 37.
38. **Ibid.,** pp. 57, 434.
39. (a). The total number of persons employed in the cutlery (restricted sense) manufacture for England was 16,167; that for England and Wales 16,214. The latter agrees closely with that derived by Lloyd from the same statistics as mine, namely, Occupation Census of 1841. Lloyd's total is 16,610, which is not supported by any detailed enumeration. See Lloyd, **op. cit.,** p. 434.

(b). The sub-branch "knife" was not given by the Census in the Summary Return for England, but in the County Return for the West Riding of Yorkshire. Evidently, "knife" was included under "cutler", the two terms being synonymous. Cf. **supra,** note 34.

(c). The sub-branch "agricultural implement" had a total of 1,111 persons for England, including 1,044 persons under "agricultural implement", and 67 persons under "shovel". The county return under "agricultural implement" was, however, 205 instead of 1,044. This discrepancy explained the fact that a detailed classification was applied to the counties in the County Returns which was not applied to England in the Summary Returns. Thus, for the West Riding of Yorkshire and Sheffield Parish, the 195 and 121 persons respectively included under the sub-branch "agricultural implement" were distributed as follows:

	West Riding	Sheffield Parish
Scythe	86	84
Spade	51	19
Shovel	34	15
Sickle	18	3
Plough	6	1
Total	195	121

(d). "Burnisher" and "chaser" were omitted, as they belonged more to the silver plate manufacture than to cutlery. Cf. Children's Employment Commission, 1843, **op. cit.,** XIV, E7 arts. 49-50.

(e). The sub-branch "saw" was not given by the Census in the Summary Return for England, but in separate County Returns. Lloyd gave a lower figure for England and Wales (819) than mine (1,019) for England alone, although he claimed to have derived it from the Census too. Besides, he seemed to suggest that "saw" was a separate branch from "tool", which was not the practice followed by the Census Authorities. Cf. Lloyd, **op. cit.,** pp. 438, 439. Lloyd's total of 4,739 persons employed in "tool" for England and Wales was incorrect; it should be 4,673 according to my enumeration of the Census statistics.

(f). The sub-branch "etc." under "tool" included different products in different counties. It is doubtful whether machine tools were included under "tools" or under "engine and machines". In my view, they were probably included under the latter, and have been treated in connection with the section on "Engine and Machine Manufacture" in Chapter VI.

40. Sheffield derives its name from the Sheaf, one of the five small streams by which the town is intersected.
41. Lloyd, **op. cit.**, pp. 100-105.
42. Development Committee of Sheffield City Council, **Sheffield: The Cutlery Capital of the British Empire**, 1920, pp. 5-6.
43. Children's Employment Commission, 1843, **op. cit.**, XIV, e36-39.
44. Allen, **History of York**, London, 1829-31, 6 vols., V, p. 54.
45. **Committee on Manufactures, Commerce and Shipping**, in P.P., 1833, VI, Q. 2906.
46. For an interpretation of the term "little master", cf. Lloyd, **op. cit.**, pp. 196-7.
47. Children's Employement Commission, 1843, **op. cit.**, XIV, E3, art. 12.
48. Lloyd, **op. cit.**, pp. 196-7. This recent description was confirmed by a contemporary account of 1841, the persistence of the older form of industrial organization being quite notable. "The economy of the country district round Sheffield does not materially differ from the town.... In these villages, orders for inferior cutlery are given by merchants termed 'factors', who reside at Sheffield and merely give orders, generally selling the steel to very small cutlers, who with the assistance of a boy or two, perform all the operations except grinding, and bring in the knives ready made on Saturday night." Children's Employment Commission, 1843, **op. cit.**, XIV, E3 art. 14. These "little masters" or "factors" had their origin in the 17th century. See Lloyd, **op. cit.**, pp. 327-8.
49. Children's Employment Commission, 1843, **op. cit.**, XIV, E3 art. 12; Lloyd, **op. cit.**, pp. 204-5.
50. Committee on Manufactures, Commerce, and Shipping, 1833, **op. cit.**, Q. 11604. "Previous to 1817 the cutlery trade was in general conducted in the manufacturing departments by persons of capital, who employed a considerable number of workmen, but the cutlery trade not requiring much capital, great numbers [of "little masters"] then commence manufacturing without any capital, who may be considered in fact the journeymen of the merchants."
51. Lloyd, **op. cit.**, p. 448.
52. Children's Employment Commission, 1843, **op. cit.**, XIV, E6 art. 36.
53. **Ibid.**, XIV, E2 art. 6; Lloyd, **op. cit.**, p. 179.
54. Lardner, **op. cit.**, I, p. 289.
55. The water-wheel.
56. Quoted by Lloyd, **op. cit.**, pp. 164-5.
57. **Ibid.**, pp. 215-6.
58. Lardner, **op. cit.**, pp. 12-3.
59. Ward, **Diary**, p. 217. Quoted by Lloyd, **op. cit.**, p. 216.
60. Faucher, **Manchester in 1844**, tr. by J. P. Culverwell, London, 1844, p. 132.
61. Lloyd, **op. cit.**, p. 215.
62. Callis, Cutlery Trade of Sheffield, in Cox (ed.), **British Industries under Free Trade**, London, 1903, p. 155.
63. Lloyd, **op. cit.**, p. 208.
64. In 1833 it was reported that there were in all a total of 50,000 to 60,000 nailers in the neighborhood of Staffordshire alone. (See Committee on Manufactures, Commerce and Shipping, 1833, **op. cit.**, p. 598) With the introduction of machine nails, the decline of hand nails was rapid. In 1866 Mr. Ball reported that there were not more than 20,000 hand nailers in the same district. See Ball, The Hand-made Nail Trade, in Timmins, **op. cit.**, p. 111.

65. Aitken, **Birmingham Trades**, in Bevan (ed.), **British Manufacturing Industries**, 15 vols., London, 1876-8, III, P. 40; Economical Statistics of Birmingham, in **Royal Statistical Society Journal**, 1839, II, pp. 436-7.
66. "In 1842 a great strike took place, several masters were made prisoners by the mob; and at Dudley the military had to be called out before its riots were suppressed." See **Report on Nail and Chain Makers in S. Staffordshire and W. Worcestershire**, by Labor Correspondent of the Board of Trade (Henry G. Calcraft), in P.P., 1888, XCI, p. 4.
67. **Committee on Payment of Wages**, in P. P., 1842, IX, Q. 1226-48.
68. Nail and Chain Makers, 1888, **op. cit.**, p. 4.
69. Aitken, **op. cit.**, p. 40.
70. **Chambers' Edinburgh Journal**, 1844, I, p. 371.
71. Children's Employment Commission, Reports, 1843, **op. cit.**, XIV, c103. See also **Ibid.**, XV, q9 no. 82; q14 nos. 60-67; Q27 art. 293.
72. Nail and Chain Makers, 1888, **op. cit.**, p. 6. The system here described in the year 1888 is, in the word of the author, "the system that has prevailed almost from the time immemorial, and it is undoubtedly old fashioned, and not to be compared with the factory system either for economy of production or efficiency of working. It is the system of a hundred years ago [1788]." For similar accounts see **Children's Employment Commission**, in P.P., 1842, XVI, p. 28; 1843, XV, q54 no. 262; 1843 XIV, b45 line 47. In the last reference the system was described by a nail factor in 1841 in the following terms. "The nail business in Dudley is carried on by the large manufacturers giving out 60 1b. bundles of iron to certain workmen, who make the nails at their own homes—men, women, boys, and girls, all work at it..... Eight days are legally allowed to the nailers to return the stock in nails, but the time in not rigidly enforced. ... The nail merchants of Dudley employ nailers both in the parish and all the parishes round about, particularly at the Lye waste near Stowbridge; has a warehouse also in the village of Sedgley."
73. Committee on Payment of Wages, 1842, **op. cit.**, Q. 1170.
74. Nail and Chain Makers, 1888, **op. cit.**, pp. 8-9. For similar accounts see Committee on Payment of Wages, 1842, **op. cit.**, Q. 1111; Children's Employment Commission, 1843, **op. cit.**, XV, Q75 art. 743, Q77 art. 754, Q88 art. 837.
75. Children's Employment Commission, 1843, **op. cit.**, XIV, b46 line 1.
76. Lardner (ed.), **op. cit.**, I, p. 193.
77. Children's Employment Commission, 1843, **op. cit.**, XV, q54 no. 262.
78. **Midland Mining Commission, Report**, in P.P. (H.L.) 1843, XXXV, p.v.
79. Committee on Payment of Wages, 1842, **op. cit.**, Q. 1112.
80. Midland Mining Commission, 1843, **op. cit.**, pp. vi, lxxxvii; **Committee on Payment of Wages**, in P.P., 1854, XVI, Q. 6-16.
81. Burritt, **Walks in the Black Country**, London, 1868, pp. 233-4. This letter was written around 1868, but also applied to conditions around 1840. For reason see **supra**, note 72.
82. Ball, **op. cit.**, p. 114.
83. Children's Employment Commission, 1843, **op. cit.**, XIV, b46 no. 146.
84. The summary return for England in the Occupation Census of 1841 gave a total of 1,711 employees in the anchor and chain manufacture. The summation of employees given in the county returns, however, resulted in a different total, viz. 1709. Of these 1,709 employees, 27 gold chain and 10 steel chain makers are not included in our analysis, as we deal only with the iron anchor and chain manufacture. Our total, therefore, is 1,672 persons.
85. Nail and Chain Makers, 1888, **op. cit.**, XCI, p. 3. See also Jones, South Staffordshire Manufactures, in Timmins, **op. cit.**, pp. 99-100; Children's Em-

ployment Commission, 1843, **op. cit.**, q55 no. 265; q83 no. 419; excellent analysis by Tawney in Ch. I of his **Establishment of Minimum Rates in the Chain Making Industry**, London, 1914. "In Cradley Heath," wrote Tawney on page 5, "the twentieth century lives side by side with the eighteenth, and the greater part of the chain manufacture is carried on as it has been from time immemorial, not in factories, but in the small workshops of the shop owners or in a room attached to the worker's cottage, the local name for which—'the brewhouse'—enshrines the domestic tradition of the industry."

86. Dodd, **British Manufactures**, London, 1844-46, Series VI, p. 222. At other times, anchor-making and shipbuilding (ship-repairing?) were carried on in one factory. For instance, Messrs. Hughes and Gerrard, "anchor smiths and ship smiths" in Liverpool, employed 32 adults and 11 children. (Children's Employment Commission, 1843, **op. cit.** XV, M48 no. 9).
87. Lardner, **op. cit.**, I, p. 184.
88. Children's Employment Commission, 1842, **op. cit.**, XVI, p. 706, no. 616.
89. The **Stranger in Liverpool**, 12th ed., Liverpool, 1838, p. 109.
90. Tildesley, Locks and Lock Making, in Timmins, **op. cit.**, pp. 85-6, 89.
91. With the exception of a few factories, the master craftsman system was prevalent. Jonah Hartill, for instance, employed 200 hands in his factory at Willenhall. See Children's Employment Commission, 1843, **op. cit.**, XV, q39 no. 175.
92. Day, **Distribution of Industrial Occupations in England, 1841-1861,** New Haven, 1927, p. 122.
93. Tildesley, **op. cit.**, pp. 85-6.
94. Children's Employment Commission, 1843, **op. cit.**, XV, q29 no. 132, Q38 art. 367.
95. **Ibid.**, Q2 art. 13.
96. **Ibid.**
97. **Ibid.**, Q44 art. 461.
98. **Ibid.**, Q44 art. 462.
99. **Ibid.**, Q42 arts. 417-8.
100. **Ibid.**, Q24 arts. 272-5.
101. **Ibid.**, Q54 art. 573.
102. **Ibid.**, Q27 arts. 299-301.
103. **Ibid.**, q9 no. 35, q13 no. 53, q27 no. 125.
104. **Ibid.**, q30 no. 135.
105. **Ibid.**, q29 no. 129.
106. **Ibid.**, Q54 art. 575.
107. **Ibid.**, Q30-31 arts. 329-32.
108. **Ibid.**, Q31-3 arts. 331-345, Q57 arts. 608-9.
109. **Ibid.**, Q33 arts. 346-8.

CHAPTER VIII. CLOTHING: I.

1. As contrasted with the Census return of 6,823 persons employed in the ribbon manufacture, the **Assistant Handloom Weavers' Commission** gave one of 17,000 to 18,000 persons. The latter return evidently included many part-time weavers, winders, fillers, and warpers. See P.P., 1840, XXIV, p. 13.
2. Factory ribbon weavers in the plain branch were reported by the Assistant Handloom Weavers' Commission, in Chester, Derby and Stafford; but not enumerated by the Census. It is possible that the Census omission was primarily due to a difference in classification. "Ribbon" weavers might have been included under "silk" weavers or "factory workers". See Assistant Handloom Weavers' Commission, 1840, **op. cit.**, XXIV, pp. 350, 337-9.

3. **Ibid., pp.** 12-13, 47.
4. The total number of looms employed was 1,862, including 598 factory looms. See **Ibid.**
5. **Committee on Petitions of Ribbon Weavers and Silk Manufacturers,** in P. P., 1818, IX, p. 97.
6. Warner, **Silk Industry of the United Kingdom,** London, 1921, p. 116.
7. Assistant Handloom Weavers' Commission, 1840, **op. cit.,** XXIV, p. 73.
8. **Ibid.,** p. 48.
9. **Ibid.,** p. 52.
10. **Committee on Handloom Weavers, Report,** in P.P., 1835, XIII, Q. 3391. Engine loom, it is plain, is a hand loom, not a ***power*** loom, as Leppington erroneously claims in her article on "The Evolution of an Industrial Town", in **Economic Journal,** 1907, XVII, p. 351.
11. Assistant Handloom Weavers' Commission, 1840, **op. cit.,** XXIV, pp. 35, 47.
12. **Ibid.,** p. 12.
13. Timmins, Coventry Ribbon and Watch Trades, in Timmins (ed.), **Industrial History of Birmingham and the Midland Hardware District,** London, 1866, pp. 181-2.
14. The new and sudden demand created by the War of 1821 was temporary, and at the conclusion of peace, its cessation threw many weavers out of employment and over-stocked the labor market with the disbanded soldiers. Large merchant employers failed, and large silk warehousemen took their place in distributing manufactured goods. "The latter naturally cared only to buy in the cheapest market, and thus the channels of trade were opened to the smaller capitalists, who had previously traded at a greater disadvantage as compared with their wealthier rivals". Timmins, **op. cit.,** pp. 183-4.
15. Assistant Handloom Weavers' Commission, 1840, **op. cit.,** XXIV, p. 36.
16. **Ibid.,** p. 37.
17. **Ibid.,** pp. 39-40.
18. **Ibid.,** p. 65.
19. **Ibid.,** p. 35.
20. **Ibid.,** p. 286.
21. **Ibid.,** p. 77. See also p. 271.
22. **Ibid.,** p. 187. This Association, however, was not the oldest, as Mr. Ambrose Moore told the Committee on Petition of Ribbon Weavers and Silk Manufacturers about "an association for preventing and punishing embezzlement and other offenses in the silk manufacture" in London, in 1818. See Committee on Petition of Ribbon Weavers and Silk Manufacturers, 1818, **op. cit.,** pp. 166-7.
23. For the apportionment of "weavers", "spinners" and "factory workers" (branch not specified) to the hosiery manufacture, see **supra,** chap. II, note 4. Felkin estimated that in 1833 there were in the hosiery manufacture 33,000 knitters and 33,500 seamers, winders and finishers (including embroiderers, menders, bleachers, dyers, dressers, etc.). See **Committee on Exportation of Machinery, First Report,** in P.P., 1841, VII, p. 237. Evidently the Census excluded seamers, winders and finishers who were partly employed.
24. **Frame-work Knitters' Commission, Report,** in P.P., (H.L.), 1845, XXIII, p. 465.
25. **Ibid.,** Appx., Pt. II (Nottingham & Derby), in P.P. (H.L.) 1845, XXV, Q. 490. See also Q. 4831.
26. "The machines [i.e. frames] for the manufacture of hosiery, are spread chiefly over upwards of 240 parishes in the three Midland counties of Nottingham, Derby and Leicester. ... This district is about 70 miles in length, and 45 in

its geatest breadth; within its limits there are 60 parishes in the county of Nottingham where are more than 6 stocking-frames at work." **Ibid.**, Q. 12 (p. 4), evidence by Wm. Felkin.

27. **Ibid.**, Appx., Pt. I (Leicester), in **P.P.** (H.L.), 1845, XXIV, Q. 965.
28. **Ibid.**, Appx., Pt. II, Q. 4831.
29. Cruft, A Nottingham Village Industry, in **Economic Review**, 1902, XII, p. 478.
30. Frame-work Knitters' Commission, Report, **op. cit.**, pp. 25-6.
31. **Ibid.**, Appx., Pt. I, Q. 3683.
32. **Ibid.**, Report, p. 503.
33. "Employed" is not equivalent to "owned". Many frames were rented from the "independent owners".
34. Frame-work Knitters' Commission, **op. cit.**, Appx., Pt. I, Q. 825, 922, 2776, 2900, 4283, 4332, 5235, 5302; Pt. II, Q. 397, 1072, 1121, 4490, 4905.
35. **Ibid.**, Appx., Pt. II, Q. 143.
36. **Ibid.**, Appx., Pt. I, Q. 29, 1822, 1915, 3664, 7351.
37. **Ibid.**, Q. 5245.
38. **Ibid.**, Appx., Pt. II, Q. 5140.
39. **Ibid.**, Q. 625.
40. **Ibid.**, Q. 626.
41. **Ibid.**, Q. 4897. See also Q. 4291.
42. **Ibid.**, Q. 861.
43. **Ibid.**, Report, p. 537.
44. **Ibid.**, pp. 549-50.
45. **Ibid.**, p. 550.
46. **Ibid.**, p. 548.
47. **Ibid.**, Appx., Pt. II, Q. 255.
48. **Ibid.**, Q. 708.
49. **Ibid.**, Q. 1092.
50. **Ibid.**, Appx., Pt. I, Q. 127.
51. **Children's Employment Commission**, in **P.P.**, 1843, XIV, F13 art. 107.
52. Frame-work Knitters' Commission, **op. cit.**, Appx., Pt. II, Q. 2604-6.
53. **Ibid.**, Appx., Pt. I, Q. 812-25, 3145, 3397, 3417.
54. "Taking in" was payment for services rendered by the middleman in fetching the raw materials from the warehouse to the shop and the finished products from the shop to the warehouse, and in assuming responsibility and supervision for the work etc.
55. Frame-work Knitters' Commission, **op. cit.**, Appx., Pts. I & II, Index, Item "shop charges".
56. **Ibid.**, Appx., Pt. II, Q. 610.
57. **Ibid.**, Report, p. 584.
58. **Ibid.**, Appx., Pt. II, Q. 1874.
59. **Ibid.**, Report, p. 496.
60. **Ibid.**, p. 516.
61. **Ibid.**, p. 447.
62. **Ibid.**, Appx., Pt. II, Q. 463.
63. **Ibid.**, Q. 467-8, 510, 972.
64. **Ibid.**, Q. 4708.
65. **Ibid.**, Appx., Pt. I, Q. 3195-6, 3205-7.
66. "The Census included 'hatter and hat-maker under one head, and it seems fair to assume that two classes with somewhat different economic functions were here merged in one. The hatters concentrated in Chester, Lancaster, Gloucester, etc., must have been largely manufacturers producing for an outside market; we know, in fact, that this was the case. How then shall

we explain the persistence of hatters in practically all the counties of England? The industry was not as yet marked by an extensive application of machinery, but it did have a somewhat elaborate technique, and considerable specialization in processes. Scattered handicraftsmen could not in great numbers have held their local markets in competition with the product of the organized manufacture. The difficulty disappears if we assume that many or most of the hatters in districts where they formed a proportion distinctly below the national average were not hat makers but were retailing shop-keepers." (Day, **Distribution of Industrial Occupations in England, 1841-1861**, New Haven, 1927, p. 104).

67. Children's Employment Commission, 1843, **op. cit.**, XIV, b42 line 3. William Waite, an eleven year old child employed by Wilson's hat factory in London, told the Commission that "his business is to take the hats from the manufacturer to the women, who are employed out of doors, to fetch the men's meals, go of (sic) errands, etc. Is running about pretty well the whole day, gets tired occasionally." **Ibid**, f275 line 14.
68. Bamford, **Walks in S. Lancashire and on its Borders**, Manchester, 1844, pp. 63-64.
69. Children's Employment Commission, 1843, **op. cit.**, XIV, b41 line 65.
70. Butterworth, **Historical Sketches of Oldham**, 1847-8, pp. 154-164.
71. **Panorama of Manchester**, Manchester, 1834, p. 165.
72. Dodd, **Days at the Factories**, London, 1843, p. 138.
73. **Ibid.**, pp. 157-8.
74. **Ibid.**, p. 139.
75. Children's Employment Commission, 1843, **op. cit.**, XIV, f273 line 16.
76. **Children's Employment Commission, 1833**, in P.P., 1833, XX, p. 1088.
77. Children's Employment Commission, 1843, **op. cit.**, XV, 01 art. 5.
78. For an excellent study on the local distribution of the glove manufacture, see Hull, **A History of the Glove Trade**, London, 1834, passim. According to Professor Day, "the city of Worcester was celebrated as the centre of a district in which glove making was a common trade, and the two counties, Somerset and Worcester, comprised a majority of the total number of workers. The census figures include dealers as well as makers, but the glovers who were merely shopkeepers were not a large enough fraction of the whole to disturb seriously the significance of the figures, **op. cit.**, p. 123.
79. Children's Employment Commission, 1843, **op. cit.**, XIV, d36-37 no. 160.
80. **Ibid.**, d37 no. 160. The above quotation is from the General Report of the Commission, in P.P., 1843, XIII, p. 129 art. 689. Another witness gave the total of sewers employed at "at least five or six thousand". **Ibid.**, p. 128 art. 688.
81. Gloves and Glovers, in **Penny Magazine**, 1843, XII, p. 102. See for similar account, Dodd, **British Manufactures**, London, 1844-46, Series V, pp. 219-220.
82. Children's Employment Commission, 1843, **op. cit.**, XIV, d39 no. 166.
83. Hull, **op. cit.**, p. 111. See also V. C. H., **Somerset**, London, 1906-11, II p. 427.
84. Children's Employment Commission, 1843, **op. cit.**, XIII, p. 129 art. 689.
85. **Ibid.**, XIV, d39 line 8.
86. The omission of 513 "slop makers and sellers" is based on the fact that the majority of them in 1841 were "sellers", while the minority were "sellers and makers", according to a separate enumeration of the county returns. There were no "makers". Furthermore, the Census enumeration is partial and incomplete, as contemporary evidence tends to show that the occupation listed under "slop makers and sellers" in 1841 applied to all makers and sellers of cheap and nasty clothes—all kinds of wearing apparel for men, women and children, especially shirts.

87. Tailors, dressmakers and seamstresses are abbreviations for the fuller but more clumsy titles of occupations listed in the Census as "tailors and breechmakers", "dressmakers and milliners", and "seamstresses and seamsters". Milliners in 1840 were proprietors of dressmaking shops—chiefly shopkeepers rather than artisans. They were, however, not numerous, constituting perhaps not more than ten percent of the whole number of persons employed as "dressmakers and milliners". "Seamstresses and seamsters"—shirtmakers—were mostly women and girls, while men constituted not more than one percent of the total.
88. Day, **op cit.**, pp. 147, 148, 149, 152, 153.
89. Outside of the Metropolis and other populous manufacturing and trading centres, it may be reasonably assumed that, because of the widespread distribution of the tailors, dressmakers and seamstresses, the wearing apparel manufacture was organized under the master craftsman as well as wage work systems. On the one hand, these artisans were shopkeeping craftsmen exercising their craft with the assistance of a few journeymen and apprentices, at their own shops, either receiving orders for making new clothes or repairing old ones from a locally limited circle of customers. On the other hand, they were wage workers working up consumers' raw materials into finished products at the consumers' homes. (Cf. Clapham, **Economic History of Modern Britain**, Cambridge, 1926, I p. 162)
90. Quoted in Galton, **The Tailoring Trade**, London, 1923, p. 90.
91. **Ibid.**, p. 103.
92. **Ibid.**, p. xciv.
93. Quoted in Galton, **op. cit.**, pp. 192-3. See also p. 184.
94. **Report**, in P. P., 1840, XXIII, p. 280.
95. Rendle (tailor, Horselydown), "Edward Smith, a narrative of the Sweating System", in **Workingman's Friend**, September 14, 1850, pp. 285, 287.
96. **Ibid.** This system of sweating by contractors, according to Kingsley, was first engineered by the Government. "Government—government?" Crossthwaite asked, "you a tailor, and not know that they first set the example, by getting the army and navy clothes made by contractors, and taking the lowest tenders? Not to know that the police clothes, the postmen's clothes, the convicts' clothes, are all contracted for on the same infernal plan, by sweaters, and sweaters' sweaters, and sweaters' sweaters' sweaters, till government work is just the very last, lowest resource to which a poor starved out wretch betakes himself to keep body and soul together? Why, the government prices, in almost every department, are half, and less than half, the very lowest living price. I tell you, the careless iniquity of government about these things will come out some day. It will be known, the whole abomination, and future generations will class it with the tyrannies of the Roman emperors and the Norman barons. Why, it's a fact, that the colonels of the regiments—noblemen, most of them—make their own vile profit out of us tailors—out of the pauperism of the men, of the slavery of the children, the prostitution of the women. They get so much a uniform allowed them by government to clothe the men with; and then—then, they let out the jobs to the contractors at less than half what government gives them, and pocket the difference. And then you talk of appealing to government." Kingsley, **Alton Locke**, 1849, ch. X. Alton Locke, although published in 1849, was composed of a series of sketches from his personal experiences in the years 1838 to 1842. (See Preface to the undergraduates of Cambridge.) In the above quotation from ch. X, Macmillan edition of 1905 is used.

97. These 'honorable' shops at the West End of London were of varying sizes; some were immature factories, some shops. In 1841, Messrs. Stultz, Houseley, and Wain employed, for instance, "a considerable number of journeymen working on the premises; in the busy season from 250 to 300; at other times about 150." (Children's Employment Commission, 1843, **op. cit.**, XIV, f237 line 37) Such a large factory, however, must have been uncommon , as it was the only instance given by the Commission for the London District.
98. Parson Lot (pseud. Charles Kingsley), **Cheap Clothes and Nasty**, London, 1850, pp. 4-5.
99. Rendle, **op. cit.**, p. 287.
100. Alton Locke, **op. cit.**, chap. X.
101. **Ibid.**, Preface to 1854 edition, addressed to the Workingmen of Great Britain.
102. Children's Employment Commission, 1843, **op. cit.**, XIV, f207 line 2.
103. **Ibid.**, f206 line 2.
104. **Ibid.**, f268-69.
105. **Ibid.**, f265 line 65.
106. **Ibid.**, f270 line 49.
107. **Ibid.**, f271 line 7.
108. **Ibid.**, f270 line 17.
109. **Ibid.**, f266 line 23.
110. **Ibid.**, f266 line 43.
111. Cf. Clapham, **op. cit.**, p. 570.
112. Children's Employment Commission, 1843, **op. cit.**, XIV, f204 line 5.
113. **Ibid.**, f225 line 60; f207 line 37; f215 line 62.
114. **Ibid.**, f211 line 13; f220 line 51.
115. **Ibid.**, F30 art. 288. For a summary of the evidences on the deplorable working conditions of the dressmakers as given before the Children's Employment Commission, See Grindrod, **The Slaves of the Needle**, London, 1844 (?), passim.

CHAPTER IX. CLOTHING: II.

1. According to the noted authority and historian of the lace manufacture, William Felkin, the embroidering and finishing processes in machine lace branch employed 180,000 persons in 1831 throughout the United Kingdom. (William Felkin, in **Committee on Exportation of Machinery, First Report**, in P.P., 1841, VII, p. 240). These persons carried on the occupation as a by-employment, practically in every town or village in the machine lace districts of the Midlands. The hand lace branch, according to a Petition to Queen Adelaide in 1830, had a total of 120,000 persons dependent upon it (Harding, **History of Tiverton**, Tiverton, 1845, p. 207).
2. The classification of the lace manufacture according to the extent of persons employed in the machine and hand lace branches is not strictly accurate, but it can be accepted as a working basis. The concentration of the machine lace branch around one region of the Midlands, and that of the hand lace branch around another, are too well known to demand further explanation. Devon, however, had a large number of persons employed in the hand lace branch in addition to those employed in the machine lace branch. For the sake of expediency they are classified together under the machine lace branch in the above table. For the apportionment of 6,144 persons from "weavers", "spinners", and "factory workers", whose branches of manufacture were not specified, see **supra**, chap. II. note 4.
3. Felkin, **History of the Machine-wrought Hosiery and Lace Manufacture**, London, 1867, p. 133.

4. Channer and Roberts, **Lace-making in the Midlands, Past and Present,** London, 1900, pp. 40-41; Special Assistant Poor Law Commissioners, **Reports on the Employment of Women and Children in Agriculture,** in P.P., 1843, XII, p. 16.
5. Jackson, A **History of Hand-made Lace,** London, 1900, p. 1.
6. Cowper, **Poetical Works,** Edinburgh, 1854, I, pp. 53-4 (Poem titled "Truth").
7. Palliser, **History of Lace,** 4th ed., 1902 (first published 1875), pp. 389-90. See also **Children's Employment Commission, Report,** in P.P., 1843, XIV, d27-30.
8. Palliser, **op. cit.,** p. 377.
9. V. C. H. **Northampton,** London, 1906, II, p. 338.
10. **Ibid.**
11. **Committee on Arts and Manufactures,** in P.P., 1836, IX, Pt. II,. p. 17, Q. 158.
12. Children's Employment Commission, 1843, **op. cit.,** XIV, d33 line 61.
13. **Ibid.,** d28, nos. 103, 111, d29 no. 115, d33 no. 140.
14. **Ibid.,** d31 line 19.
15. Report, 1836, **op. cit.,** Q. 167.
16. Children's Employment Commission, 1843, **op. cit.,** XIV, f9 line 66. See also **Ibid.,** f14 line 21.
17. The family members were chiefly employed in the preparatory processes such as winding, warping, threading, etc., even in a date as late as 1861. See **Report on the Lace Manufacture,** by H. S. Tremenheere, in P. P., 1861, XXII, pp. 466, 472, 476-7.
18. Children's Employment Commission, 1843, **op. cit.,** F3 art. 16; F8 art. 65.
19. **Ibid.,** f36 line 25, f41 line 40.
20. These two system, as described by a Nottingham lace-runner (that is, lace embroiderer) before the Children's Employment Commission in 1841, worked as follows: The witness, "at the time of the turnout of the lace runners, saw many of the mistresses; thinks there are in Nottingham from 80 to 100 mistresses who take lace to embroider; these mistresses employ young women at their own houses, and on an average about 5 to 10 each; they also employ a number of women at Nottingham, Leicester, Derby, Sutton and the neighboring villages; the women to whom the work is thus given, employ themselves children and young persons, and also again deliver out a part of the lace to other parties, so that some times the work goes through three, and occasionally four hands; is of opinion that each of the first hand mistresses to whom the lace is delivered from the warehouse, employs directly or indirectly 300 hands. Of the total number of lace runners in this neighborhood, thinks that something less than half are under 18 years of age." (Report, 1843, **op. cit.,** XIV, f42 line 60 to f43 line 4). The shops or houses of these mistresses, or agents,—the larger ones at least—might be called immature factories, where one could find some concentration of labor under the employer's supervision. The number of children employed per house varied from two or four to ten or twenty according to the Children's Employment Commission. Mr. Felkin told the latter that the number of Children employed per house averaged about ten. **(Report,** XIV, f48 line 20, F13 art. 109).
21. **Report,** in P. P., 1833, XX, p. 547.
22. Grindrod, **Slaves of the Needle,** London, 1844, pp. 12-3.
23. Children's Employment Commission, 1843, **op. cit.,** f43 lines 29-43. See also f40 lines 41-50.
24. **Ibid.,** f44 lines 12-19.
25. **Committee on Sweating System, First to Fifth Reports,** in P.P., 1888, XX, XXI; 1889, XIII, XIV; 1890, XVII; passim.

26. Hopkinson, Home Industries in Nottingham, in **Economic Review,** 1903, XIII, p. 337.
27. Children's Employment Commission, 1843, **op. cit.,** XIV, f40 lines 53-55.
28. **Factory Inspector's Report for Saunders' District ending January, 1840,** in P. P., 1840, XXIII, p. 21. The factories that were included by the factory inspectors were not all factories in the strict sense, for some of them were "tenement factories".
29. The factory act of 1833, 3 and 4 will., IV, c. 103, applied to "any cotton, woolen, worsted, hemp, flax, tow, linen, lace or silk mill or factory wherein steam or water or any other power is or shall be used to propel or work the machinery in such mill or factory, either in carding, roving, spinning, piecing, twisting, winding, throwing, resting, doubling, netting, making thread, dressing, weaving or finishing of cotton, wool, worsted, hemp, flax, tow or silk, either separately or mixed, in any such mill or factory, situated in any part of the United Kingdom of Great Britain and Ireland." (In P.P., 1833, II) Lace factories (i.e. those employing power driven machinery), therefore, came under the legislative interference, and in all probability, were included in the Factory Returns of 1838. In the Returns, however, lace factories were not mentioned, and might have been merged together with the cotton, woolen, worsted, silk or flax and hemp factories. This view is confirmed by an analysis of the Factory Returns of 1838 and the Census Returns of 1841 for the worsted (restricted sense) branch of the worsted (broad sense) manufacture. In this analysis it is found that in several of the counties where lace factories were located, Leicester and Nottingham for instance, the factory employees exceeded all employees as returned by the Census by a fairly large margin. In this study, no attempt is made to sift from the Factory Returns of 1838 any approximate data on the lace factories, as such attempt will be futile as well as confusing,—futile because we have not sufficient facts at our disposal.
30. Committee on Exportation of Machinery, First Report, 1841, **op. cit.,** p. 251.
31. Children's Employment Commission, 1833, **op. cit.,** p. 968.
32. **Ibid.,** pp. 981, 982.
33. Glover, **History of Derby,** Derby, 1829, 2 vols., I, pp. 246-7. "The lace manufactures of Derbyshire are in number 40, and they employ 800 persons, besides giving employment to between 3 and 4 thousand females, who figure or run the net when it is taken from the loom."
34. Children's Employment Commission, 1833, **op. cit.,** p. 368.
35. Children's Employment Commission, 1843, **op. cit.,** XIV, f2 line 43.
36. Committee on Exportation of Machinery, First Report, 1841, **op. cit.,** Q. 2676.
37. Felkin, History, **op. cit.,** pp. 340-1.
38. Children's Employment Commission, 1833, **op. cit.,** p. 369.
39. **Committee on Regulation of Mills and Factories,** in P.P., 1840, X, p. 111, no. 9994.
40. Felkin, in Exportation of Machinery, First Report, 1841, **op. cit.,** pp. 242, 251.
41. Children's Employment Commission, 1833, **op. cit.,** p. 368.
42. Glover, **op. cit.,** II, p. 426.
43. Committee on Regulation of Mills and Factories, 1840, **op. cit.,** X, Q. 9977.
44. Humphrey, **Great Britain, France and Belgium—a Short Tour in 1835,** New York, 1838, p. 140.
45. Felkin, Lace and Hosiery Trades of Nottingham, in **Royal Statistical Society Journal,** 1866, XXIX, p. 540.
46. Inwards, **Straw Hats: Their History and Manufacture,** London, 1922, pp. 10, 15; Davis, **Luton: Past and Present,** Luton, 1874, p. 17.

47. Tansley, On the Straw Plait Trade, in **Royal Society of Arts Journal,** 1860-1, IX, p. 69; V.C.H.,**Essex,** London, 1907, II, p. 376.
48. **Report,** in P.P., 1843, XIV, a50, no. 289.
49. Tansley, **op. cit.,** p. 71. This description, although written in 1860, was largely true of the earlier as well as of the later period. The persistence of the plaiting school as well as of the industrial organization in the straw plait manufacture has always been an interesting feature.
50. Children's Employment Commission, 1843, **op. cit.,** XIV, a49, a56.
51. Taylor, **Scenes of Wealth,** London, 1826, p. 21.
52. Tansley, **op. cit.,** p. 72; Davis, **op. cit.,** p. 23.
53. Austin, **The Straw Plaiting and Straw Hat and Bonnet Trade of the Luton District,** Luton, 1871, p. 63.
54. Quoted by Davis, **op. cit.,** pp. 26-7; also by Austin, **op. cit.,** p. 76.
55. Davis, **op. cit.,** p. 10; Austin, **op. cit.,** p. 33.

Annotated Bibliography

I. PARLIAMENTARY PAPERS

For bibliographical aids[1] consult the Annual Index for respective year and other indexes published at irregular intervals and listed below. Of the latter class of indexes the most helpful is **A Catalogue of Parliamentary Reports and a Breviate of their Contents arranged under Heads according to the Subjects, 1696-1834,** which gives a brief summary of each report.

In the following bibliography Roman numerals give the numbers to the bound volumes of the **Sessional Papers** (more commonly known as **Parliamentary Papers)** for the respective years. Unless otherwise stated, the following reports are presented to the House of Commons by the Select Committees, and printed by the House of Commons.

A. Indexes

1829,	unnumbered:	**General**	**Index,**	1801-1826.
1834,	L:	„	„	1696-1834.
1837,	LII:	„	„	1835-1837.
1840,	XLIX:	„	„	1832-1838.
1845,	XLIV:	„	„	1832-1844.
1850,	XLVII:	„	„	1844-1850.
1854,	LXX:	„	„	1801-1852.
1862,	LXV:	„	„	1852-1861.
1870,	LXXI:	„	„	1852-1869.
1880,	LXXXIII:	„	„	1870-1879.
1889,	LXXXIX:	„	„	1880-1889.
1904,	CXII:	„	„	1890-1899.
1911,	CIV:	„	„	1900-1909.

B. Reports, etc.

1802-3, V. **Petitions from the Yorkshire Woolen Manufacturers.**
On the state of the manufacture and the mode of conducting it under the "domestic" and factory systems.

1806, III. **State of Woolen Manufacture of England.**
A valuable document giving for the first time a clear-cut analysis of the industrial systems in the manufacture, viz. the "small clothier" system in the West Riding and the "master clothier" system in the West of England. The term "domestic system" was first introduced. Opposition to the factory system and machinery refuted.

1808, II. **Cartwright's Loom.**
Opposition to the introduction of power loom and burning down of Cartwright's factory described.

1812, II. **Crompton's Mule.**
Report on the petition respecting the famous invention.

1812, II. **Framework Knitters' Petitions.**
Subjects for petitions included abuses from truck, frame-rent, etc. See also other reports cited below; 1819, V; 1845, XXIII-XXV (H.L.); 1854-55, XIV.

1816, III. **State of the Children Employed in the Manufactories of the United Kingdom.**
Peel's Committee report, chiefly on cotton factories. Aside from the information given on the factory labor conditions, earliest statistics with reference to factory

1. For indexes not published in the Sessional Papers Series, consult King, **Catalogue of Parliamentary Papers, 1801-1900,** London, 1904. For general information on sessional papers, consult the excellent pamphlet by Lees-Smith, H. B.: **A Guide to Parliamentary and Official Papers,** Oxford, 1924.

sizes in cotton and woolen manufactures, although incomplete, give a fairly general picture. Half of the English factories included, and all Scotch factories except one at New Lanark. These Statistics were for the first time analyzed by Clapham, in "Some factory statistics of 1815-16". **Economic Journal**, 1915, XXV, pp. 475-9.

1817, VI. **Coventry Watchmakers' Petitions.**
Evidence only, no report.

1818, IX. **Laws relating to Watchmakers.**
This and the preceding reports supply ample information with respect to the progress and state of the watch manufacture from 1796 to 1816. Factory system in Coventry finishing branch complained of by witnesses as cause for distressed conditions. 1818 report contains a digest or index to the Evidence printed in 1817, and a list of acts relating to clocks and watches.

1818, IX. **Petitions of the Ribbon Weavers and Silk Manufacturers.**
Complains distress of workmen on account of inadequate wages, and recommends extension of the "Spitalfields Acts" to the London workmen. Evils of half-pay apprenticeship system also given.

1819, V. **Framework Knitters' Petition.**
Complaint against making of cut-up (worsted) stockings in Leicester, which caused redundant output of inferior quality, with reduced wages.

1822, VII. **Laws regulating the Linen Trade of Ireland.**
On the Linen Board, laws for the regulation of the trade, and premium payments for exportation of linen cloth.

1824, V. **Artizans and Machinery.**
Six reports on laws relating to export of machinery, emigration of artizans, and combinations. Useful information on the factory system in metal manufactures, especially machine-making.

1825, V. **Laws relating to the Export of Tools and Machinery.**
Deplores prohibition hindering development. Good on textile and metal industries. See also other reports cited below: 1841, VII.

1825, V. **Linen Trade of Ireland.**
Conditions of the trade; claims to encouragement. Contains good description of the merchant employer system in northern England (West Riding), Scotland and Ireland. This and the 1822 reports give a very complete view of the history and state of the manufacture.

1826-7, V. **Emigration from the United Kingdom.**
Three reports, with evidence. Alleges unemployment of Lancashire cotton weavers to the introduction of power looms as a primary cause.

1828, VIII. **State of the British Wool Trade.**
Chiefly on commercial aspect; but scattered information on merchant employer system. A Report from the House of Lords Committee.

1830, X. **Means of lessening the Evils arising from the Fluctuation of Employment in Manufacturing Districts.**
States causes and effects. Plan for improvement, its objections and advantages. No evidence printed.

1831-2, XV. **Bill for Regulation of Factories.**
A highly controversial report from the famous Sadler's Committee, the extensive scope of which was partly responsible for the early death of its author in 1835. Baines, Engels, Greg, Plener on the one hand, and Hammond, Hutchins and Harrison on the other, held different views on the value and truthfulness of the report. For an excellent recent review of this report and 1833-4 reports on

Children's Employment see Hutt, Factory system of the early 19th century, in **Economica**, 1926, VI, pp. 78-93.

1831-2, XIX. **Silk Trade.**

Covers ribbon and silk, especially full on silk throwing mills. Complains smuggling and foreign competition, especially from the French, and distressed conditions of the handloom weavers. This report summarized in **Westminster Review**, 1833, XVIII, pp. 1-30.

1833, V. **Agriculture.**

Describes allotment system. Contains scattered information on different manufactures. Principal purpose was to survey the state of agriculture and of persons employed in it. Report with minutes of evidence, appendix and index.

1833, VI. **Manufactures, Commerce and Shipping.**

Report, with minutes of evidence, appendix and index, on capital and wages, machinery, industrial and commercial methods. Good on cotton manufacture.

1833, XX, XXI; 1834, XIX, XX. **Children's Employment [in Factories] Commission.**

First and second reports published in 1833; supplementary report, in two parts, in 1834. The Commission, appointed as a result of severe criticisms on the partisan character of the report of Sadler's Committee, was thought to be biassed; but in two months' time brought out their reports distinctly in favor of legislation, though not as far as Sadler and Ashley would have desired. The appointment of factory inspectors to enforce legislation, a result of the recommendations moved by this Commission, constituted a "turning point" in the English factory legislation, and brought about the official compilation for the first time in 1835 the most valuable set of data from the viewpoint of industrial organization, viz. the "Factory Returns". The supplementary report contained returns on employees, motive power, child and woman labor, etc., by all English factories individually. These individual returns, unforunately, were not analyzed or summarized by the Commission. It is hard to believe that such valuable statistics have been overlooked by the historians of textile industries.

1834, XLIII; 1835, XL; 1836, XLV; 1837, XXXI; 1837-8, XXVIII, XLV; 1839, XIX; 1840, XXIII; 1841, X; 1841 Sess. 2, VI; 1842, XXII; etc. **Factory Inspectors' Reports.**

To quote from Hutchins and Harrison, **History of Factory Legislation,** 3rd ed., 1926 (first published 1903), p. 71, "these reports were to be sent in twice a year to one of the Secretaries of State, or oftener if he desired it. From 1836 to 1844 they were sent in every quarter, but only printed every half year. Sometimes special reports on particular subjects were made by the inspectors, by order of Parliament or of the government. With the object of rendering the law as uniform in application as possible, the four inspectors were required to meet twice a year in London and agree on a common plan of action as to their official proceedings and policy, and a joint report of these meetings was to be compiled and forwarded to the Secretary of State along with the single reports." In 1834, LXIII, a map of the inspectors' districts was inserted. In 1842, XXII, **Inspector Horner's Report** was the most valuable of all, giving the actual sizes of all textile factories in his district including West Riding, Lancashire and other Northern counties. The Factory Inspectors' Reports continued to be issued until 1878, at which year they were replaced by the **Annual Report of H. M. Chief Inspector of Factories and Workshops.**

1834, XLIII; etc. **Factory Returns.**

In 1834, XLIII, Inspector Horner was the only one to give factory returns for his district; for the years 1835, 1838, 1847, 1850, 1856, 1862, 1868, etc., factory returns were given by each of the four inspectors for cotton, woolen, worsted, silk and flax factories in the United Kingdom. 1835 report is incomplete, and in many respects unreliable, due to the difficulty encountered in first collecting data from the factories in a systematic manner. (See 1840, X, **Second Report on Mills and Factories**, p. 119, appendix no. 2; 1857, Sess. 1, III, p. 11, **Factory Inspectors' Reports ending October, 1856**). Since 1838 the returns have become reliable. Statistics on the number of factories, employees and motive power were given in returns for all years mentioned above; power loom statistics in 1835, 1850, 1856 and subsequent returns, spindle statistics in 1850, 1856 and subsequent returns. The reference to the returns of respective years is as follows.

1834 in H.C., 1834, XLIII.
1835 in H.C., 1836, XLV.
1838 in H.C., 1839, XLII.
H.C., 1843, LVI.
1847 in H.C., 1847, XLVI.
H.C., 1849, LIV.
1850 in H.C., 1850, XLII.
1856 in H.C., 1857, sess. 1, XIV.
H.C., 1858, LVII.
1862 in H.C., LV or LX.
1868 in H.C., LXIV.
Etc.

Secondary authorities such as Baines, Ure, James, McCulloch, Porter, Poole, Page, etc. quoted the factory returns with varying degree of completeness and slight changes to suit their purposes. V. C. H., Lancashire, quoted 1835-1901 returns for that country.

1834, X; 1835, XIII. **Petitions from Handloom Weavers.**
State of handloom weavers and causes of their distress, covering both England and Scotland. Causes included power looms, war taxation, currency contraction, yarn exportation, low wages, corn laws, competition among manufacturers and commission houses, want of combination. Remedies included tax on power looms and limitation of hours for the working of such looms, compulsory apprenticeship rules, central and local boards to fix minimum wages, etc. Gives valuable information on the merchant employer system.

1835, V; 1836, IX. **Arts and Manufactures.**
Pt. I in 1835, V, or 1836, IX; pt. II in 1836, IX. Useful information on metal industries, especially machine-making. Reports, minutes of evidence, and appendix.

1835, XXXI. **Commissioners of Inquiry into Excise Establishment, etc.**
A part of reports in 1835, XXXI, dealing with glass, paper, etc. The whole series of reports, 19 in all, began in 1833 and ended in 1837. For complete references to these 19 reports, consult Indexes referred to above.

1837, XX. **Fourdrinier's Patent.**
Chiefly on advantages of machine vs. hand made papers. Purpose of report is to find out if the patent deserves public compensation because of its usefulness.

1839, XLII; 1840, XXIII, XXIV; 1841, X. **Commissioners on Handloom Weavers.**
1839 and 1840 reports were presented by the Assistant Commissioners; 1841 report by the commissioners. The contents included:

1839, XLII, (pt. I):
Symons, J.C.: **South of Scotland, Switzerland, Austria, France, Belgium.**
Harding, J.D.: **East of Scotland.**

1840, XXIII, (pts. II and III):

Mitchell, J.: **East of England.**
Austin, A.: **Southwest of England.**
Keyser, S.: **Yorkshire (W.R.), Macclesfield, etc.; Germany.**
Chapman, H.S.: **Yorkshire (W.R.).**
Otway, C.G.: **Ireland.**
Muggeridge, R.M.: **Ireland.**

1840, XXIV, (pts. IV and V):

Fletcher, Joseph: **Midland Districts of England, including Coventry, Banbury, Congleton, Macclesfield, Leck, Derby, Hosiery Districts.**
Miles, W.A.: **Gloucester, Worcester; N. and S. Wales.**
Muggeridge, R.M.: **Lancaster, Westmoreland, Cumberland, Yorkshire, W.R. (part of).**
Symons, J.C.: **Note to Report on Scotland.**
Hickson, W.E.: **Report on the Condition of Hand-loom Weavers.**

1841, X.

Summary report by the Commissioners, Nausau William Senior, Samuel Jones Lloyd, William Edward Hickson and John Leslie. Commissioners originally appointed to consider the condition of the *unemployed* handloom weavers, but went beyond the terms of the appointment and considered the condition of *all* handloom weavers. After describing the state of affairs and exposing the fallacies involved in most of the popular remedies suggested, the commissioners proposed more free trade, protection of workmen against combinations, more education and emigration, encouragement of good designs and cheap means of copyrighting them. These reports, including those of the commissioners and assistant commissioners, give most valuable information on industrial organization in textile manufactures, especially on merchant employer system and handloom factory system. **Hickson's Report,** in 1840, XXIV, embodied conclusions by one who had travelled the whole of the United Kingdom in order to obtain a full picture, and on that account is the most valuable of all reports.

1840, VI. **Copyright of Designs.**

1840, X; 1841, IX. **Act for the Regulation of Mills and Factories.**
Six reports with index in 1840, X; and a final concluding report in 1841, IX. Give valuable information on factory system; also factory statistics as furnished by the factory inspectors and published in 1840, X, pp. 118, 127, 158. These inspectors pointed out the difficulty in defining a factory as employed in the Factory Returns.

1841, VII. **Laws affecting the Exportation of Machinery.**
Two reports with index and appendix. Contains valuable information and statistics on machine-making factories, chiefly in the appendices.

1842, IX. **Payment of Wages in Goods.**
Truck system analyzed in regard to history, development, abuses and remedies. Report with minutes of evidence, appendix and index.

1842, XV, XVI, XVII; 1843, XIII, XIV, XV. **Children's Employment Commission.**
Most comprehensive reports on industries not regulated by the Factory Acts, including mining and manufacturing. Gives first-hand information on industrial organization. Detailed references are as follows:

1842, XV: **First Report on Mines.**
On iron, copper, tin, lead and coal mines; on mining, smelting and other finishing processes of metal industries. Analysis of truck system.

1842, XVI, XVII: **Appendices I and II** to above Report.

1843, XIII: **Second Report on Trades and Manufactures.**

1843, XIV, XV: **Appendices I and II** to above report, containing evidence taken in 1840 and 1841. The detailed references are:

1843, XIV, Appendix I, A—F, a—f.

Burns, J.G.: **Paper mills, S.E. counties of England.**

Kennedy, J.L.: **Print grounds etc., Lancashire.**

Scriven, S.S.: **Staffordshire Potteries, etc.**

Stewart, L.: **Manufactures in Dorset, Devon, and Somerset, and in the city of Norwich.**

Symons, J.C.: **Trades of Sheffield; ironworks, etc. in Leeds and West Riding of Yorkshire.**

Grainger, R.D.: **Manufactures of Nottingham, Derby, Leicester, Birmingham, and London.**

1843, XV, Appendix II, G—S. i—t, 1-112.

Roper, F.: **S. Ireland.**

Tancred, T.: **W. Scotland.**

Franks, R.H.: **E. Scotland.**

Leifchild, J.R.: **Glass houses, potteries, paper mills and miscellaneous manufactories in Northumberland and Durham North of the River Wear.**

Austin, A.: **West and North of Lancashire.**

Martin, T.: **N. Ireland.**

Waring, E.: **S. Gloucestershire.**

Fellows, J.M.: **Derbyshire potteries.**

Horne, R.H.: **Iron and other manufactures of S. Stafford and neighborhood of Worcester and Salop.**

Fletcher, J.: **Detached manufactures in Hampshire, Surrey, and Kent.**

Barham, Charles: **Iron foundries in the West of England.**

Jones, H.H.: **Wales.**

Index to above Report on Trades and Manufactures.

1843, VII. **Allotments of Land.**

1843, XII. **Special Assistant Poor Law Commission on the Employment of Women and Children in Agriculture.**

Contains excellent short account of the family craftsman system in shirt button making; also scattered references to other manufactures.

1843, XXXV, H.L. **Midland Mining Commission.**

On S. Staffordshire iron and coal mines.

1844, XXVII; 1845, XXXIV 1846, XXXVI; 1847, XXXVII; 1847-8, XLII; 1849, XLII; 1850, XLI; etc. All these refer to H.L. Sessional Papers. **Commissioner of Mining Population, Report.**

Annual report published since 1844, with scattered statistics on iron output and information on ironworks in England and Scotland. Valuable though not systematic.

1844, VII. **Joint Stock Companies.**

First report, with minutes and evidence taken in 1841 and 1843. Appendix and index. Contains **Statements respecting the Establishment of Joint Stock Woolen Mills in the West Riding of Yorkshire** in the appendix, a valuable document on the history and organization of joint stock woolen mills.

1844, XV, H.L. **An Account of the Quantities of the Several Articles Charged with Duties of Excise in the Former Years ended 5th April, 1844.**

On paper, glass, etc.

1844, XXVII. **Occupation Abstract.**

Contains the second Census of Occupation for Great Britain, taken in 1841. A summary of it may be found in an article on the "Occupations of the People of

Great Britain", **Chambers' Edinburgh Journal**, 1844, II, pp. 262-84; and an excellent criticism of its defects in Day, **Distribution of Industrial Occupations in England, 1841-1861**, New Haven, 1927, pp. 90-92. Evidently, in the Census enumeration the part-time outworkers organized under the different industrial systems, such as craftsman and merchant amployer systems, were not included.

1845, XXIII-XXV, H.L. **Framework Knitters' Commission.**
Report in 1845, XXIII, H.L.; evidence, appendices, and indexes in 1845, XXIV-XXV, H.L. On framework knitters in the hosiery districts of Leicestershire, Nottinghamshire, and Derbyshire. Most comprehensive and detailed analysis of the middleman system for one single industry yet understaken, not excepting the monumental reports on the **Sweating System** in the late 'nineties. Commission appointed in response to a Petition presented to the House of Commons in 1843 by upwards of 25,000 framework knitters in the counties of Nottingham, Derby and Leicester, etc., which centered its inquiry on the following subjects: low wages, fraudulent work, frame-rent system, truck payment, delivery of work tickets, etc. Conclusions were: depressed conditions of workers, prevalence of truck, injury of frame-rent system, competition of child and woman labor, excess of labor supply. Remedies included birth control through abstinence, extension of manufacture with a simultaneous improvement in quality of output.

1854, XV. **Payment of Wages.**
Truck system in mining districts.

1854-55, XIV. **Stoppage of Wages (Hosiery).**
On the abuses of the middleman system in hosiery manufacture. A revival of 1845 inquiry.

1861, XXII. **Report by Mr. H. Seymour Tremenheere to the Home Secretary upon the Expediency of subjecting the Lace Manufacture to the Regulation of the Factory Acts, with evidence.**

1863, XVIII; 1864, XXII; 1865, XX; 1866, XXIV. **Children's Employment Commission.**
Five reports on industries not yet regulated by the Factory Acts, with appendices. A second attempt to investigate industrial conditions similar to the first one in 1842-3. Interesting side-lights on the persistence of industrial organization in those industries organized under the outwork systems.

1876, XXIX, XXX. **Factory and Workshop Acts.**
Contains interesting analysis of outwork industries organized primarily under the merchant employer system. Persistence of industrial organization in such industries as lace, hosiery, straw plait, glove, nail and chain, etc.

1884, V. **Report on the English Silk Industry**, by Thomas Wardle, of Leek, Stafford.
Excellent analysis of history and industrial organization of silk manufacture in leading centres.

1888, XCI. **Report on Condition of Nail and Chain Makers in S. Staffordshire and W. Worcestershire**, by the Labor Correspondent of the Board of Trade (Henry G. Calcraft).
Best analysis of the industry organized under the merchant employer system, with historical notes. Gives only exhaustive bibliography on accounts published in the **Sessional Papers** in 1842, IX; 1854, XV; 1854-5, XIV; 1875 and 1876.

1888, XX, XXI; 1889, XIII, XIV; 1890, XVII. **Sweating System.**
Five reports with indexes to evidence from the House of Lord's Committee. First important inquiry on the system, chiefly confined to the London industries.

1907, VI; 1908, VIII. **Home Work.**

II. NON-OFFICIAL PUBLICATIONS

All contemporaneous and non-contemporaneous works that have been either cited or consulted in writing up the present treatise are listed, classified, and annotated.

As far as possible, the writer has attempted to classify each reference under the section to which it properly belongs, but cross-references, in a few cases, are hard to avoid. Each section is preceded by a brief survey of the bibliographical aids.

Most of the references may be found in the Yale Library, but a small portion are available only in the Harvard University Library, Columbia University Library, New York Public Library and Boston Public Library.

A. Industrial organization

1. General

For bibliographical aids consult Schmoller, Bücher, Sombart, Schwiedland, Unwin and Usher. Usher's is annotated, and deals mostly with literature taking a hostile attitude towards Bücher's theory.

Allen, G. C.

The Industrial Development of Birmingham and the Black Country, 1860-1927, London, 1929.

Methods of Industrial Organization in West Midlands, 1860-1927, in **Economic Journal Supplement (Economic History Series No. 4),** January, 1929, pp. 535-553.

Ashley, W.J.:

Early History of English Woolen Industry, 1887, pp. 72-73.

Introduction to English Economic History, I, pt. ii, 1892, p. 219ff.

Economic Organization of England, London, 1914, pp. 35-6, 198.

In these works, Ashley introduced for the first time the systematic classification of industrial systems from the German literature, chiefly from Thun. Similar attempt is now made by Unwin (1904), and Lloyd (1913).

Brodnitz, Georg: Betriebskonzentration und Kleinbetrieb in der englischen Industrie, in **Jahrbuch für Nationalökonomie,** 1908, XC, pp. 173-218; 1909, XCII, pp. 51-86, 145-84.

Analysis of contemporary English industrial organization, chiefly from the **Factory and Workshop Returns** and the Reports of Parliamentary Committees on **Home Work** (1907-1908). Contrasts large and small scale industries.

Bücher, Karl:

Article on "Gewerbe", in Conrad (ed.), **Handwörterbuch der Staatswissenschaften,** 1st ed., 1890-94, 6 vols., III, pp. 922-50; or 4th ed., 1923-27, 8 vols., IV, pp. 966-99. Bibliography, p. 950 (1890-94 ed.), pp. 998-9 (1923-27 ed.).

Die Entstehung des Volkswirtschaft, Tübingen, 1893 (tr. fr. 3rd German ed., 1900, as **Industrial Evolution,** New York, 1901 by S. M. Wickett), ch. IV.

Summarizes previous researches and results of own investigation in a masterly fashion, presenting a scheme of industrial systems that has been frequently quoted, but criticized by Below, Gerlich, Liefmann, and many others.

Clapham, J.H.: **Economic History of Modern Britain, 1820-1850,** Cambridge, 1926.

Chapter V on "Industrial Organization" around 1840 gives a comprehensive summary of most manufactures covered in the present treatise, grouping together a mass of information with little attempt at a consistent analysis of the different industrial systems.

Gerlich, Fritz: **Geschichte und Theorie des Kapitalismus,** München und Leipzig, 1913.

Chapter VII (pp. 299-374) on the "Middle Ages and the Present" gives an excellent survey of the relationship between technique and industrial organization, pointing out many defects in Bücher's scheme in classifying industrial systems in chronological terms.

Hasbach, W.: Zur Characteristik der englischen Industrie, in **Jahrbuch für Gesetzgebung,** etc., 1902, XXVI, pp. 1015-1062.

Contrasts large and small scale industries of England in respect to industrial organization and other aspects, and proposes a scheme based on the nature of income received from industrial employments.

Hazard, Blanche E.: **The Organization of the Boot and Shoe Industry in Massachusetts before 1875, Cambridge, 1921.**

An able analysis on the evolution of industrial systems.

Lloyd, G.I.H.: **Cuttery Trades,** London, 1913.

Excellent summary of industrial systems in chapter I. A first-rate study of the "little master" system in the cutlery trades, with comparative analysis of industrial systems in cotton, woolen and worsted, linen, ribbon, hosiery, nail and chain manufactures.

Marshall, Alfred: **Principles of Economics,** 1st ed., 1892; 6th ed., 1910.

Devotes four chapters (1910 ed., pp. 240-313) to a thorough analysis of the advantages of large-scale production and factory system.

Marx, Karl: **Capital,** Chicago, 1912-13 (Kerr edition), 3 vols. (Date of first publication in German, I, 1867; II, 1885; III, 1894).

Socialistic interpretation of industrial systems.

Palgrave's Dictionary of Political Economy, new edition by Henry Higgs, 3 vols., 1923-26.

Short articles with bibliography on **Domestic System** (L. L. Price), **Factory System** (R.W. Cooke-Taylor), **Home Industries** (Mark G. Davison), and **Industries, Large and Small** (L. L. Price).

Riekes, Hugo: Der Fabrikbegriff und die Handwerksorganisation, in **Jahrbuch für Nationalökonomie,** 1902, LXXIX, pp. 185-214.

Theoretical and practical distinction, with author's own scheme of classification.

Schmoller, Gustav:

Die geschichtliche Entwicklung des Unternehmung, in 12 pts., pts. 4-5, in **Jahrbuch für Gesetzgebung,** etc., 1890, XIV, pp. 1043-76.

Grundriss der allgemeinen Volkswirtschaftslehre, 2 vols., 1st ed., Leipizig, 1900, I, pp. 417-34. Bibliography, I, pp. 411-2.

Results of extensive researches by the distinguished German economic historian. Concept of social classes under different industrial systems taken over by Sombart.

Schwiedland, Eugen: Der Wettkampf der gewerblichen Betriebsformen, in **Grundriss der Sozialökonomik,** Tübingen, 1914,VI, pp. 24-53.

Bibliography and summary of industrial systems based on Bücher. Analysis of the factory system and large undertakings.

Sombart, Werner:

Die gewerbliche Arbeit und ihre Organisation, in **Archiv für Sozialegesetzgebung und Statistik,** 1899, XIV, pp. 1-52, 310-368. Bibliography, pp. 1-16.

Der moderne Kapitalismus, 1st ed., Leipzig, 1902, 2 vols., 4th ed., 1922. Chs. I and II (1902 ed.).

Both references contain same substance with slight changes in arrangement, presenting Sombart's own socialistic interpretation of industrial systems.

Das Wirtschaftsleben in des Hochkapitalisums, München, 1928.

Thun, Alphons: **Industrie am Niederrhein,** 2 vols., Leipzig, 1879.

His classification of industrial systems into housework, handicraft, house industry, and factory (pp. 240-250) was first introduced into the English economic history by Ashley, in 1887.

Unwin, George: **Industrial organization in the sixteenth and seventeenth centuries, Oxford, 1904.**

Introductory chapter gives an excellent summary of industrial systems, and an original diagram on the evolution of the "industrial pedigree". Bibliography, pp. 253-70.

Usher, A.P.:

Generalizations of economic history, in **American Journal of Sociology, 1916, XXI,** pp. 474-91.

Industrial History of England, Boston, 1920. Chapter I on "Forms of industrial organization", pp. 1-23, and bibliography for the same, pp. ii-iii.

The first reference is an independent classification of economic stages based on the extent of market in different periods of world economic history; the second gives an able but controversial discussion on the classification of industrial systems, directing severe criticism on Bücher's theory.

Weber, Max: **Wirtschaftsgeschichte,** ed. by Hellmann, S. and Palyi, Dr. M., 2nd ed., 1924, München und Leipzig (tr. by Frank H. Knight as **General Economic History,** New York, 1927).

Stimulating discussion on the classification of industrial systems by a distinguished German sociologist and economist. The present volume is a posthumous edition of author's lectures given in 1919-20. See especially chs. VII, VIII, XI, XII.

Wielden, Josef: Handwerk und Industrie, in **Volkswirtschaftliche Zeitfragen, 1912,** XXXIV, no. 270.

Theoretical and practical distinction. Illustrates difficulty and gives own conclusions.

2. Craftsman system.

This and the following sections on the merchant employer and factory systems deal with literature restricted to each of the three industrial systems. The references given under "1. General" are equally valuable, and must not be overlooked. For bibliographical aids, consult Sieveking and Sombart.

Below, George von: **Territorium und Stadt,** München und Leipzig, 1900.

Contains important criticism on Bücher's distinction between wagework and handicraft systems in chronological terms (pp. 321ff.).

Sieveking, Heinrich: Geschichte der gewerblichen Betriebsformen und der zünftigen, städtischen und statlichen Gewerbepolitik, in **Grundriss der Sozialökonomik,** 1914, VI, pp. 1-23.

On medieval craftsman system, especially useful for discussion on the relation of craftsman to market. Bibliography, pp. 2-3.

Sombart, Werner: **Der moderne Kapitalismus,** 1st ed., Leipzig, 1902, vol. I.

Chapters IV, V and VI devoted to the concept, organization and conditions for the existence of the craftsman system. Bibliography, I, p. 79.

3. Merchant employer system

The recent literature printed after the publication of the Parliamentary Committee Reports on the **Sweating System** (1888-90) and **Home Work** (1907-8) are omitted because of limited space. For a thorough understanding of the persistence of the small-scale industries organized under the merchant employer system in the nineteenth and twentieth centuries, these reports must remain the chief source of information. For bibliographic aids, consult Sombart and Bräuer. Stieda's bibliography (1889) is exhaustive, but out of date. Liefmann's is critical, and useful as a guide to advanced study of the subject.

Agricultural Economics Research Institute, **Rural Industries of England and Wales,** Oxford, 1926-27, 3 vols.

Concludes that "the only basis upon which rural industries can be firmly established is that of a high standard of technical knowledge and skill, suitable machinery, and commercial organization." (I, p. vi) Field investigation into the industrial organization of small-scale industries: causes, persistence and suggestions for future development.

Booth, Charles: **Life and Labor of the People in London,** London, 1892-97.

Only extensive study of the old industrial system prevalent in London industries undertaken by a private individual,—a mine of information comparable with the Parliamentary Committee Reports on the **Sweating System** (1888-90) and **Home Work** (1907-8).

Bräuer, K.: Hausindustrie und Heimarbeit mit besonderer Rücksieht auf das rheinisch-mainische Wirtschaftsgebiet, in **Zeitschrift für Sozialwissenschaft**, 1915, VI, pp. 25-39, 87-98.

Theoretical consideration and discussion on the classification of elements in the merchant employer system. Author's own scheme proposed. Bibliography.

Furger, Fridolin.

Zum Verlagssystem als Organisationsform des Frühkapitalismus im Textilegewerbe, Stuttgart, 1927.

Gaskell, P.: **The Manufacturing Population of England**, London, 1833.

Chapter I gives an idyllic picture of the "domestic system" before the industrial Revolution.

Gonner, E.C.K.: Survival of domestic industries, in **Economic Journal,** 1893, III, pp. 23-32.

Excellent brief summary of factors for decline and survival of the "domestic" industries.

Grant, S.F.: The Survival of the Small Unit in Industry, in **Economic Journal,** 1922, XXXII, pp. 489-505.

An analysis of the recent parliamentary reports, covering many industries still organized under the merchant employer system.

Kropotkin, P.:

Articles on "The breakdown of our industrial system", "The coming reign of plenty", "Industrial villages of the future", "The small industries of Britain," in **Nineteenth Cenutry,** 1888, XXIII, pp. 497-516, 817-37; XXIV, pp. 513-30; 1900, XLVIII, pp. 256-71.

Fields, factories and workshops, Boston and New York, 1899.

Studies on the persistence of small scale industries and their organization by a Russian anarchist—an opponent to the large scale production under the factory system. His **Fields** and **Small industries of Britain** contain a careful analysis of the factory and workshop statistics around 1900.

Liefmann, Robert: **Über Wesen und Formen des Verlags (der Hausindustrie)**, Freiburg i.B., 1899.

A penetrating theoretical analysis of the merchant employer system, with adverse criticism on most of the well-known theorists. His distinction between independence of outworker (Hausindustriellen) under the merchant employer system and the dependence of factory worker is the chief contribution.

Sombart, Werner: Article on "Hausindustrie", in Conrad (ed.), Handwörterbuch, **op. cit.,** 1890-94 ed., IV, pp. 1138-69; or 1923-27 ed., V, pp. 179-204.

Theoretical and factual. Covers England, United States and most European nations including Germany, France, etc. Extensive bibliography in 1890-94 edition (pp. 1158-69).

Stieda, W.: **Litteratur, heutige Zustände und Entstehung der deutschen Hausindustrie,** Leipzig, 1889.

Has exhaustive bibliography, with critical discussion; also an account of the rise and present status of the merchant employer system in German industries.

Westerfield, Ray B.: **The Middleman in English Business,** 1660-1760, New Haven, 1915.

Woods, K.S.: **Rural Industries round Oxford,** Oxford, 1921.

First volume of a series of studies made on behalf of the Agricultural Economics Research Institute, referred to above.

4. Factory system

For bibliographical aid, consult Williams, Sombart, Hutchins, etc. Williams, in her **Guide** (referred to below, section C. 1.) has an exhaustive bibliography on contemporary literature.

Babbage, Charles: **Economy of Machinery and Manufacture,** London, 1832.

A well-known summary of economic advantages resulting from large scale production and factory system.

Cooke-Taylor, William: **Factories and the factory system, from parliamentary documents and personal examination,** 1844.

First chapter defines and analyzes the factory system, which is defended in the subsequent chapters. His definition of a factory is accepted *verbaitm* by Carroll D. Wright, the leading American writer on factory system.

Cooke-Taylor, R. Whateley: **Introduction to a history of the factory system,** London, 1886.

Traces the system from the ancient times to date, interesting but at times superficial.

Cooke-Taylor, R. Whateley: **The Modern Factory System,** London, 1891.

A thoughtful but favorable analysis of the system by a factory inspector, chiefly devoted to a description of the social effects and resultant legislation. Contains a brief survey of literature on factory controversy in the 'forties (pp. 223-27).

Cooke-Taylor, R. Whateley: **The Factory system and the factory acts,** 1894.

A short but popular summary of his larger works.

Gaskell, P.: **The Manufacturing Population of England,** London, 1833.

An adverse analysis of the factory system by a medical authority. The same book was published with minor changes in 1836 under the title **Artisans and Machinery.**

Heffter, Werner: Fabrik und Werkstatt; ihre Merkmale und ihre gesetzlichen Beschränkungen in der Beschäftigung von Arbeiterinnen und jugendlichen Arbeitern, in **Zeitschrift für Sozialwissenschaft,** 1901, IV, pp. 313-18; 383-93.

Hutchins, B.L. and Harrison, A.: **A History of Factory Legislation,** 1st ed., London, 1903; 3rd ed., 1926.

Standard work, with excellent bibliography on primary and secondary sources.

Sombart, Werner: **Die moderne Kapitalismus,** 1st ed., 1902, 2 vols.

Critical review of writers on factory system including Babbage, Gaskell, Ure, Marx, etc. appears on pages 1 to 17 in volume I.

Ure, Dr. Andrew: **Philosophy of Manufactures: or, an exposition of the scientific, moral and commercial economy of the factory system of Great Britain,** 2nd ed., 1835, 3rd ed., 1861.

A systematic analysis of the system and its effects by a contemporary scientist. Favorable conclusions.

Wright, Carroll D.: Report on the Factory System of the United States, 1882, in **Tenth Census of the United States,** 1880, Manufactures.

Takes over William Cooke-Taylor's definition of a factory, and gives a favorable picture of the system. Bibliography.

Wright, Carroll D.: Factory System as an element of civilization, in **Journal of Social Science** (formerly **American Journal of Social Science**), 1882, XVI, pp. 101-26.

A popular summary of the Report cited above.

B. Individual industries

1. Cotton

A majority of the references here listed have been consulted in writing up the present treatise, but a few are included because of their merits and omission by Chapman. As a bibliographical aid, Chapman's **Select Bibliography** is by far the most helpful despite its early date of publication (1904).

Ashworth, Henry: Statistics of the present Depression of Trade at Bolton, in **Royal Statistical Society Journal,** 1842, V. 74-81.

By a leading cotton manufacturer and member of the Manchester Chamber of Commerce.

Baines, Edward, Jr.: **History of the cotton manufacture in Great Britain,** London, 1835.

Standard work on the subject for early years, with valuable statistics. Intended to be judicial, but colored by the author's views in favor of the factory system and laissez faire. "With the exception of Smith's **Memoirs of Wool,** published so far back as 1747, it is the only work that gives a clear and copious account of the rise, progress, and present condition of any of the great branches of manufacture carried on in the Kingdom." (McCulloch, in **Edinburgh Review,** 1835, LXI, 471.)

Burns, Richard: **Statistics of the cotton trade,** London, 1847.

Statistics on exports, prices, imports, factories, employees, arranged in convenient tabular form; also a chronological history of various inventions, improvements, etc. A reprint from the **Commercial glance.**

Butterworth, Edwin: **Historical Sketches of Oldham,** 1847-48.

Probably the only work giving statistics on factory sizes of Oldham and other important cotton manufacturing centres for 1847 and previous years,—one of the most valuable small volumes on early local industrial history.

Butterworth, Edwin: **Historical Account of the Towns of Ashton-under-Lyne, Stalybridge and Dukinfield,** Ashton, 1842.

A worthy companion to the above volume, but with fewer details and statistics.

Chapman, Sydney J.: **The Lancashire Cotton Industry, a Study in Economic Development,** Manchester, 1904.

The best account on the industrial organization of cotton manufacture before the publication of recent source studies of Daniels and Unwin, portraying the transition from the "domestic" to the factory system. Scholarly, and thoroughful. Contains a "Select Bibliography" (pp. 277-304) on parliamentary papers and secondary works, partly annotated and compiled with considerable skill. It may be regretted, however, that such valuable primary sources as the reports of the several parliamentary committees and commissions on the employment of children in factories from 1816 to 1841 are omitted from Chapman's list, in which there is also a misquotation of the Handloom Weavers' Report of 1840 (p. 282).

Chapman, Sydney J.: **The Cotton Industry and Trade,** London, 1905.

A pendant to the author's larger work cited above, and "intended to be an elementary introduction to the economics of the cotton industry and trade" (Preface). Not primarily historical, but containing information on industrial organization.

Chapman, S.J. and Ashton, T.S.: The sizes of businesses, mainly in the textile industries, in **Royal Statistical Society Journal,** 1913-1914, LXXVII, pp. 467-555.

Elaborate statistical analysis based upon trade directories, covering chiefly textile industries in Yorkshire, Lancashire, Ireland, Scotland and Wales, United States and Canada; Europe; India and Japan. Illuminating conclusions on representative size of a business, and causes for its variation.

Clapham, J.H.: Some factory statistics of 1815-16, in **Economic Journal,** 1915, XXV, 475-9.

An interpretation of the early factory sizes in cotton and woolen manufactures, based on the **Report of the Minutes of Evidence taken before the Select Committee on the State of the Children Employed in the Manufactories of the United Kingdom** (in P.P., 1816, III). This Report, according to Clapham, was long neglected by the industrial historians of cotton manufacture including Guest, Baines, Ellison, and Chapman. The writer, however, misquoted the term "manufactories" as "manufactures".

Cooke-Taylor, William: **The Handbook of Silk, Cotton, and Woolen Manufactures,** 1843.
A popular account, including history, processes, conditions of workers.

Daniels, G.W.: Early Records of a Manchester Cotton Spinning Firm, in **Economic Journal,** 1915, XXV, 178-83.
Records of Messrs. M'Connel's cotton factory giving size in terms of persons employed and other information.

Daniels, G.W.: **The Early English Cotton Industry,** Manchester, 1920.
Primarily a source study of Messrs. M'Connel's records and unpublished letters of Samuel Crompton, bringing the history of cotton industry down to the end of the eighteenth century, with an excellent analysis of the transition of industrial organization from the "domestic" to the factory system. Professor Daniels promises another volume to bring the history down to date.

Donnell, E.J.: **Chronological and Statistical History of Cotton,** New York, 1872.
A voluminous reference containing 650 pages arranged chronologically. Poor but useful.

Ellison, Thomas: **The Cotton Trade of Great Britain, including a history of the Liverpool Cotton Market and of the Liverpool Cotton Brokers' Association,** London, 1886.
A commercial and industrial history with full statistical tables, written by one who possesses a first hand knowledge acquired from practical personal experience.

Faucher, Léon: **Manchester in 1844; its present condition and future prospects** (tr. by J.P. Culverwell from a part of the French work **Etude sur l'Angleterre,** Paris, 1845), London, 1844.

French, Gilbert J.: **The Life and Times of Samuel Crompton,** 2nd ed., 1860.
Manuscript study. Contains in the Appendix, R. Cole, **Some account of Lewis Paul, and his inventions of the machine for spinning cotton-wool by rollers.**

Guest, Richard: **A Compendious History of the Cotton Manufacture, with a disproval of the claim of Sir Richard Arkwright to the invention of its ingenious machinery, Manchester,** 1823.
Contains numerous illustrations. A much quoted and highly controversial book.

Montgomery, James: **The Cotton Manufacture of the United States of America, contrasted and compared with that of Great Britain, with comparative estimates of the Cost of Manufacturing in both countries,** Glasgow, 1840.
That the author under-estimates the cost of manufacturing for cotton factory in Great Britain and over-estimates the same for cotton factory in the United States has been pointed out by the writer of a pamphlet called **Strictures on Montgomery,** Newburyport, 1841 (Anonymous).

Montgomery, Robert: Economic and Social Conditions of Manchester and the Surrounding Districts in 1834 and 1884, in **Manchester Statistical Society Transactions,** 1884-5 session, pp. 1-30.

Radcliffe, William: **Origin of the New System of Manufacture, commonly called "Power Loom Weaving",** Stockport., 1828.
Valuable for information on industrial organization from the end of the eighteenth century and written by the inventor of the famous "dressing machine". Reminiscences.

Schulze-Gavernitz, J. von: **Der Grossbetrieb: Ein wirtschaftliche und sociales Fortschrift: Eine Studie auf dem Gebiete der Baumwollindustrie,** 1892 (tr. by Oscar S. Hall as **The Cotton Trade in England and on the Continent,** London, 1895; tr. by Gueroult as **La Grande Industrie,** Paris, 1896).
Shows the growth of centralization in the English cotton industry and its resulting economy as the main thesis, with illuminating analysis accompanied by a good deal of statistical proofs.

Senior, Nassau William: **Letters on the Factory Act, as it affects the Cotton Manufactures,** London, 1844, 2nd ed.

On the handloom weavers, written by the noted economist and a member of the Handloom Weavers' Commission, 1838-1841.

Unwin, George: The Transition to the Factory System, in **English Historical Review,** 1922, XXXVIII, pp. 206-18, 383-97.

Now superseded by his **Samuel Oldknow** referred to below.

Unwin, George; Hulme, A.; and Taylor, G.: **Samuel Oldknow and the Arkwrights,** Manchester, 1924.

Traces the industrial revolution at Stockport and Marple, mostly from recently discovered records of Oldknow's cotton mill in the eighteenth century. A realistic picture of the great economic transition.

Ure, Andrew: **The Cotton Manufacture of Great Britain,** London, 1836, 2 vols.

A poorly arranged work, historical as well as technical. Quotes valuable factory statistics from **Supplementary Report of the Children's Employment Commission,** in P.P., 1834, XIX, pp. 312-314.

Wood, L.S. and Wilmore, A.: **The Romance of the Cotton Industry in England,** Oxford, 1927.

A good popular account of growth and development, with excellent illustrations.

Woodbury, Charles J. H.: **Bibliography of the Cotton Manufacture,** 1909.

See especially section on "History and Economics" of the cotton manufacture (pp. 102-160). Covers all countries and periods, but poorly classified and not annotated. Includes magazine articles.

2. Woolen and worsted

For bibliographical aids, consult Lipson and Heaton. Lipson's bibliography contains unannotated references classified under no less than nine heading, for "while the book is based on the original authorities, detailed references have been omitted in view of the nature of the Series to which it belongs" (Author's Preface). Heaton's bibliography gives full references, also not annotated.

Ashley, W.J.: Early History of the English Woolen Industry, in **American Economic Association Publication,** 1887..

A pioneer work, tracing the development of the industries in their transition from the handicraft to the "domestic" system. Somewhat revised in his later work on **Economic History,** 1892-93, Pt. II, chap. III.

Baines, Edwards: An Account of the Woolen Manufacture of England; with special reference to the Leeds Clothing District, in **Royal Statistical Society Journal, 1859,** XXII, pp. 1-34.

An excellent short article covering all aspects of the woolen and worsted manufactures—technical, industrial and commercial. Reprinted in his larger work, **Yorkshire, Past and Present,** 1870, 2 vols. (bound in 4 vols. with consecutive paging every two volumes), II, pp. 629-96, and brought up to date with revisions.

Bischoff, James: **A Comprehensive History of the Woolen and Worsted Manufactures,** London, 1842, 2 vols.

A useful but biassed book modelled on Smith's **Memoirs of Wool** published in 1747, from which it was partly taken. Valuable tables of statistics, but uncritical quotations from books, pamphlets, reports, petitions, and documents.

Bonwick, James: **Romance of the Wool Trade,** London, 1887.

A popular but poorly arranged work, with frequent citations.

Burnley, James: **History of Wool and Wool-combing,** London, 1889.

Standard history, but mediocre. Chiefly technical.

Clapham, J.H.:

Industrial Organization in the Worsted and Woolen Industries of Yorkshire, in **Economic Journal**, 1906, XVI, pp. 515-22.

Transference of the Worsted Industry from Norfolk to the West Riding, in **Economic Journal**, 1910, XX, pp. 195-210.

Woolen and Worsted Industries, London, 1907.

The two articles are suggestive and helpful, being closely related to the industrial organization of woolen and worsted manufactures in 1840 and thereabouts. The book on the **Woolen and Worsted Industries** is the first of its kind, giving contemporary account with scattered historical references.

Deschesne, Laurent: **L'Evolution économique et sociale de l'industrie de la laine en Angleterre**, Paris, 1900.

A historical sketch dealing with technical, industrial and commercial development. Only author on the textile manufactures to treat the Factory Returns on a plan somewhat similar to the present writer's. It is to be regretted, however, that Deschesne did not go to the sources directly, and instead, resorted to secondary works such as Porter's **Progress of the Nation, Statesman's Yearbook, etc.. Deschesne's** analysis differs from the present writer's as the former deals with the United Kingdom and England-and-Wales, while the latter touches upon England only.

Forbes, Henry: History of the Worsted Manufacture in England, in 1851 **Exhibition Lectures**, II, pp. 301-31.

Fox, Joseph H.: **The Woolen Manufacture at Wellington, Somerset**, London, 1914.

Compiled from the records of an old family business, covering a period of two and half centuries. Illustrative details.

Greenwood, W.: Fifty Years of British Industry from the Workman's Point of View, in **Eonomic Review**, 1900, X, pp. 323-32.

A vivid recollection of the Yorkshire "domestic" system.

Heaton, Herbert: The Leads White Cloth Hall, in **Thoresby Society**, 1913, XXII, pp. 131-71.

A source study based on the manuscripts of the White Cloth Hall, with brief sketch of the "small clothier" system in West Riding.

Heaton, Herbert: **Yorkshire Woolen and Worsted Industries, From the Earliest Times up to the Industrial Revolution**, Oxford, 1920.

A scholarly sketch based chiefly upon information gleaned from printed materials of the last two centuries, and from manuscripts covering the whole period from the thirteenth century to the eighteenth. Chiefly economic, with first-hand information on the "domestic" system from the fifteenth to the eighteenth centuries (chaps. III and IX, pp. 89-123, 282-321).

James, John: **History of the Worsted Manufacture in England**, London, 1857.

A useful work poorly compiled. Cites Factory Returns, Excise Returns, Reports of the Handloom Weaver's Commission of 1838, and other volumes of Parliamentary Papers. Valuable reprints of factory statistics, and a good deal of first-hand information inaccessible elsewhere.

James, John: **History of Bradford**, 1841.

Chiefly on woolen and worsted manufacture, expanded in his revised work entitled **Continuations and Additions to the History of Bradford and its Parish**, 1866.

Lipson, E.: **History of the Woolen and Worsted Industries**, London, 1921.

First volume in the "Histories of English Industries Series" edited by the author, the purpose of which is to "help to kindle an interest in industrial history among those who follow the paths of industry" (General Editorial Preface). An original study, including a rewritten and extended chapter on the medieval woolen industry in Vol. I of his **Economic History of England**, 1915, pp. 391-443. Valuable up to the eighteenth century but brief and incomplete on the nineteenth century. The author's failure to analyze the factory statistics since 1835 constitutes a serious omission.

Lohmann, F.: **Die Staatliche Regelung der englischen Woolindustrie vom XV bis zum XVIII Jahrhundert,** Leipzig, 1900.

Serious and documentary study, containing good accounts of the merchant employer system and its abuses.

Morris, G. W.: **The Golden Fleece. An Introduction to the Industrial History of England,** Oxford, 1922.

A skillful textbook designed to meet the needs of those students desiring a more realistic and concrete survey of English economic history. Popular.

Samuel Brothers: **Wool and Woolen Manufactures of Great Britain,** London, 1859.

A historical sketch.

Sprague, Oliver M. W.: **English Woolen Industry in the 17th and 18th Centuries,** Cambridge, 1897.

An unpublished doctoral dissertation at Harvard.

Turner: Bradford Piece Halls, in **Bradford Antiquary,** 1888, I, pp. 135-39.

On history and organization.

Wadsworth, A. P.: The History of the Rochdale Woolen Trade, in **Rochdale Lit. and Sci. Soc. Trans.,** 1925, XV, pp. 90-110.

Good local sketch from the Middle Ages to the nineteenth century.

3. Silk, flax and hemp

Literature on these manufactures, with the exception of writings by Warner, Rawlley and Gill, very seldom touches on the subject of industrial organization falling under our period of investigation. Gill is the only writer furnishing us with a critical bibliography,dealing chiefly with the Irish sources on linen manufacture.

Combe, John: **Suggestions for Promoting the Prosperity of the Leeds Linen Trade,** Leeds, 1865.

A pamphlet on the introduction of power looms and the difficulties involved.

Dodd, George: Spitalfields, in Knight, Charles (ed.): **London,** 1842, II, pp. 385-400.

A local survey of the industrial organization in the leading centre for silk handloom weaving, by a well-known writer on factories.

George, N. Dorothy: **London Life in the Eighteenth Century, London,** 1925.

Contains good description of the industrial organization in the Spitalfields silk handloom weaving (pp. 176-194).

Gill, Conrad: **The Rise of the Irish Linen Industry,** Oxford, 1925.

Worthy to rank with studies for other textile industries by Chapman, Unwin, Daniels, Lipson, Heaton, etc. Its main interest, as the author states in the Preface, is "to trace the change from domestic to factory production". A thoroughgoing and scholarly contribution based on original sources. Critical bibliography (pp. 344-49).

Hertz, G.B.: English Silk Industry in the 18th Century, in **English Historical Review,** 1909, XXIV, pp. 710-27.

A short, excellent sketch on early industrial organization.

Horner, John: **The Linen Trade of Europe during the Spinning Wheel Period,** Belfast, 1920.

A useful work, with frequent quotations and summaries taken from the chief writers of earlier times and official sources. For England, see pages 215 to 266.

Porter, G. R.: **A Treatise on the Origin, Progressive Improvement, and Present State of the Silk Manufacture,** in vol. XLIV of Lardner's Cabinet Cyclopaedia Series, London, 1831.

An early work written by a noted author, but too technical.

Rawlley, Ratan C.: **Economics of the Silk Industry, a Study in Industrial Organization,** London, 1919.

An excellent analysis, more contemporary than historical. The work is intended to be a comparative survey, dealing chiefly with conditions in Great Britain and France.

Warden, Alexander: **The Linen Trade, Ancient and Modern,** London, 1864; 2nd ed., 1867.

A standard work by a Scotch manufacturer, with careful use of primary sources. Includes all countries and periods, but only gives a slight chronological account of developments before the nineteenth century. For England see pages 381 to 388.

Warner, Sir Frank: **The Silk Industry of the United Kingdom: Its Origin and Development, London,** 1921.

An expansion of his article written 17 years ago on "The British Silk Industries", in **Society of Arts Journal,** 1904, LII, pp. 123-39. Analyzes the industrial organization of leading centres such as Spitalfields, Macclesfield, Manchester, Derby, Leek etc. A pioneer work.

Whiter, James S.: **Silk Industry of Great Britain and Its Revival,** London, 1882.

A pamphlet attributing the decline of English silk industry to the rise of commercial warehousemen. Wishes to see the industry marketed again by small drapers of the early nineteenth century.

4. Paper, pottery and glass

Many works are written either from technical or artistic viewpoints, of which only a few of exceptional merits are listed below. For bibliographical aids, consult the monumental work of Solon on pottery, Great Britain Patent Office Library's list on pottery and glass, Dillon's selected bibliography on glass. There is no bibliography on paper, except that of Hunter's covering the period 1390-1800.

Anonymous: A day at the Darwen Paper Mills, Lancashire, in **Bradshaw's Manchester Journal,** 1842, III, pp. 33-9, 50-55.

Description of processes, machines and different departments of a paper factory. Sidelights on factory size and extent.

Anonymous: A peep at the Staffordshire Potteries, in **Chambers' Edinburgh Review,** 1839, VIII, pp. 345-6.

Picturesque account of a visit to the famous district by a contemporary.

Anonymous: Paper making, in **Penny Magazine,** 1833, II, pp. 378-384.

An account of Magnay's Albury Mill near Guildford.

Bladen, V. W.: The Potteries in the Industrial Revolution, in **Economic Journal Supplement (Econ. Hist. Series No. 1),** January, 1926, pp. 117-130.

Short, interesting article on the emerging factory system.

Boch, S. von: **Geschichte der Töpferarbeiter von Staffordshire im XIX. Jahrhundert,** Stuttgart, 1899. Collection des Münchener volkswirtschaftliche Studien, XXXI.

A careful treatise.

Boyle, J., An account of strikes in the Potteries in 1834-6, in **Royal Statistical Society Journal,** 1839, I, pp. 37-44.

Supplies valuable statistics on number of factories and employees, with an estimate of capital and wages.

Dillon, Edward: **Glass,** London, 1907.

First general history of glass in English language written from an artistic viewpoint, from the earliest period to the present. Covers all countries. Select bibliography, pp. xxii-xxviii, not annotated.

Great Britain Patent Office Library: **Subject list of works on the silicate industries (ceramics and glass) in the Library of the Patent Office,** London, 1914.

Not annotated. Covers all countries and all aspects. See especially sections "historical and descriptive" (pp. 10-11, 39-40) and "local" (pp. 23-4, 29, 48).

Hunter, David: **The literature of paper making, 1390-1800,** Chillicothe (Ohio), 1925.

Jewitt, Llewellynn: **The Ceramic Art of Great Britain from the Prehistoric Times down to the Present Day,** London, 1878.

A standard work treating each pottery by localities. A mine of information poorly arranged.

Méteyard, Eliza: **Life of Josiah Wedgwood,** 2 vols., London, 1865-66.

Written from Wedgwood's private correspondence and family papers, conscientious and exhaustive, but not critical.

Munsell, J.: **A chronology of paper and paper making,** Albany, 1857.

Compiled from numerous English and foreign sources. Has merits and defects common to works of this nature, giving no references to sources.

Owen, Harold: **The Staffordshire Potter, London, 1901.**

A study of trade union history by a sympathetic student. Quotes statistics from Boyle on factory size and extent.

Porter, G. T.: **A treatise on the progressive improvement, and present state of the manufacture of porcelain and glass,** London, 1832.

A volume in Lardner's Cabinet Cyclopaedia series. Chiefly technical, with historical notes. Covers England and other countries.

Powell, Harry J.: **Glass Making in England,** 1923.

Contains two chapters on glass houses before 1840.

Royal Statistical Society: An approximate statement of the present extent and value of the manufacture of glass in the United Kingdom, in **Royal Statistical Society Journal,** 1839, I, pp. 539-40.

Contains six tables.

Shaw, S.: **Staffordshire Potteries,** London, 1900 (first published in Hanley, 1829).

A mine of information for later writers, written chiefly from oral accounts of contemporary witnesses in the district.

Solon, M. L.: **Ceramic Literature an analytical index to the works published in all languages on the history and technology of the ceramic art,** London, 1910.

Carefully compiled, classified and annotated by an extensive writer on the subject. Comprehensive, selective, and critical.

Solon, M. L.: **A brief history of old English porcelain and its manufactories,** London and Derby, 1903.

A history of leading porcelain manufactories up to 1820.

Spicer, Albert D.: **The paper trade, a descriptive and historical survey of the paper trade from the commencement of the nineteenth century,** London, 1907.

A study of parliamentary papers and other sources. Excellent description of paper mills and their history in leading centres throughout the United Kingdom. Valuable statistical tables on machines, output and duties paid.

Ward, John: **History of Stoke-on Trent,** London, 1843.

A work started by Shaw, but failing in financial resources, completed by Ward. Exhaustive treatment of the famous pottery district by localities.

Wedgwood, Joseph C.: **Staffordshire Pottery and Its History,** London, 1913.

Compiled from standard works and MSS of his great-great-grandfather. Treats evolution of industrial organization in the pottery manufacture, rise of capital and coming of steam machinery.

5. Metals

There has been no bibliography devoted exclusively to the metal manufactures considered in the present treatise. Special sections given in the general bibliographies such as Williams' and Power's, however, furnish safe guide. In many cases, literature on the metal manufactures is technical in nature, and it is only during recent years that studies such as those of Ashton, Lloyd, Lord and Tawney begin to give sufficient attention to industrial organization. Standard works are quoted for the manufactures of iron and steel, engine and machine, and cutlery. For the minor iron and steel products most information is drawn directly from the Parliamentary Papers, with occasional references to the articles and chapters devoted to each particular product in the composite industrial series by Timmins, Armstrong, Bevan, Dodd, and Lardner. (The general bibliographies and composite industrial series referred to here are listed and annotated in the section on 'General Works'.)

Anonymous: Two Days in Birmingham, in **Chamber's Edinburgh Journal**, 1844, I, pp. 369-73.

A contemporary account of visits to the button, pin, nail, and screw factories in Birmingham.

"Artifex" and "Opifex", **Causes of Decay in a British Industry**, London, 1907.

A pioneer treatise written by two leading Birmingham gun manufacturers, attributing the decline of the industry chiefly to the unwise governmental legislation. Tables, charts, and statistics, but practically nothing on industrial organization. Chapters I and II are devoted to a historial summary.

Ashton, Thomas S.: **Iron and Steel in the Industrial Revolution**, Manchester, 1924.

A manuscript study of leading firms such as Boulton and Watt, Darby, Crowley, Huntsman etc., covering the ironworks of South Yorkshire and the Midlands during the period 1700 to 1815. Describes epoch-making inventions, with frequent references to industrial organization. Thorough, readable, and well-illustrated, although the tuthor's depreciation of Parliamentary Papers as the "principal quarry" for economic historians under-estimates the value of such sources to the students of industrial organization.

Ashton, Thomas S.: Domestic System in the Early Lancashire Tool Trade, in **Economic Journal Supplement (Economic History Series No. 1)**, January, 1926, pp. 131-140; The Records of a Pin Manufactory 1814-21, **Economica**, 1925, V, pp. 281-92.

A study of the records of two firms, Peter Stubbs from 1777 to 1803, and Stubbs, Wood & Co. from 1814 to 1821. An interesting analysis of the transition from the "domestic" to the factory system.

Bore, Henry: **The Story of the Invention of Steel Pens**, New York, 1890.

Includes a description of the manufacturing processes and the enterprise of the famous steel pen manufacturer, James Perry. Frequent quotations from various sources,—poems, magazines, cyclopaedias, local and industrial histories, etc. A sketchy but popular account.

Bume, John T.: **Josiah Mason**, London, 1890.

A biography of the noted steel pen manufacturer, with two chapters on the history of steel pen and electroplating trades rarely found elsewhere.

Dodd, George: Clerkenwell, in Knight, Charles (ed.), **London**, 1841-6, 6 volumes, III, pp. 129-44.

Includes good description of industrial organization in watchmaking, besides topographical notes on the District.

Fairbairn, Sir William: **Lectures on the Rise and Progress of Engineering, Derby**, 1859.

By a famous pioneer in engineering.

Fairbairn, Sir William: **Iron, its History, Properties, and Processes of Manufacture**, Edinburgh, 1861.

First published in 1856 in the 8th edition of the **Encyclopaedia Britannica** and now revised in book form. (2nd ed., 1865, 3rd ed., 1869. Tr. into French, with additions by M. Gustave Maurier) Authoritative work in its day, including history and trade statistics, but chiefly technical.

Farey, John: **Historical Account of the Steam Engine,** London, 1827.
Written by an engineer who had the command of first-hand information from the inventor and the advantage of observation as a contemporary witness. Useful chapter on the application of steam engine to different industries.

Forrest, Noah: The Chain and Trace Makers of Cradley Heath and its Vicinity, in **Trans. National Association for the Promotion of Social Science,** 1859, pp. 654-6.
Notes on the truck system.

Goodman, J.D.: On the Progress of the Small Arms Manufacture, in **Royal Statistical Society Journal,** 1865, XXVIII, pp. 494-506.
An excellent analysis of the industrial organization, although touching mainly on conditions in 1865. Same article appeared in Timmins (ed.), **Industrial History of Birmingham and the Midland Hardware District,** London, 1866.

Hollond, John: **A Treatise on the Progressive Improvement and Present State of Manufactures in Metal,** 1st ed., 1831-4, 3 vols.; new ed., 1853, rev. by Robert Hunt.
Vols. I and II on "Iron and Steel".

Holland, George Calvert: **An Inquiry into the Condition of the Cutlery Manufacture,** Sheffield, 1842.
A pamphlet deploring the increasing gap between rich and poor, the lack of a living wage, the overproduction of inferior articles, especially of cast metal, fraudulent marks.

Jeans, James Stephen: **Steel, its History, Manufacture, Properties, and Uses,** 1880.
First general work on steel industry by an extensive and authoritative writer.

Leppington, Miss C. H. E.: The Evolution of a Modern Industrial Town, in **Economic Journal,** 1907, XVII, pp. 345-57.
A contemporary survey of social and economic conditions in Coventry—the watch-making and ribbon weaving centre. Scattered references on history and industrial organization.

Lloyd, G.I.H.: **The Cutlery Trades, an Historical Essay in the Economics of Small-scale Production,** London, 1913.
A brilliant analysis of the "little master" system in the famous Sheffield industry, especially of those "rapidly disappearing features which appear to be survivals from an earlier industrialism". Valuable appendices containing data on water power, forges and warehouses, employees, rules and regulations of trade associations in the early nineteenth century.

Lord, John: **Capital and Steam Power 1750-1800,** London, 1923.
Largely based upon the papers of the engineering firm of Boulton and Watt. Author points out the close correlation between the rise of capital and the coming of steam power, and gives invaluable compilations from original records on the distribution of steam engines erected by Boulton and Watt, by industries as well as by localities.

Meade, Richard: **The Coal and Iron Industries of the United Kingdom,** London, 1882.
A painstaking compilation by the Assistant Keeper of Mining Records for the government, condensing all available information relating to coal, iron and steel. Each iron making district and its iron output were considered separately, beginning with the year 1740.

Morrall: **Short Description of Needle-making,** 1866.
A prize essay and only treatise on the subject.

Mushet, David: **Papers on Iron and Steel,** 1840.
One of the earliest manufacturers to make scientific experiments gives the results of his efforts, with valuable statistics on output, and a few articles devoted to history.

Pole, William (ed.): **The Life of Sir William Fairbairn,** London, 1877.
Chapters I and II give valuable summary on the history of engineering and a historical interpretation of the term. Appendix is an exhaustive bibliography on the works and papers written by Fairbairn, chronologically arranged (pp. 475-92).

Roe, J.W.: **English and American Tool Builders,** New Haven, 1916.
Only work on the subject, except the writings of Smiles. In the opinion of the author, "the history of English and American tool building... covers substantially the entire history of the art".

Royal Statistical Society: Steam power in Lancashire and its immediate vicinity, in **Royal Statistical Society Journal,** 1839, I, pp. 315-6.
Statistics on increase since 1835 by industries as well as by localities.

Royal Statistical Society: Steam power of Birmingham, in **Royal Statistical Society Journal,** 1839, II, pp. 436-7.
Gives the dates when steam power was first introduced into the different Birmingham manufactures, such as nail, screw etc., and also the amounts of steam power employed in 1839.

Rylett, Rev. Harold: Nails and Chains, in **English Illustrated Magazine, 1889-90,** VII, pp. 163-75.
A popular account of the technique and industrial organization from the viewpoint of a social reformer. Well illustrated.

Scrivenor, Harry: **History of the Iron Trade, from the Earliest Records to the Present Period,** London, 1841 (new ed. 1854).
Long a standard work, chiefly on Great Britain (chs. I to VI, XIII to XIV), but with brief references to foreign countries also (chs. VII to XII). Historical, technical and commercial. Statistics on blast furnaces and pig iron outputs since 1740.

Sheffield: The Cutlery Capital of the British Empire, issued by the Development Committee of the Sheffield City Council, 1920.
On the rise and present position of Sheffield.

Smiles, S.: **Industrial Biography: Iron Workers and Tool Makers,** 1863.
A popular and interesting book, based on original sources; but now partly replaced by Roe's.

Smiles, S.: **Lives of Boulton and Watt,** Philadelphia, 1865.
Concise and readable; partly replaced by T. Marshall's **James Watt (1736-1819),** 1925. Compiled from the original Soho MSS.

Taylor, Leslie B.: **A Brief History of the Westley Richards Firm 1812-1913,** 1913.
On one of the earliest gun manufacturing factories.

Tawney, R. H.: **Establishment of Minimum Rates in the Chain-making Industry under the Trade Boards Act of 1909,** London, 1914.
Excellent analysis of the merchant employer system in Chapter I, although a contemporary account.

Wilkie, George: **The Manufacture of Iron in Great Britain; with Remarks on the Employment of Capital in Ironworks and Collieries,** Edinburgh, 1857.
Historical and descriptive. Clear analysis.

6. Clothing

Literature on the clothing manufactures, with the possible exception of lace and hosiery, is extremely scarce. Nor does any bibliography exist that gives a fairly satisfactory survey of the subject. Parliamentary papers, therefore, must be the chief source of information for detailed study.

Anonymous: Gloves and Glovers, in **Penny Magazine,** 1843, XII, pp. 94-6, 101-3.
Early history and anecdotes; excellent description of merchant employer system.

Austin, Thos. Geo.: **The Straw Plaiting and Straw Hat and Bonnet Trade**, Luton, 1871
Chiefly a digest of the Occupation Census of 1861 for the metropolis of the industry, the Luton District. Only treatise giving sufficient attention to industrial organization except Tansley's short article referred to below.

Channer, C.C. and Roberts, M.E.: **Lace Making in the Midlands, Past and Present**, London, 1900.
Historical anecdotes on hand made lace, with interesting sidelights on industrial organization.

Cobbett, J.M.: **On the converting of English grass into straw for making plat for hats and bonnets**, London, 1823.
Pamphlet no. 8 in the Cobbett's **Cottage Economy Series**, a plea for home industry.

Felkin, William: **A History of the Machine Wrought Hosiery and Lace Manufactures**, London, 1867.
Still a standard work on the subject, although poorly arranged, by the well-known statistician of the industry. A witness before various Parliamentary Inquiries. For reference see **Children's Employment Commission, Report**, in P. P., 1833, XX, pp. 513-26; **Committee on Export of Machinery, First Report**, in P.P., 1841, VII, pp. 231-56; **Framework Knitters' Commission, Report**, in P.P. (H.L.), 1845, XXIII, pp. 18-19. Incorporates first-hand knowledge of technique and inventions with his personal analysis of statistics on capital, wages, employees, machines, raw materials, finished products etc.

Galton, F.W. (ed.): **The Tailoring Trade**, London, 1923.
No. 1 of series on the **Select Documents illustrating the History of Trade Unionism.** A thorough and pains-taking piece of research by an artisan, with valuable extracts on the outwork system (vi, lxi-lxii, xciv, 90, 103, 184, 192, 220, 222). Bibliography, but not annotated (pp. 224-37).

Gould, I. Chalky: Straw Plaiting—A Lost Essex Industry, in **Essex Naturalist, 1906**, XIV, pp. 184-92.
An interesting local study, chiefly technical.

Grindrod, Ralph Barnes: **The Slaves of the Needle; an Exposure of the Distressed Condition, Moral and Physical, of Dressmakers, Milliners, Embroiderers, Slop-workers, etc.**, London, [1844?]
Extracts from the **Reports of the Children's Employment Commission** (in P.P., 1843, XIV) for the London District, by a learned social reformer.

Gutteridge, Joseph: **Lights and Shadows in the life of an artisan**, Coventry, 1893.
Life story of a ribbon weaver, born in 1816.

Holford, Sir Henry: **A plea for the framework knitters**, 1847.
Bitter against extreme laissez-faire policy towards the distressed conditions of the framework knitters arising chiefly from the middleman system.

Hopkinson, A.W.: Home Industries in Nottingham, in **Economic Review** (quarterly published by the Christian Social Union, now discontinued), 1903, XIII, pp. 334-9.
Contemporary survey of industrial organization in lace and hosiery industries from a broad and human viewpoint.

Hull, William: **History of the Glove Trade**, London, 1834.
A study of leading glove centres, with scattered notes on industrial organization.

Jackson, Mrs. F. Neville: **A History of Hand Made Lace**, London, 1900.
Dealing with the origin of lace, the growth of the great lace centres, the mode of manufacture, the methods of distinguishing and handling various kinds of lace. Sketchy for England. Bibliography.

Kingsley, Charles: **Alton Locke, Tailor and Poet**, 1849. Edition used in this treatise is by Macmillan, London, 1905.

An autobiography drawn from recollections in 1838-1842. Intimate though biassed portrayal of the "sweating" system in chaps. X and XXI. The present edition contains reprints of his pamphlet listed below.

Kingsley, Charles (published under pseudonym "Parson Lot"): **Cheap Clothes and Nasty,** London, 1850.

A denunciation of the "sweating" system.

Moody, A.P. and Tompkins, A.E.: **Devon Pillow Lace: Its History and How to Make it,** London, 1907.

An analysis of the famous Honiton hand lace industry.

Nottingham Free Public Libraries, **A Contribution towards a Bibliography of Hosiery and Lace,** Nottingham, 1896. 16pp.

Includes encyclopaedia and magazine articles.

Palliser, Mrs. Bury: **History of Lace,** 1st ed., 1875; 4th edition, 1902, entirely revised, rewritten and enlarged under the editorship of M. Jourdain and Alice Dryden.

A standard work on hand lace, comparable with Felkin's on machine lace. Covers European countries, especially England (chaps. XXII toXXXII, pp. 285-417). Uses primary sources; has many quotations and illustrations.

Rendle, E.: Edward Smith, a Narrative of the Sweating System, in **Workingman's Friend,** Sept. 14, 1850, pp. 281-7.

A brief analysis of the system by a working tailor.

Shaw, William: **An affectionate pleading for England's oppressed female workers, 1850.**

An account of a meeting of ragged, underpaid needle-women, revealing terrible poverty.

Tansley, A. J.: On the Straw Plait Trade, in **Royal Society of Arts Journal,** 1860-61, IX, pp. 69-77.

An excellent article on industrial organization, etc. of the manufacture, highly commended by the Society.

Tawney, R.H.: **The Establishment of Minimum Rates in the Tailoring Industry under the Trade Boards Act of 1909,** London, 1915.

Chapter I on the "Organization of the Tailoring Industry" (pp. 1-23) is the best short analysis of the "sweating" system still existing.

Wright, Thomas: **The Romance of the Pillow Lace; being the history of lace-making in Bucks, Beds, Northants and neighboring counties, together with some account of the lace industries of Devon and Ireland, Olney** (Bucks), 2d ed., 1924.

Much more connected with hand-made lace. Describes methods and designs. Bibliography.

C. General works

1. Bibliography

Besides the special bibliographies here noted, see also others of a more general nature, especially Cunningham, Mantoux, Held, Knowles, etc., as listed in section 3.

Beales, H.L. and Cole, G.D.H.: Select list of books on economic and social history, 1700-1850, in **Journal of Adult Education,** Vol. I, no. 2, March, 1927.

McCulloch, J.R.: **Literature of Political Economy—a classified list,** London, 1845.

Incomplete but useful, giving partial comments from the viewpoint of a Ricardian economist. Occasionally gives historical and biographical notes.

Power, Eileen: The Industrial Revolution, 1750-1850, in **Ecenomic History Society, Bibliographies series, no. 1,** London, 1927.

A carefully selected and annotated bibliography, giving most of the important works recently published.

Williams, Judith Blow: **A guide to the printed materials for English social and economic history, 1750-1850,** New York, 1926, 2 vols.

A monumental work begun some ten years ago as the author's doctoral dissertation at Columbia, involving patient research in the leading libraries of America and Great Britain. Introductory comments to each section are especially helpful.

2. Cyclopaedias

Cyclopaedias of a strictly technical nature are omitted, the following list including only those that give general and economic information on the manufactures studied in the present treatise, with sufficient technical references for an understanding of the industrial organization.

Bevan, G.P.: **British Manufacturing Industries,** 1876, 15 vols.

Comprehensive survey of all industries, edited by Bevan with a two-volume summary under the title **The Industrial Classes and Industrial Statistics** (vols. XIII and XIV). Contemporary, but throws light on the earlier period.

Lardner, Rev. D. (ed.): **Cabinet Cyclopaedia,** 1831-4.

Covers all subjects of general interest, but for our purpose consult 3 volumes on **Manufactures in Metals** by J. Holland, 1831-4; 2 volumes by G.R. Porter on **Silk** and **Porcelain and Glass,** 1831-2.

McCulloch, J. R.: **Dictionary of Commerce,** 1st ed., 1832; 2nd ed., 1834.

McCulloch, J.R.: **A Statistical Account of the British Empire,** 2nd ed., 2 vols., London, 1839.

See sections on "manufactures" in vol. I, pp. 612-733. The paucity of materials available for such a compilation is pointed out by the editor, "notwithstanding the magnitude of the materials from which we have drawn, we have had to regret the extreme scarcity, and sometimes, indeed, the total want, of all authentic and trustworthy information on various important points. In many instances we have had to build our conclusions on reasonings from analogy, or on a few detached facts; and we have not infrequently been compelled, from the total want of experience and information, to decline drawing any conclusions". (I, p. viii).

Palgrave, H.: **Dictionary of Political Economy,** new ed., by Henry Higgs, 3 vols., 1923-26.

Contains short articles on important manufactures, with useful bibliography.

Poole, B.: **Statistics of British Commerce, being a compendium of the productions, manufactures, imports and exports of the United Kingdom,** London, 1852.

Porter, G.R.: **The Progress of the Nation,** London, 1836,3 vols.;new ed. by F. W. Hirst, 1912.

Chiefly statistical.

Society for the Diffusion of Useful Arts, **Penny Cyclopaedia,** London, 1833, 27 vols., with index.

Ure, Andrew: **A Dictionary of Arts, Manufactures and Mines,** 1839.

Technical but valuable. Other editions, 1843, 1853, 1860, 1867, 1875-8.

3. Industry and industrial revolution

For bibliographical aids, consult Cunningham, Webb, Mantoux, Held, Knowles, Hammond, etc.

Armstrong, Sir W.G. (ed.): **Industrial Resources of the Northern District,** 2nd ed., London, 1864.

Excellent articles by experts on various industries, including hemp, carpet, paper, pottery, glass, iron, engine and machine, and hat. Each article deals more with contemporary than historical conditions, but is still useful for our purpose.

Ashley, Sir W.J. (ed.): **British Industries,** 1903.

A series of general reviews for business men and students, with historical references, by experts. Covers cotton, woolen, worsted, flax, iron and steel, etc.

Bowden, Witt: **Industrial Society in England Towards the End of the Eighteenth Century,** New York, 1925.

Has good, annotated bibliography. For our purpose, consult heading under "Non-contemporaneous accounts and bibliographies" (pp. 305-11).

Cox, Harold (ed.): **British Industries under Free Trade,** London, 1903.
Essays by experts, throwing light on history and industrial organization of many industries, including silk, linen, engine and machine, cutlery, etc.

Cunningham, W.: **The Growth of English Industry and Commerce in Modern Times,** 1st ed., 1882, 2 vols. (vols. II and III).
Vol. III was reissued under the title **The Industrial Revolution,** containing an extensive bibliography. The various editions of this pioneer work bear the following dates: 1st ed., 1882; 2nd ed., 1890- 92; 3rd ed., 1896-1903; 4th ed., 1903-5; 5th ed., 1910.

Cunningham, W.: Localization of Industry, in **Economic Journal,** 1902, XII, pp. 501-6.
On causes of localization.

Day, Clive: Distribution of industrial occupations in England, 1841-1861, in **Connecticut Academy of Arts and Sciences, Transactions,** 1927, XXVIII, pp. 79-235.
Stimulating study on the extent of market covered by each English manufacture, based on the Occupation Censuses of 1841 and 1861.

Dodd, George: **British Manufactures,** London, 1844-46.
Chiefly accounts of his visit to factories in different manufactures, describing technique rather than industrial organization. Contains six series on chemicals, metals, textiles etc., a large part of which appeared previously in the **Penny Magazine, 1841-44,** X-XIII; and the author's **Days at the Factories,** London, 1843.

Fay, C.R.: **Life and Labor in the Nineteenth Century,** Gambridge, 1920.
Able and clear analysis of the parliamentary reports on the Handloom Weavers andFramework Knitters in chs. XIV, XVI, XVII.

Hammond, J. L. and Barbara:
The Town Laborer, 1760-1832, 1st ed., 1917; now ed., 1925.
The Skilled Laborer, 1760-1832, 1st ed., 1919; new ed., 1926.
The Rise of Modern Industry, London, 1925.
These studies are distinguished works on the great transition in English industries, although colored by the authors' philosophical outlook and sympathy for labor. **Town Laborer** contains a good bibliography of primary and secondary sources; **Skilled Laborer** has excellent summaries of important parliamentary reports on the cotton, woolen, worsted and silk handloom weavers and the hosiery knitters (chs. IV-VIII, pp. 47-256); and finally, **Rise** is the best short summary of industries such as engine, iron, pottery, and cotton for the period (chs. VIII-XI).

Held, Adolf: **Zwei Bücher zur socialen Geschichte Englands,** Leipzig.
Takes over Thun's (1879) classification of industrial systems and stages, and contains a short critical account of some works relating to the coming of the factory system and its effects. A standard German work on English industrial revolution, comparable with Mantoux's.

Knowles, Mrs. Lillian C.A.: **Industrial and Commercial Revolution in Great Britain during the Nineteenth Century,** London, 1920; 3rd (revised) ed., 1924.
Contains excellent account of industrial changes in Pt. II, "Industrial Revolution caused by Machinery" (pp. 15-109). Useful and select bibliography for each chapter.

Mantoux, Paul J.
La Revolution industrielle au XVIIIe Siècle, Paris, 1906.
The Industrial Revolution in the Eighteenth Century, rev. ed. transl. by Marjorie Vernon, London, 1928.

A classic on the subject, especially useful because of its analysis of the early industrial system in woolen industry and extensive annotated bibliography on primary and secondary sources.

Moffit, Louis W.: **England on the Eve of the Industrial Revolation, 1740-1760,** London, 1925.

Contains good bibliography on non-contemporaneous sources.

Page, William (ed.): **Commerce and Industry, 1815-1914,** 1919, 2 vols.

Vol. II contains valuable statistics compiled from the parliamentary papers, on population, revenue and taxation, imports and exports, shipping, railway, production, wages, prices, factories, banks, etc.

Pinchbeck, Ivy. **Women workers and the industrial revolution, 1750-1850, London,** London, 1930.

A source study covering women in agriculture, industry and trade. Industries included are textiles (cotton, woolen, worsted, silk, linen), lace, straw plait, glove, hosiery, button and nail, etc.

Redford, A.: **Labor Migration in England, 1800-1850,** Manchester, 1926.

Contains good account .of the early development of factory system in textile industries.

Symons, J.C.: **Arts and artisans at home and abroad,** Edinburgh, 1839.

Includes "Two letters on economy of British Cotton Mills" (pp. 230-39).

Timmins, S. (ed.): **Industrial History of Birmingham and the Midland Hardware District,** London, 1866.

Similar to Armstrong's, but covers a more important industrial district. Treats manufactures such as iron, clock and watch, lock and key, gun, nail, screw, pin, steel pen, button, wire needles, and glass.

Traill, H.D. (ed.): **Social England,** London, 1895-1901, 6 vols.

Vol. VI contains short articles, with bibliographies, on textiles, metals, and pottery.

Webb, Sidney and Beatrice: **History of Trade Unionism,** 1st ed., 1894; 2nd ed., 1902.

Includes extensive bibliography (pp. 499-543, 1902 ed.) not embodied in 1920 rev. ed.

Weber, Alfred: Industrielle Standortslehre, in **Grundriss der Sozialökonomie,** 1914, VI, pp. 54-82.

A thorough analysis of factors affecting localization of industry.

4. Local history, travels and description

Local histories varied in fullness for different counties throughout the nineteenth century, of which the compilations for the period around 1840 deplorably deficient in giving information on social and economic subjects—particularly industrial organization of different manufactures. The present bibliography makes no pretense to completeness, including only those works that have been cited or consulted, but not referred to in our bibliography on "Individual Industries". For references not here cited, consult the bibliographical works of Anderson and Humphrey.

Accounts of tourists gave scattered references to industrial organization, of which the most valuable were those by Faucher, Kohl, Cooke-Taylor, and Head.

Allen, Thomas: **History of the County of York,** London, 1829, 6 vols.

Anderson, J. Parker: **The Book of British Topography, a classified alphabetical catalogue of the topographical works in the Library of the British Museum relating to Great Britain and Ireland,** London, 1881.

Includes 14,000 books, and "by far the most comprehensive hitherto published of books relating to the history and antiquities of Great Britain and Ireland" (author's preface, viii). A standard work compiled by a member of the National Library.

Anonymous (an American Lady): **Change for the American Notes, in Letters from London to New York,** New York, 1843.

Gives accounts of the distressed conditions of the London dressmakers.

Ashton, Joseph: **A Picture of Manchester,** Manchester, 1816.
Contains description of the new Cotton Exchange in 1809 (pp. 204-14).

Bamford, Samuel: **Walks in S. Lancashire and on its borders,** Blackley (near Manchester), 1844.
Includes interesting statistics on hat factories and employees in Oldham.

Burton, John R.: **A History of Kidderminster,** London, 1890.
A study of the famous carpet weavıng centre.

Cooke-Taylor, William: **Notes of a Tour in the Manufacturing Districts of Lancashire,** London, 1842.
Keen but favorable observations on the factory system in different manufacturing districts by a noted authority of the period, in a series of letters.

Davis, Frederick: **Luton, Past and Present,** Luton, 1874.
Contains excellent short account of the rise and progress of the straw plait and hat and bonnet manufactures (pp. 17-26).

Dent, Robert K.: **The Making of Birmingham,** Birmingham, 1894.
Includes short account of the steel pen manufacture (pp. 453-5).

Fletcher, J.S.: **The Making if Modern Yorkshire, 1750-1914,** London, 1918.
A local history, political, social and economic, by a voluminous and noted authority.

Glover, Stephen:**The History of the County of Derby,** Derby, 1829, 2 vols.
Poorly compiled, but contains a wealth of information on lace and hosiery manufactures.

Granville, A.B.: **The Spas of England,** London, 1841, 2 vols.

Hardwick, Charles: **History of the Borough of Preston,** Preston, 1857.

Head, Sir George: **Home Tour Through the Manufacturing Districts of England,** London, 1836.
A desultory ramble through the manufacturing districts, giving accounts of manufacturing establishments and the processes carried therein.

Humphrey, H.: **Great Britain, France and Belgium, a Short Tour in 1835,** New York, 1838.

Humphreys, A.L.: **A handbook to County Bibliography, being a bibliography of bibliographies relating to the counties and towns of Great Britain and Ireland,** London, 1917.
"Result of many years' vigilance in taking notes, followed by four years of active work in amplifying, and preparing for the press". A monumental work covering all subjects, classified alphabetically according to county divisions.

Kohl, J. G.: **England and Wales,** Philadelphia, 1846.
A keen and critical account of the industrial conditions by a German tourist.

Liverpool, The Stranger in, 12th ed., Liverpool, 1838.

Manchester Historical Recorder: being an analysis of the municipal, ecclesiastical, biographical, commercial and statistical history of Manchester from the earliest period, chronologically arranged, Manchester, 1874.

Manchester, Panorama of, and Railway Companion, Manchester, 1834.

Nowrojee, J. and Merwanjee, H.: **Journal of a residence of two years and a half in Great Britain,** London, 1841.

Parsons, Edward: **History of Leeds etc.,** Leeds, 1834, 2 vols.

Poole, Benjamin (compiler): **Coventry: Its History and Antiquities,** London, 1870.
Contains good account of ribbon and watch manufactures (pp. 359-65).

Shaw, William A.: **Manchester Old and New,** London, date? 3 vols.

Snell, F.J. (ed.): **Memorials of Old Devonshire,** London, 1904.

Contains a first-hand account by Miss Alice Dryden on the "Honiton Lace", describing industrial organization and attempts at revival of the famous centre for hand lace.

Taylor, Rev. Isaac: **Scenes of wealth, or views and illustrations of trades, manufactures and commerce...**, London, 1826.

A series of queries and replies comparable to the popular publications such as "Conversations on Political Economy", written in dialogue form.

Tupling, G.H.: **Economic History of Rossendale,** Manchester, 1927.

One of the few local histories written from modern standard and viewpoint. Scholarly analysis of the industrial systems in cotton and woolen manufactures before the middle of the nineteenth century.

Victoria County History (elsewhere abridged as V.C.H. in this treatise), ed. by William Page, 1900-date.

The bulk of the work varies with individual counties, ranging from eight volumes for **Lancaster** to one volume for many small counties. The volumes for **Chester, Northumberland, Westmoreland** and **Wilts** have not yet appeared, while for other counties only the first or first few volumes have been published. Each county, however, has a section devoted to the industries written by specialists and students in their respective fields, with the exception of such counties as **Devon, Hereford, Huntington, Kent, Leicester, Norfolk,Northampton,** and **Stafford.** The following articles are recommended for their special merit or usefulness.

Textiles:

Cotton

Lancaster, II,379-92, by S. J. Chapman.

Woolen and worstod

Essex, II, 380-403, by Miller Christy.

Gloucester, II, 193-99, by Miss R. F. Butler and Miss C. Violet Butler.

Suffolk, II, 254-71, by George Unwin. This article is reprinted in Unwin, Studies **in Economic History,** ed. by R. T. Tawney, London, 1927, pp. 262-301.

Worcester, II, 281-97 (woolen), 297-99 (carpet), by C. H. Vellacott.

York, II, 406-29, by Miss Maud Sellers.

Silk

Essex, II, 462-69, by Miss C. Fell Smith and Miller Christy.

Middlesex, II, 132-8, by Charles Welch.

Flax and hemp

Dorst, II, 344-53, by Miss M. M. Crick.

Paper, pottery and glass:

Paper

Buckingham, II, 111-12, by Miss C. Jamison.

Hertford, IV, 256-8, by Lewis Evans.

Lancaster, II, 407-8, by Douglas Knoop.

Middlesex, II, 195-7, by Charles Welch.

Surrey, II, 418-20, by M. S. Giuseppi.

Pottery and glass

Derby, II, 374-76, by Mrs. J. H. Lauder.

Durham, II, 309-14, by Miss Maud Sellers.

Lancaster, II, 403-6, by Douglas Knoop.

Middlesex, II, 141-58, by Charles Welch.

Salop, I, 434-40, by John Randall.

Surrey, II, 281-305, by M. S. Giuseppi.

Warwick, II, 244-49, by Miss Ethel M. Hewitt.

Worcester, II, 276-81, by J. W. Willis-Bund and Miss Marjorie V. Taylor.

York, II, 429-49, by Miss Maud Sellers.

Metals:

Iron, engine and machine, and cutlery

Derby, II, 356-62, by Mrs. J. H. Lauder and C. H. Vellacott.

Durham, II, 278-92, by Miss Maud Sellers.

Lancaster, II, 360-65, 367-74, by Douglas Knoop.

Salop, I, 460-79, by John Randall.

Surrey, II, 263-76, 410-18, by M. S. Giuseppi.

Worcester, II, 267-71, by Miss Marjorie V. Taylor.

York, II, 387-406, by Miss Maud Sellers.

Button, etc.

Buckingham, II, 127-8 (needle), by Miss C. Jamison.

Gloucester, II, 206-7 (pins), by Miss R. F. Butter.

Warwick, II, 234-40 (buttons, needles and pins) by Miss Ethel M. Hewitt.

Worcester, II, 273-4 (needles), by Miss Marjorie V. Taylor.

Nail, anchor and chain

Salop, I, 479-81 (chain), by John Randall.

Worcester, II, 271-3 (nail and chain), by Miss Marjorie V. Taylor.

Clock and watch, gun

Lancaster, II, 366-7 (watch), by Douglas Knoop.

Middlesex, II, 158-65 (clock and watch) by Charles Welch.

Warwick, II, 226-32 (gun), 242-44 (clock and watch). by Miss Ethel M. Hewitt.

Clothing:

Ribbon

Warwick, II, 257-62, by Miss Ethel M. Hewitt.

Hosiery, lace

Bedford, II, 122-4 (lace), by Arthur Ransom.

Buckingham, II, 106-9 (lace), by Miss C. Jamison.

Derby, II, 367-72 (hosiery and lace), by Mrs. J. H. Lauder.

Northampton, II, 336-39 (lace) by T. J. George.

Nottingham, II, 352-63 (hosiery and lace), by Miss Ethel M. Hewitt.

Hat and glove

Lancaster, II, 393-4 (hat), by Douglas Knoop.

Oxford, II, 255-9 (glove), by Reginald W. Jeffery.

Somerset, II, 427-9 (glove), by Miss Ethel M. Hewitt.

Surrey, II, 259-63 (hat), by M. S. Giuseppi.

Straw plait and straw hat and bonnet

Bedford, II, 118-22, by Arthur Ransom.

Buckingham, II, 113-14, by Miss C. Jamison.

Essex, II, 375-79, by Miller Christy.

Hertford, IV, 251-6, by Miss Ethel M. Hewitt.

Wheeler, James: **Manchester: its political, social and commercial history,** 1836.

Wylie, William H.: **Old and New Nottingham,** London, 1853.
Contains history of the hosiery and lace manufactures, compiled from the works of Deering, Blackner, Henson and Felkin.

INDEX

INDEX

INDEX

INDEX

INDEX

INDEX

INDEX

INDEX

INDEX

INDEX